ENERGY LEVEL ALIGNMENT OF SOLID INTERFACES

ENERGY LEVEL ALIGNMENT OF SOLID INTERFACES

Raymond T Tung
City University of New York, USA

World Scientific

NEW JERSEY • LONDON • SINGAPORE • BEIJING • SHANGHAI • HONG KONG • TAIPEI • CHENNAI • TOKYO

Published by

World Scientific Publishing Co. Pte. Ltd.
5 Toh Tuck Link, Singapore 596224
USA office: 27 Warren Street, Suite 401-402, Hackensack, NJ 07601
UK office: 57 Shelton Street, Covent Garden, London WC2H 9HE

British Library Cataloguing-in-Publication Data
A catalogue record for this book is available from the British Library.

ENERGY LEVEL ALIGNMENT OF SOLID INTERFACES

ISBN 978-981-98-2574-5 (hardcover)
ISBN 978-981-98-2575-2 (ebook for institutions)
ISBN 978-981-98-2576-9 (ebook for individuals)

For any available supplementary material, please visit
https://www.worldscientific.com/worldscibooks/10.1142/14659#t=suppl

Desk Editor: Rhaimie Wahap

PREFACE

Modern technologies advance at a dazzling pace. Nearly all aspects of our personal lives make use of electronics enabled by advanced solid-state devices already miniaturized in size and getting even smaller. Our smart phones, computers, and other devices process information by manipulating the flow of electrons across materials interfaces. At the writing of this book, the state-of-the-art Si ULSI circuits are at a technology node of 2.0 nm, with 1.6 nm node devices being actively developed. As the size of the individual device decreases, the number of various interfaces between materials per area skyrockets. An interesting consequence is that nearly the entire volume of material actively involved in the operation of each individual device is within a very short distance of an interface and may be considered as a part of that interface. In essence, each device may be viewed as a structure assembled with various interfaces rather than built from different materials! And the operation of any integrated circuit may be regarded more conveniently as a merger of transport properties of different interfaces than an interaction among the properties of difference materials. The importance of interfaces to modern devices cannot be overstated. Needless to say, a thorough understanding of the relationship between atomic structure and interface properties is essential to continued advancements in technologies. Of various properties of a solid interface, the difference in the energy band positions across that interface is particularly important for electronic transport. For metal-semiconductor (MS) interfaces, which are omnipresent in electronic and optoelectronic semiconductor devices, the Schottky barrier height (SBH) controls electronic transport properties. Optimized SBHs can significantly improve the performance of devices, while devices may fail with inadequate SBHs! At a semiconductor heterojunction, the difference between the energy band edges of two semiconductors, called the band offset (BO), controls its transport properties as well as the overall performance of devices in many photonics and energy applications. Obviously, device engineers should know the theory behind the SBH and the BO well enough to be able to control or tune the SBH/BO to optimize device performance. Indeed, band alignment problems have been studied from the perspective of technology development for many decades. Scientifically speaking, the relative position of the energy bands of two solids at an interface between them is a fundamental problem in condensed matter physics that presents a challenge for research scientist to explain. Many theories and models were proposed in the last ~50 years with the intention and hope to explain the SBH and the related BO.

With so much from the technology side riding on the control of the SBH/BO and just as much interest in solving the SBH/BO mystery from the academic side for so long, we certainly expect to hear a positive answer when a question like "do we know everything now?" is asked. However, a quick glance of the current literature finds that the field of SBH/BO is still in a state of confusion, or at least in a state of transition. In the vast majority of current publications, empirical theories that have been around a long time are used to explain the experimentally measured SBH. Being empirical and based on conjectures, these theories have little connection with scientific principles but, by design, are able to agree with some of the experimental data. At the same time, empirical models fail for a significant portion of the MS interfaces studied. When an empirical model fails for an interface, it's not known why, and it's not known how to correct the mistakes of empirical models. On the brighter side, a scientifically based approach also emerged in the literature in recent years. An ability to quantitatively explain the SBH/BO at all interfaces studied so far has been demonstrated. Because band alignment is a direct result of the charge distribution of an interface, the most straightforward way to explain/predict the SBH/BO is through charge distribution modeling. Taking advantage of the similarity between charge densities at solid interfaces to those of bulk solid crystals, a neutral polyhedra theory (NPT) reasonably modeled the charge distributions and the SBH/BOs of various solid interfaces in recent years. Therefore, a scientifically sound theory that quantitatively explain the SBH/BOs of solid interfaces is available at the present. However, having come into light only within the last <10 years, this interface-charge-distribution-based method and its demonstrated ability to widely explain/predict the SBH/BO are not covered in any of the standard textbooks and comprehensive reviews on band alignment. As a result, the majority of researchers in the field are not aware of the existence of this scientific alternative to empirical models! To help researchers unfamiliar with the latest developments in this field make the transition from empirical models to this charge-density-based approach is the main reason for the writing of this book. From personal experience, this is not an easy task. It took the author ~40 years to understand band alignment at solid interfaces!

After a thesis in experimental surface science, my first job was a postdoc at Bell Labs working on silicide reaction, which was a technologically important topic at the time because silicides were used as metallic contacts and interconnects in Si IC circuits. While playing with the reaction of ultrathin layers of metal with Si in an UHV chamber that belonged to John Poate, I stumbled across a fascinating phenomenon, which was that the orientation of ultrathin epitaxial $NiSi_2$ layers grown on Si(111) could be controlled to be entirely one of the two possible orientations. High resolution electron microscopy conducted by Murray Gibson and others determined their atomic structures. The SBHs of these two interfaces, called type A and type B, were obviously something that might interest people. With instructions and equipment from Kwok Ng, I discovered that the SBHs for the two epitaxial $NiSi_2$/Si(111) interfaces were different. (Sec. 7.1) That's when I started reading about Schottky barriers. There were quite a few theories on SBH formation at the time, but none could explain the experimentally observed SBH difference between the two epitaxial interfaces. "Fermi level pinning" (FLP) dominated the experimental scene at the time, which referred to the observation that the experimental SBH was nearly constant for a semiconductor and independent of the metal. All SBH theories at the time were focused on just one thing: the explanation of the FLP phenomenon. And every theory used the same thing, namely interface states, to explain FLP. Even though it was clear that interface states could be part of the SBH explanation, I couldn't understand scientifically why theories always put them on the side with the semiconductor and assumed the metal to be unaffected by the formation of the interface. (Sec. 4.2) Apparently, doing so allowed FLP to be explained, but it made no sense to me, fresh out of graduate school and still gung-ho that quantum mechanics should explain everything. It seemed that empirical theories had to be wrong because they conflicted with quantum mechanics. (Sec. 7.2)

My own experience with current-voltage measurement of SBH introduced me to a mysterious parameter, the ideality factor. This parameter was not in the thermionic emission theory that described current transport of Schottky barriers but needed to be included artificially to fit the experimental data obtained from polycrystalline MS interfaces. There was no self-consistent explanation of this factor from various models based largely on interface states at the time. While reading an article by John Freeouf et al (JVST 1982) one day, I stared at a figure for a while. I almost fell off my chair! Right there was the answer to the ideality factor mystery! I quickly wrote an analytic theory on current transport to cover the more generalized case that the SBH might be inhomogeneous. John Sullivan, who was doing his Ph.D. at Bell Labs did a thorough job of numerical simulation with a PADRE program that Mark Pinto developed. (Sec. 6.1.2) Various dependencies of the ideality factor experimentally observed were found to agree with SBH inhomogeneity. (Sec. 6.4) This shocking discovery had implications on many levels. It showed that the SBHs of polycrystalline MS interfaces were pervasively inhomogeneous. Therefore, the interfaces employed decades ago to demonstrate the FLP effect in fact had Fermi level positions that were "unpinned" and spatially varying in the band gap! The phenomenon known as "Fermi level pinning" turned out not to be a fundamental mechanism that kept the FL constant in the band gap for each semiconductor independent of the metal, based on which all the empirical theories were invented. The actual SBH mechanism, being responsible for the locally varying SBH, clearly depended on the interface structure, in conflict with empirical models. "Fermi level pinning" was a misnomer. In reality, it described a macroscopic observation that the average of a range of SBHs at a polycrystalline MS interface was insensitive to the metal. Even though empirical models were incompatible with the observed SBH inhomogeneity, they continued to be used for SBH explanation. At that time, there just was no other theory that could quantitatively explain the experimental SBH, albeit only some of it.

Not long after changing profession from industrial research to academia, I also switched research activities from mainly experimental work to full-time theoretical calculations. With a generous arrangement and technical help from Leeor Kronik, I started to calculate electronic structures of semiconductor heterojunctions and MS interfaces. During these calculations, an intriguing general property of the interfaces and the bulk crystals of zincblende semiconductors practically forced itself to be noticed. The charge distribution within a volume in the immediate vicinity of an individual atom, either anion or cation, was found to be always approximately the same no matter in what semiconductor or interface this volume, called "neutral polyhedron", was located. (Sec. 2.3) Reflecting a "nearsightedness" of the electron density, this interchangeability, or transferability, of charge distribution suggested a viable scientific approach to predict the SBH/BO quantitatively. It seemed that charge distributions of bulk semiconductors could be used to assemble, neutral-polyhedron-by-neutral-polyhedron, the charge distribution of any interface approximately and, from there on, an SBH/BO could be produced for that interface! This approach turned out to be very successful in practice. The SBH/BOs of not only epitaxial MS and heterojunction interfaces with known atomic structures but also some polycrystalline interfaces were quantitatively explained by this method. (Secs. 5.3, 5.4, 5.5, and 7.3) Detailed analysis of electronic structures of MS interfaces with different atomic structures also quantitatively explained the apparent Fermi level pinning phenomenon, i.e. the metal-independent average SBH at polycrystalline interfaces. (Sec. 7.4) A scientific method to quantitatively explain the entire spectrum of seemingly disparate dependencies of SBH/BO is thus in place. Instead of requiring the reader to piece together different aspects of this method from publications scattered in the literature, this book describes in a coherent manner the entire principles and procedures of this method based on charge distribution. It is intended for researchers and graduate students with basic knowledge of quantum mechanics who want to have a thorough grasp of how interface band alignment is decided. Because the basic idea of this problem can also be understood without quantum mechanics, a narrative with illustration(s) is provided at the end of each chapter to summarize the physics just discussed without elaborate math. Undergraduate students and others who are not planning to research in this area but would like to understand this topic should be able to do so from reading these summaries.

Any energy band position of a solid, such as its valence band maximum (VBM), is an energy level. "How the energy bands of two solids are aligned at a solid interface" is essentially a question on the difference between energy levels on two separate systems. This book begins with a definition of what an energy level is for an isolated finite system. This finite system, such as a molecule, is known as the "host" of its discrete energy levels. When this host is no longer isolated but is in the company of, and weakly coupled to, other sub-systems, called "bath", its energy levels are affected. Through the use of a semiclassical theory, how the energy levels of a host are rigidly shifted by the electric potential of the bath can be expressed explicitly. This definitive host-bath relationship is essentially the foundation on which energy level alignment problems of any kind can be understood. In the remainder of Chapter 1, an exercise is undertaken to illustrate the host-bath relationship through discussions of the energy levels of redox species in a solvent. In this exercise, the free energy required to change the solvent configuration (the bath) such that the HOMO of an acceptor (one host) is raised or lowered by a certain amount with respect to the LUMO of a donor (another host) is worked out. It is worth noting that the alignment of the energy levels is controlled by the charge distribution of the bath. The energy levels and electronic structures internal to the bath do not matter. In Chapter 2, the groundwork is laid for the host-bath approach to be applied to analyze band alignment at solid interfaces. The tricky part actually comes at the very beginning, when appropriate hosts for energy bands on both sides of a solid interface need to be identified. Energy bands are derived from the periodic atomic structures of infinite crystals and the latter should act as the host ordinarily. However, at an interface, the periodic structure has to end. It is thus necessary to define a finite volume with the charge distribution of a bulk crystal as the host of the crystal's energy bands. An important question is how the bulk charge distribution should be terminated into a "model solid" that acts as the host of the energy bands. The methods to properly define a model solid are discussed in detail, with the convenient choices identified as the neutral polyhedra theory (NPT) and the atoms-in-molecules method (AIM). Common crystalline metals, semiconductors, and oxides are analyzed with these two methods and the properties of their model solids are listed at the end of Chapter 2. With discussions of the atomic structures and the electronic states of the surfaces of crystalline solids in Chapter 3, we also get an opportunity to witness how the energy band alignment is analyzed through the distribution of charge at the surface. A brief foray into the explanation of the work functions experimentally observed from the surfaces of elemental metals from modeled charge distribution of the surface demonstrates how the host-bath concept is applied to a solid interface. In this case, the surface is the "interface" between a solid and vacuum.

Beginning in Chapter 4, we focus on solid interfaces. After a dependence of the SBH/BO on the interface charge distribution is demonstrated explicitly, traditional theories for interface band alignment are discussed. Because of the use of surface quantities to represent bulk crystals, these traditional theories contain inherent errors from surface dipoles and are unable to explain the dominant experimental observation of FLP. Empirical models based on metal-induced gap states (MIGS) that can explain some but not all the experimental data are discussed. A general discussion of the formation of electronic states at solid interfaces then follows and shows an explicit dependence of the MIGS on the metal, contrary to the assumption crucial to empirical MIGS models. With empirical models evidently unscientific, it becomes imperative to adhere to charge distribution as the scientific basis in the explanation of the entire dependences of experimental SBH/BO. The rest of Chapter 4 presents an overall plan to proceed in that direction. With model solids acting as the hosts to the energy bands of the crystals, the bath for band alignment problems in general is identified and an approximate treatment of its effect is described. Essentially, the physics of band alignment is fully contained in this overall host-bath plan for solid interfaces. These procedures are applied successfully to explain, through modeled charge distributions, the VBOs of semiconductor heterojunction interfaces and the SBHs of MS interfaces, respectively, in Chapters 5 and 7. The ability to model the interface charge distribution hinges on the transferability of charge distribution between the interface and bulk crystals, which is a consequence of the nearsightedness principle of electron density. In Chapter 5, the extremely local dependence of the electron distribution is demonstrated through direct comparisons. Chapter 6 describes SBH measurement. Experimental measurement of anything is ordinarily a boring subject matter unworthy of primary billings. However, in the case of the SBH of polycrystalline MS interfaces, it is anything but boring. Potential pinch-off and saddle-point formation in the space charge regions of MS interfaces with inhomogeneous SBH lead to widely different transport properties that depend on the spatial distribution of the SBH. The demonstration that polycrystalline MS interfaces have pervasively inhomogeneous SBH necessitates a reinterpretation of the FLP phenomenon, which is addressed in Chapter 7. A historic and personally nostalgic discussion of a sharp SBH dependence on the atomic structure for single crystal $NiSi_2$/Si interfaces begins that chapter. Theoretical calculations of these interfaces and other epitaxial interfaces between elemental metals and zincblende semiconductors are then described. These calculations show an expected dependence of the MIGS distribution on the metal and also perhaps unexpectedly that the full extent of the interface dipole lies largely outside of the semiconductor! Cut-and-stitch numerical experiments show that the metal-cation bonded and metal-anion bonded interfaces behave differently because of a difference in the location of metallicity boundaries. This discovery justifies the employment of different model solids in successfully modeling all the calculated SBHs. It also leads to a quantitative and logical explanation of the FLP mystery that has intrigued scientists for more than 50 years. The FLP phenomenon can be understood from the interface charge distribution after all, just like all other experimental observations on SBH/BO can.

The author thanks his many collaborators in SBH/BO research, listed here in roughly chronological order: Drs. J. M. Poate, J. M. Gibson, A. F. J. Levi, K. K. Ng, J. Werner, J. P. Sullivan, H. Fujitani, H. Ishiwara, S. Oda, S. M. Yalisove, S. Ohmi, Y. Li, W. Long, L. Kronik, H. Haick, and M. Hong. He is grateful to John Poate and Len Feldman for encouraging him to write this book and for critical reading of the manuscript, and to Drs. S. Chambers, H. Efeoglu, and V. LaBella for kind review of a proposal for this book. Permissions to show figures from various authors in this book are greatly appreciated. The author also thanks Drs. S. Oda, H. Ishiwara, L. Kronik and D. Cahen for hosting his stays during sabbatical and short leaves.

This preface was written in September 2025 while traveling through Xi'an, Guilin, and Taipei.

Ray Tung (董梓則)
Brooklyn College, CUNY

FOREWORD

There is another story behind this important monograph on semiconductor and materials physics. It is the personal scientific odyssey of Ray Tung. He completed his PhD at Penn with Professor Bill Graham on single atom diffusion on surfaces and joined me at Bell Labs in 1980 as a Post Doctoral Fellow. It was an exciting time at the Labs with so many scientific thrusts. We were investigating the challenges facing the processing of silicon for the next generation integrated circuits. In particular, scientists and engineers at the labs had pioneered the use of metallic silicides such as platinum silicide for stable rectifying contacts with well-defined barrier heights. We were researching the formation of these contacts under atomically clean conditions when Hiroshi Ishiwara of the Tokyo Institute of Technology showed the formation of single crystal epitaxial silicides under normal processing conditions. This stimulated the field and Ray and John Bean formed epitaxial Nickel and Cobalt silicides under atomically clean conditions.

Ray then in a classic series of experiments changed the understanding of metallic contacts to semiconductors. He and colleagues demonstrated the growth of different epitaxial phases of Ni and Co silicides. Importantly the barrier heights depended on the interface structure. This was in contradiction to the theoretical understanding of Schottky barriers and caused much consternation in the field. This was a challenge for Ray, and he took up the daunting task of theoretically understanding the electronic properties of semiconductor interfaces. He left the Labs in 2002 for CUNY and devoted his time to this task. He and collaborators have developed a quantum mechanical model that embodies the physical structure of interfaces to explain their electronic properties. This treatise which describes the whole gamut of the physical and electrical properties of semiconductor interfaces is an important addition to the scientific literature.

John Poate
Emeritus Vice President of Research
Colorado School of Mines

CONTENTS

ACRONYMS

Acronym	Full Expression	Explanation/ First occurrence
AC	anion-centered	
AC-AIM	anion-centered AIM model solid	2.2.3
AES	Auger electron spectroscopy	3.1.3
AFM	atomic force microscope	3.1.3
AIM	atoms in molecules theory	2.2.3
AWS	anion-centered WSC	2.2.2
BEEM	ballistic electron emission microscopy	6.3.3
BO	band offset	4.1
CBM	conduction band minimum	2.1
CBO	conduction band offset	4.1
CC	cation-centered	
CC-AIM	cation-centered AIM model solid	2.2.3
CNL	charge neutrality level	3.4.1
CPE	chemical potential equalization	4.2
C-V	capacitance-voltage	6.2.2
CWS	Cation-centered WSC	2.2.2
DFT	Density Functional Theory	1.2.3
EAR	Electron Affinity Rule	4.2
EBIC	electron beam induced conduction	6.3.3
FLP	Fermi-level pinning	3.4.3
ISR	interface specific region	4.1

(*Continued*)

(*Continued*)

Acronym	Full Expression	Explanation/ First occurrence
HF	Hartree-Fock	1.2.2
HOMO	highest occupied molecular orbital	1.2.4
HREM	high resolution electron microscopy (HREM)	7.1.2
I-V	Current-voltage	6.2.1
KS	Kohn-Sham	1.2.3
LEED	low energy electron diffraction	3.1.3
LEEM	low energy electron microscope	3.1.3
LEGO	local electron densities generated in oligo-cells	5.3.1
LUMO	lowest unoccupied molecular orbital	1.2.4
MIGS	metal-induced gap states	4.3
MS	metal-semiconductor	
MSC	model solid cell	2.2
NDBP	net density between consecutive planes	7.2.4
NPT	neutral polyhedra theory	2.3.3
PES	photoemission spectroscopy	6.2.3
PR	photoresponse	6.2.3
RHEED	reflective high energy electron reflection	3.1.3
SBH	Schottky barrier height	2.2
SEM	scanning electron microscopy	6.3.3
SMT	Schottky-Mott Theory	4.2
SSR	surface specific region	3.2
STM	scanning tunneling microcopy	3.1.3
TF	Thomas-Fermi	1.2.3
TFE	thermionic-field-emission	6.2.1
VBM	valence band maximum	2.1
VBO	valence band offset	4.1
WF	work function	3.3.1
WSC	Wigner-Seitz cell	2.1

Chapter 1

ENERGY LEVELS: "HOSTS" AND "BATHS"

1.1 Introduction

Our daily lives are increasingly enriched by, and dependent upon, advanced technologies surrounding us. As the semiconductor industry continues to push for more advanced functionalities in integrated circuits that are getting smaller and smaller, the role played by material interfaces in semiconductor devices, already extremely important at the present, can only get more dominant. The properties of interfaces essentially control the operation and performance of advanced devices. Therefore, interfaces of various kinds need to be understood thoroughly in order for the semiconductor industry to continue its relentless pursuit of smaller and better devices. Particularly important for an interface, from the perspective of device transport properties, is the difference in the energy band positions across that interface. How the energy levels of one material are positioned with respect to those of another is also a scientific problem that has attracted much attention in condensed matter physics.

For both scientific and technological reasons, there is a need to know all the factors that contribute to the difference in the energy levels of two materials across an interface. Unfortunately, this involves quantum mechanics. From a quantum mechanical description of two materials forming an interface, we could in principle extract information about the energy level alignment from the coupled equations. The messy results from such an exercise wouldn't help us much toward an understanding of the energy level alignment problem. Our comprehension of the physics world is based largely on well-established classical ideas of how things work. To understand the various dependences of the alignment of energy levels on the properties of the interface, it is helpful and necessary to put complicated quantum mechanical results in the language commonly used to describe science classically. This is what this chapter attempts to do. Take the energy level of a material, for example. Its full quantum mechanical definition already contains information about how it varies from coupling with other materials and, therefore, information about how energy levels on different materials are aligned. However, the physics of energy level alignment is not easily appreciated from the complicated quantum mechanical equations until we make a few semiclassical approximations. These approximations break a complicated and coupled system into separate sub-systems that may be treated individually and have only classical interactions between them. This move allows the energy level difference at an interface to be understood in familiar language. Unfortunately, in order to establish the validity of such a semiclassical approach and identify the factors that control energy levels, it is necessary to delve into hardcore quantum mechanics. This is an unavoidable step that needs to be taken for the entire book to make sense. Fortunately, this only appears in the first chapter. Being somewhat boring and not directly affecting our ability to comprehend the physics of energy level alignment, this chapter may be skipped on first reading and perhaps should be a reference to return to if specific questions about the semiclassical picture arise later in the book. Before moving on from this chapter, however, readers should read a summary written in layman's language at the end of this chapter, which explains the basic idea behind the present semiclassical approach without the bother of quantum mechanics.

Even though readers are probably more interested in energy levels of solids, any solid is a complicated electronic system with energy levels that are definable only after considerable elaboration. Since the factors that determine the alignment of energy levels between two solids are the same as those controlling the alignment of discrete energy levels in small systems, e.g. molecules, we will deal with small systems in this chapter, from which these factors are more clearly identified. Our first order of business is to define energy levels for an isolated electronic system, known as the "host" of these energy levels, with respect to a global reference, such as the vacuum level. Discussions then broaden to cover the situation when this host is not in isolation but in the company of other electronic systems, called "bath". The expression for an energy level of the host is now modified due to the presence of the bath. Although seemingly dealing with the energy levels of only one electronic system, this semiclassical "host–bath" description contains all the physics needed to explain the alignment of isolated energy levels. When the alignment of energy bands at solid interfaces is explained later in this book, the same host–bath framework will be used.

1.2 Total Energy of an Isolated System

1.2.1 *The Hamiltonian*

The energy level of a system describes how the total energy of the system varies with the addition or removal of an electron. Our discussion of the energy level has to begin with the total energy of an isolated finite system that contains a number, N_H, of nuclei and a fixed number, n_H, of electrons. This is a well-known problem that has been treated in detail in excellent textbooks in quantum chemistry. We only summarize the results here. The Hamiltonian for our system can be written as

$$H_N^{\text{el}} = -\frac{\hbar^2}{2m}\sum_i^{n_H}\nabla_i^2 - \frac{e^2}{4\pi\varepsilon_0}\sum_{i,\alpha}\frac{Z_\alpha}{|\vec{r}_i - \vec{R}_\alpha|} + \frac{e^2}{4\pi\varepsilon_0}\sum_{i<j}^{n_H}\frac{1}{|\vec{r}_i - \vec{r}_j|} + \frac{e^2}{4\pi\varepsilon_0}\sum_{\alpha<\beta}^{N_H}\frac{Z_\alpha Z_\beta}{|\vec{R}_\alpha - \vec{R}_\beta|}, \tag{1.1}$$

where $\vec{r}$'s are the coordinates of the electrons, and $\vec{R}$'s and Z's are, respectively, the positions and charges of the ion cores. Within the Born–Oppenheimer approximation, the motions of the nuclei, being much slower than the electrons, are ignored when the motions of the electrons are considered. The electrons are assumed to adjust their motion to any shift in the nuclear positions almost instantaneously and be able to always maintain their density optimized for the instantaneous nuclear positions. The electronic Hamiltonian is used to solve for the wave functions that describe the motion of N electrons in the system, through the Schrödinger equation

$$H_N^{\text{el}}\psi_N(\vec{x}_1, \vec{x}_2, \ldots, \vec{x}_N) = E_N^{\text{el}}\psi_N(\vec{x}_1, \vec{x}_2, \ldots, \vec{x}_N), \tag{1.2}$$

where $\vec{x}_i$ represents both the spatial and the spin coordinates ($\vec{r}_i$ and s_i) of the ith electron. E_N^{el} is the total energy associated with the N-electron system. Since the motion of the nuclei is not part of the electronic Hamiltonian, the positions of the nuclei enter into the Hamiltonian only parametrically. Therefore, the Coulomb potential from the nuclear charges can be viewed as an "external potential" to the many-electron system, and the Hamiltonian can be broken down into a part that operates on one electron at a time (T stands for the kinetic energy operator) and a part H_{ee} that governs electron–electron interaction:

$$H_N^{\text{el}} = \{T + H_{\text{ext}}\} + H_{\text{ee}}. \tag{1.3}$$

The total number of electrons of the Hamiltonian operator is fixed and should match the total number of electrons of the wave function that it operates on. The wave function solution to Schrodinger's equation satisfies the usual ortho-normal conditions. Since electrons are indistinguishable Fermions, the wave function of a multi-electron system must be anti-symmetric. A natural consequence of the anti-symmetry of electronic wave functions is that no two electrons with the same spin can have the same spatial coordinate. The lowest expectation value of the Hamiltonian is the ground-state energy, $E^{\text{el}}_{N,0}$,

$$E^{\text{el}}_{N,0} = \langle \psi^{+}_{N,0} | H^{\text{el}}_{N} | \psi_{N,0} \rangle, \tag{1.4}$$

which is assumed to be non-degenerate and is produced when the true ground-state wave function $\psi_{N,0}$ is inserted. Two well-known and popular methods to compute the total energy of an isolated system are the Hartree–Fock approach and the density functional theory (DFT).

1.2.2 *Hartree–Fock approximation*

Calculations conducted with the Hartree–Fock (HF) theory usually give reasonable results on the electronic structure of material systems, even though HF does not provide an "exact" solution. Since the physical interpretation of its equations is clear and its limitations well understood, the HF method sets an excellent foundation for the one-electron energy level concept used throughout this book. HF begins with a trial N-electron wave function in the form of a single "Slater determinant" of N individual spin orbitals, φ_i:

$$\psi_{\text{HF},N} = \frac{1}{\sqrt{N!}} \begin{vmatrix} \varphi_1(\vec{x}_1) & \varphi_2(\vec{x}_1) & \cdots & \varphi_N(\vec{x}_1) \\ \varphi_1(\vec{x}_2) & \varphi_2(\vec{x}_2) & \cdots & \varphi_N(\vec{x}_2) \\ \vdots & \vdots & \ddots & \vdots \\ \varphi_1(\vec{x}_N) & \varphi_2(\vec{x}_N) & \cdots & \varphi_N(\vec{x}_N) \end{vmatrix} \tag{1.5}$$

The meaning of the absolute value is just the determinant of a matrix in the usual sense. Since electrons have spins, a proper wave function for any electron necessarily contains a spin component, of which two states, "up" and "down", span the entire spin coordinate. Each single-particle wave function $\varphi_i(\vec{x}_j)$ is a simple product of a spatial orbital $\phi_i(\vec{r}_i)$ and a spin wave function $\sigma_i(s_i)$. Since spin orbitals are normalized and orthogonal to unlike spins, the HF wave function can be verified to be normalized and anti-symmetric. Now, the job of the HF theory is to look for the set of orbitals that minimizes the total electronic energy of the system. After plugging in the HF wave function, Eq. (1.4) becomes

$$\begin{aligned}[E_N^{\mathrm{el}}(\psi_{\mathrm{HF},N})] &= \langle \psi_{\mathrm{HF},N}^{+} | H_N^{\mathrm{el}} | \psi_{\mathrm{HF},N} \rangle \\ &= \sum_i^N \int d\vec{r} \phi_i^*(\vec{r}) \left(\frac{-\hbar^2 \nabla^2}{2m} - eV_{\mathrm{ext}}(\vec{r}) \right) \phi_i(\vec{r}) \\ &\quad - \frac{e^2}{8\pi\varepsilon_0} \sum_{i=1}^N \sum_{j=1}^N \int d\vec{r} \int d\vec{r}\,' \frac{\phi_i^*(\vec{r})\phi_i(\vec{r})\phi_j^*(\vec{r}\,')\phi_j(\vec{r}\,')}{|\vec{r}-\vec{r}\,'|} \\ &\quad - \frac{e^2}{8\pi\varepsilon_0} \sum_{i=1}^N \sum_{j=1}^N \int d\vec{r} \int d\vec{r}\,' \frac{\phi_i^*(\vec{r})\phi_j(\vec{r})\phi_j^*(\vec{r}\,')\phi_i(\vec{r}\,')}{|\vec{r}-\vec{r}\,'|} \delta_{\sigma_i,\sigma_j}\end{aligned} \tag{1.6}$$

As a "functional" of the N-electron wave function, the expectation value of the electronic energy, $[E_N^{\mathrm{el}}(\psi_{\mathrm{HF},N})]$ in Eq. (1.6), has a physical meaning only when the chosen wave function $\psi_{\mathrm{HF},N}$ actually minimizes the total energy. Since Eq. (1.6) has to remain stationary against variations in $\delta\phi_i^*(\vec{r})$, one arrives at the following equation for $\phi_i(\vec{r})$:

$$\begin{aligned}&\frac{-\hbar^2}{2m}\nabla^2\phi_i(\vec{r}) - eV_{\mathrm{ext}}(\vec{r})\phi_i(\vec{r}) + \frac{e^2}{4\pi\varepsilon_0}\int d\vec{r}\,' \frac{\rho(\vec{r}\,')}{|\vec{r}-\vec{r}\,'|}\phi_i(\vec{r}) \\ &\quad - \frac{e^2}{4\pi\varepsilon_0}\sum_{j=1}^N \int d\vec{r}\,' \frac{\phi_j^*(\vec{r}\,')\phi_i(\vec{r}\,')}{|\vec{r}-\vec{r}\,'|}\phi_j(\vec{r})\delta_{\sigma_i,\sigma_j} = \varepsilon_i\phi_i(\vec{r}),\end{aligned} \tag{1.7}$$

where ε_i, the "Lagrange's multiplier" or the eigenvalue, is a fixed real number and the local density of electrons is defined as $\rho(\vec{r}) = \sum_{i=1}^N \phi_i^*(\vec{r})\phi_i(\vec{r})$. The N equations, one for each electronic orbital, are known as the "Hartree–Fock equations". A collection of N spin orbitals obtained self-consistently and simultaneously from the HF equations leads to the lowest possible total electronic energy for the system and, therefore, appears to be the ground state of the system. One should remember, however, that this is only the energy-minimized solution among all wave functions expressible as a single Slater determinant. Nevertheless, Hartree–Fock theory produces useful approximations and sets an upper bound for the true ground-state energy. Methods to improve the HF results, such as by using trial wave functions of a more complex form, a la "configuration interaction", can be found in quantum chemistry books.

It is worthwhile to examine the physical meaning of the eigenvalue for each Hartree–Fock equation. The total energy of the isolated system calculated by the HF method, expressed in terms of the eigenvalues, is

$$\langle \psi^{+}_{\mathrm{HF},N} | H^{\mathrm{el}}_{N} | \psi_{\mathrm{HF},N} \rangle = \sum_{i}^{N_e} \varepsilon_i - \frac{1}{2}\frac{e^2}{4\pi\varepsilon_0} \sum_{i,j}^{N_e} \iint d\vec{r}\,' d\vec{r} \frac{\phi_j^*(\vec{r}\,')\phi_j(\vec{r}\,')}{|\vec{r}-\vec{r}\,'|} \phi_i^*(\vec{r})\phi_i(\vec{r})$$
$$+ \frac{1}{2}\frac{e^2}{4\pi\varepsilon_0} \sum_{i,j}^{N_e} \iint d\vec{r}\,' d\vec{r} \frac{\delta_{\sigma_i,\sigma_j}}{|\vec{r}-\vec{r}\,'|} \phi_i^*(\vec{r}\,')\phi_i(\vec{r})\phi_j^*(\vec{r})\phi_j(\vec{r}\,'). \tag{1.8}$$

Multiplying Eq. (1.7) by ϕ_i^* and integrating over space, ε_i is shown explicitly to be

$$\varepsilon_i = \int d\vec{r}\phi_i^*(\vec{r}) \left(\frac{-\hbar^2}{2m}\nabla^2 - eV_{\mathrm{ext}}(\vec{r}) + \frac{e^2}{4\pi\varepsilon_0} \sum_{j\neq i}^{N} \int d\vec{r}\,' \frac{\phi_j^*(\vec{r}\,')\phi_j(\vec{r}\,')}{|\vec{r}-\vec{r}\,'|} \right) \phi_i(\vec{r})$$
$$- \frac{e^2}{4\pi\varepsilon_0} \int d\vec{r}\phi_i^*(\vec{r}) \sum_{j\neq i}^{N} \int d\vec{r}\,' \frac{\phi_j^*(\vec{r}\,')\phi_i(\vec{r}\,')}{|\vec{r}-\vec{r}\,'|} \phi_j(\vec{r})\delta_{\sigma_i,\sigma_j}. \tag{1.9}$$

The energy of the electron in the electric potential created by the positive ions and all other electrons is the first line on the right-hand side of Eq. (1.9). Known as the Hartree energy, this is what the energies would be for (fictitious) electrons that were distinguishable from each other. In other words, if only one of the $N!$ terms of Eq. (1.5) were used as the trial wave function, the Hartree energy would be found as the total energy. The second term of Eq. (1.9) has no classical analog and is the electrostatic interaction between one electron and the density of electrons with the same spin. This is the attractive interaction between this electron and the "exchange hole" created in its proximity from the density of electrons with the spin. The form suggests that Eq. (1.9) is the energy needed or released when one electron is moved from the vacuum level to a system with $N - 1$ electrons.

1.2.3 *Density functional theory and the Kohn–Sham equation*

As just illustrated for the Hartree–Fock method, all observable information about an electronic system may be extracted from its wave function through the expectation values of specific operators. Looking at the huge number of coordinates for a common electronic system and the very limited number of observables it delivers, one may question whether all the details of the wave function are really necessary in order to fully define a quantum mechanical state. Indeed, the idea emerged that the electron density of the system, which has only three spatial coordinates, may be a function that contains all the information of a system and thus is capable of representing the system. The earliest attempt to directly employ the electron density in total energy calculation was the Thomas–Fermi (TF) model, which assumed that the kinetic energy associated with a particular electron density, per volume, was the same as the kinetic energy for a free electron gas of the same density. Assuming a constant shape of the exchange–correlation hole, the TF functional of electron density was used to calculate the energy of small molecules with, however, very poor results. A well-known unphysical prediction of the TF model was that no molecule was stable against dissociation! From this inauspicious beginning, theories based on electron density would evolve dramatically and rise to the preeminence they enjoy today. Crucial to the success of density functional theory were two theorems of Hohenberg and Kohn (HK).[1] The first of these demonstrated that the non-degenerate ground-state density of a system indeed contained all the information of the system and the second showed that the energy functional was minimized when the ground-state density was used, and thus the solution may be sought through the variational principle.

While working with wave functions, the total energy of a system can be written down as Eq. (1.6). Now, with the electron density as the input, the form of the total energy, although not known immediately, is clearly a functional of the density $E = E[\rho(\vec{r})]$. Since the electronic Hamiltonian is known to consist of three different operators, one can individually express these three contributions as

$$E[\rho(\vec{r})] = T[\rho(\vec{r})] + U_{\text{ext}}[\rho(\vec{r})] + U_{\text{ee}}[\rho(\vec{r})], \tag{1.10}$$

where T, U_{ext}, and U_{ee} are, respectively, the kinetic energy functional, the functional for the interaction with external potential, and the functional for electron–electron interaction. Of these three, only the form for U_{ext} is known exactly:

$$U_{\text{ext}}[\rho(\vec{r})] = -e \int d\vec{r}\, V_{\text{ext}}(\vec{r})\rho(\vec{r}). \tag{1.11}$$

The sum of the other two functionals is the HK functional $F_{\text{KS}}[\rho(\vec{r})]$, which is universal but of unknown form. One therefore needs to propose and justify the form of the HK functional in order to proceed. Here, it is helpful to understand the reason that the TF theory performed so poorly before. The failure to produce chemical bonding in TF calculations could be largely attributed to the inadequacy of its kinetic energy functional. In view of this, Kohn and Sham (KS) proposed an approach to proceed within the outlines of DFT, while minimizing the impact of our ignorance on the kinetic energy functional.[2] KS assumed that the kinetic energy for a particular electron density can be approximated by the kinetic energy of a non-interacting electron gas with the same density ($= \sum_{i=1}^{N} |\phi_{\text{KS},i}(\vec{r})|^2$). Writing the wave function of this model system as a Slater determinant of N orthogonal "KS spin orbitals" ($\xi_i(\vec{x}_j) = \phi_{\text{KS},i}(\vec{r}_j)\sigma_i(s_j)$) and plugging it into the Schrodinger equation, the kinetic energy part of the total energy can be expressed as

$$T_{\text{KS}} = -\frac{\hbar^2}{2m}\sum_{i=1}^{N}\int d\vec{r}\phi^*_{\text{KS},i}(\vec{r})\nabla^2\phi_{\text{KS},i}(\vec{r}). \tag{1.12}$$

Even though the entire functional for electron–electron interactions is not known, the classical Coulomb interaction between electrons has to be part of it. Writing that part out explicitly leaves us with just one unknown functional $U_{\text{XC}}[\rho(\vec{r})]$ that covers effects due to the exchange and correction interactions between electrons and also contains the errors associated with the form of the KS kinetic energy term:

$$\begin{aligned} E[\rho] = &-\frac{\hbar^2}{2m}\sum_{i=1}^{N}\int d\vec{r}\phi^*_{\text{KS},i}(\vec{r})\nabla^2\phi_{\text{KS},i}(\vec{r}) + U_{\text{XC}}[\rho] \\ &+\sum_{i=1}^{N}\int d\vec{r}|\phi_{\text{KS},i}(\vec{r})|^2\left\{-eV_{\text{ext}}(\vec{r}) + \frac{e^2}{8\pi\varepsilon_0}\int d\vec{r}'\frac{\rho(\vec{r}')}{|\vec{r}-\vec{r}'|}\right\}. \end{aligned} \tag{1.13}$$

Since the energy is minimized when the correct density is inserted, the functional is stationary against all possible ways of varying the electron density. When each orbital wave function is varied individually, a Kohn–Sham equation can be written as

$$\begin{aligned} &\left\{-\frac{\hbar^2}{2m}\nabla^2 - eV_{\text{ext}}(\vec{r}) - eV_{\text{XC}}(\vec{r}) + \frac{e^2}{8\pi\varepsilon_0}\int d\vec{r}'\frac{\rho(\vec{r}')}{|\vec{r}-\vec{r}'|}\right\}\phi_{\text{KS},i}(\vec{r}) \\ &\quad = \varepsilon_{\text{KS},i}\phi_{\text{KS},i}(\vec{r}), \end{aligned} \tag{1.14}$$

where $\varepsilon_{\mathrm{KS},i}$ is an eigenvalue and the effective potential is given by

$$-eV_{\mathrm{XC}}(\vec{r}) = \frac{\delta U_{\mathrm{XC}}[\rho(\vec{r})]}{\delta \rho(\vec{r})}. \tag{1.15}$$

Several aspects of the KS approach should be pointed out. First of all, because the total density appears in the operator of the KS equation, the solution can be obtained only self-consistently, similar to how the HF equations can be solved. One has to remember, however, that the Slater determinant product of KS orbitals is just a computational tool and is not the wave function of the interacting electron system. Second, there is no approximation involved in arriving at the correct electron density through the KS method, once the functional is precisely known. The KS approach is in principle "exact" because the XC functional, through its definition in Eq. (1.16), accounts for the full quantum mechanical effect. There is nothing that has been dropped from the rigor of the HK theorems. Any imprecision, beyond computational errors, made in real calculations through the KS approach should be attributed to an imperfect knowledge of the functional rather than to the method itself. Along the same vein, a density that leads to a lower total energy is, by definition, closer to the ground-state density but only if the exact functional is used. The energies rendered by imperfect functionals have no real physical significance. With inaccuracies suspected in the functional used, DFT calculations cannot be used to set an "upper-bound" on the total energy, unlike the HF approximation.

1.3 Energy Levels of Isolated System

The total electronic energy of a finite electronic system, obtained through either the HF or the KS formalism, is a collective property of all the electrons and ions in the system. Since electrons are indistinguishable and interact with each other, it does not seem meaningful to speak of the energy of any individual electron in a multi-electron system. Technically, a single-electron energy operator that commutes with the total Hamiltonian does not exist. It is in the context of how the total energy of a multi-electronic system is changed when an electron is removed or added that such a single electron energy level can be defined. We will use a neutral molecule with N electrons in its ground state as an example. In this discussion, there is also a reservoir of electrons at infinity at our disposal. The least amount of energy required to remove an electron from this molecule and place it at rest at infinity is known as the ionization potential, I, of the molecule:

$$-I = E_0^N - E_0^{N-1}, \tag{1.16}$$

where E_0^N and E_0^{N-1} are the ground-state energies, respectively, of the neutral molecule with N electrons and the ionized molecule with $N-1$ electrons. Within the HF theory, these energies may be written out explicitly in terms of the electron orbitals, as in Eq. (1.9). One now assumes that the orbitals for the systems with N and $N-1$ electrons are identical, except for the missing of the orbital with the highest eigen energy for the N-electron system, ϕ_{HOMO}, for the ionized molecule. Taking the difference between the total energies for these two systems leads to

$$\begin{aligned} E_0^N - E_0^{N-1} \\ &= \int d\vec{r}\phi^*_{\text{HOMO}}(\vec{r})\left\{-\frac{\hbar^2}{2m}\nabla^2 - eV_{\text{ext}}(\vec{r}) + \frac{e^2}{8\pi\varepsilon_0}\int d\vec{r}'\frac{\rho_N(\vec{r}')}{|\vec{r}-\vec{r}'|}\right\}\phi_{\text{HOMO}}(\vec{r}) \\ &\quad - \frac{e^2}{4\pi\varepsilon_0}\int d\vec{r}\phi^*_{\text{HOMO}}(\vec{r})\sum_{j=1}^{N}\int d\vec{r}'\delta_{\sigma_{\text{HOMO}},\sigma_j}\frac{\phi_j^*(\vec{r}')\phi_{\text{HOMO}}(\vec{r}')}{|\vec{r}-\vec{r}'|}\phi_j(\vec{r}). \end{aligned} \tag{1.17}$$

A comparison with Eq. (1.9) above shows that the right-hand side of Eq. (1.17) is exactly the definition of the eigenvalue for the occupied orbital with the highest energy, namely,

$$-I = \varepsilon_{\text{HOMO}}. \tag{1.18}$$

This is Koopmans' theorem,[3] which is an exact result within the HF theory, provided that the orbitals of the remaining $N-1$ electrons do not relax when the highest lying electron is removed from the N-electron molecule. This assumption is unlikely as atomic positions and electronic distribution relax in general. The error introduced by the assumption of frozen electronic configuration in Eq. (1.19) is approximately offset by the correlation energy, which is the error associated with the assumption of a single Slater determinant form for the HF wave function, for most systems. Despite its possible errors, we shall assume that Koopmans' theorem is accurate enough to be used for our analysis of one-electron energy levels.

Within the HF theory, the definition of an unoccupied energy level is also straightforward. We can consider the retrieval of an electron in reverse. The maximum amount of energy that can be released by moving an electron from the reservoir at the vacuum level to the molecule is known as the electron affinity, A, of the molecule. This level is called the lowest unoccupied molecular orbital (LUMO) and is the difference between the ground-state energies of the neutral molecule with N electrons, E_0^N, and the negatively ionized molecule with $N+1$ electrons, E_0^{N+1}. Writing these energies out according to Eq. (1.9) and taking the difference between them allows the electron affinity to be identified as the eigenvalue of the LUMO of the system:

$$-A = \varepsilon_{\text{LUMO}} = E_0^{N+1} - E_0^N. \tag{1.19}$$

A more transparent way to grasp the idea of LUMO is to identify it with the HOMO of the ionized molecule with $N+1$ electrons.

The identification of the occupied and unoccupied one-electron energy levels of an N-electron system with the HF orbital energies has the advantage of having explicit expressions for these levels, although such an identification is understood to be only approximate numerically. One is naturally interested in whether there are similar features in the calculation based on the KS theory of the DFT, which is the mainstay of modern calculation of electronic structure, that may also be used to represent the one-electron energy levels. Strictly speaking, the KS orbitals are just a tool of the calculation, as they belong to a fictitious non-interacting system. Therefore, it would seem that the eigenvalues and orbitals of the individual KS equations should not have meaningful physical significance. Fortunately, this is not quite right. For a finite N-electron system, for which the effective potential and the electron density vanish at infinity, the decay of the electronic density is dominated by the KS orbital with the highest eigenvalues. At the same time, the decay of the electronic density is known to be dependent on the ionization energy of the system. Thus, it has been shown that the energy eigenvalue for the HOMO of the KS orbital is exactly the negative of the first ionization energy of the system. Therefore, the highest occupied one-electron energy level is the largest KS orbital energy:

$$-I_{\text{HOMO}}^{\text{op}} = \varepsilon_{\text{KS,HOMO}}, \tag{1.20}$$

where $\varepsilon_{\text{KS,HOMO}}$ is the highest eigenvalue.

Furthermore, when the correct functionals are used, this relationship is exact. As for other KS orbital energies and wave functions, there is no direct proof that they have physical significance. However, it has been argued that in fact they have great semiquantitative value, much like HF energies and wave functions, because they are consistent with the exact physical density.

They may be argued to have even more significance than the HF energies because the electron correlation may be properly treated in KS. A common practice in current research is that the KS orbital energies are being used widely to compare with experimental results on many physical phenomena in various fields. The identification of KS energies as the occupied one-electron energy levels of a system is very much in keeping with the spirit of this book of capturing the basic physics behind various concepts without being fixated on numerical accuracy. How an energy level of a system is affected by its chemical environment is straightforward to understand and handle through the equation for eigenvalues, while it would be mired in the expression for the total energy. Therefore, the discussions of energy level alignment in this book will be based on the eigenvalues of the systems.

1.4 "Host" and "Bath": Semiclassical Treatment

Our discussions of the total energy of an electronic system and its one-electron energy levels have thus far dealt with a system in isolation. Since we want to understand the alignment of different energy levels at material interfaces, the electronic systems to be analyzed are definitely not in isolation. To make the energy level concept relevant, it is necessary to put a system in an environment surrounded by other electronic systems. However, these systems should not interact or couple so strongly that the individuality of energy levels for sub-systems is lost. The overlap in the electron densities between two systems essentially determines the degree of interaction. Lightly coupled electronic systems without significant overlaps can be treated semiclassically and form the subject of the present discussion. The exchange interactions between different sub-systems are due to the cross terms, e.g. $\varphi^*_{j,\alpha}(\vec{r}\,')\varphi_{k,\beta}(\vec{r})$, from using the anti-symmetric Slater determinant for the wave functions of all electrons on different sub-systems. In the semiclassical approach, we assume that the overall system can be distinguished into P different sub-systems and that the numbers of electrons on the sub-systems are, respectively, n_1, n_2,..., and n_P, with $N = \sum n_i$. Each sub-system can be thought of as an atom, a molecule, or larger. Not all sub-systems need to be electrically neutral. Exchange interactions between electrons on different sub-systems can be eliminated by using a trial wave function of the form

$$\begin{aligned}\Psi_N &= \psi_1(\vec{x}_{1.1}, \vec{x}_{1.2}, \ldots, \vec{x}_{1,n_1}) \times \psi_2(\vec{x}_{2.1}, \vec{x}_{2.2}, \ldots, \vec{x}_{2,n_2}) \\ &\quad \times \cdots \times \psi_P(\vec{x}_{P.1}, \vec{x}_{P.2}, \ldots, \vec{x}_{P,n_P}), \end{aligned} \tag{1.21}$$

where ψ_i is the single Slater determinant wave function for n_i electrons on the ith sub-system, according to Eq. (1.5). When the trial wave function Eq. (1.21) is used to evaluate the expectation value of the energy of the entire system, one gets

$$E_N^{\text{el}}(\Psi_N) = \langle \Psi_N | H_N^{\text{el}} | \Psi_N \rangle = \sum_{i=1}^{P} \langle \psi_i^* | H_i^{\text{el}} | \psi_i \rangle + \sum_{i \neq j}^{P} \langle \psi_i^* \times \psi_j^* | V_{j,i} | \psi_i \times \psi_j \rangle, \tag{1.22}$$

where H_i^{el} is the electronic Hamiltonian for the isolated ith sub-system, i.e.

$$H_i^{\text{el}} = \frac{-\hbar^2}{2m} \sum_{\alpha=1}^{n_i} \nabla_{i,\alpha}^2 - \frac{e^2}{4\pi\varepsilon_0} \sum_{\beta=1}^{m_i} \sum_{\alpha=1}^{n_i} \frac{Z_{i,\beta}}{|\vec{r}_{i,\alpha} - \vec{R}_{i,\beta}|} + \frac{e^2}{4\pi\varepsilon_0} \sum_{\alpha<\gamma}^{n_i} \frac{1}{|\vec{r}_{i,\alpha} - \vec{r}_{i,\gamma}|}, \tag{1.23}$$

where m_i is the number of nuclei on the ith sub-system and $V_{j,i}$ is the Coulomb interaction between the electrons on the ith sub-system and all charges on the jth sub-system:

$$V_{j,i} = -\frac{e^2}{4\pi\varepsilon_0} \sum_{\beta=1}^{m_j} \sum_{\alpha=1}^{n_i} \frac{Z_{j,\beta}}{|\vec{r}_{i,\alpha} - \vec{R}_{j,\beta}|} + \frac{e^2}{8\pi\varepsilon_0} \sum_{\gamma=1}^{n_j} \sum_{\alpha=1}^{n_i} \frac{1}{|\vec{r}_{i,\alpha} - \vec{r}_{j,\gamma}|}. \tag{1.24}$$

We now examine how a specific individual energy level of sub-system #1 is affected by the presence of other sub-systems in this semiclassical approach. Setting the total energy of Eq. (1.22) stationary against variations in the complex conjugate of this specific orbital, $\varphi_{1,k}^*$ on the first sub-system, subject to the constraint that this orbital is normalized, one gets the equation

$$H_1^{\text{el}} \varphi_{1,k}(\vec{r}) + \left\{ \sum_{j=2}^{P} \frac{e^2}{4\pi\varepsilon_0} \sum_{\beta=1}^{m_j} \frac{-Z_{j,\beta}}{|\vec{r} - \vec{R}_{j,\beta}|} + \sum_{j=2}^{P} \frac{e^2}{4\pi\varepsilon_0} \right.$$
$$\left. \times \sum_{\alpha=1}^{n_j} \int d\vec{r}\,' \frac{\varphi_{j,\alpha}^*(\vec{r}\,')\varphi_{j,\alpha}(\vec{r}\,')}{|\vec{r} - \vec{r}\,'|} \right\} \varphi_{1,k}(\vec{r}) = \varepsilon_{1,k}\varphi_{1,k}(\vec{r}), \tag{1.25}$$

with which to solve for $\varphi_{1,k}$. It is important to note the absence of cross terms in the integrands of the above equation. Also, note that the terms in the wiggly parentheses represent the electrostatic potential energy of this electron with the density (and not the wave function) of negative and positive charges on all other sub-systems. Under the present assumption of large but finite separations between sub-systems, that potential has a weak spatial dependence over the volume of sub-system #1. Therefore, this is approximately a constant contribution to all electrons of the #1 sub-system. We call this the "empty (MT) potential" energy

$$-e\bar{V}_{\mathrm{MT}}^{\#1} = \frac{-e}{4\pi\varepsilon_0}\int d\vec{r}\,'\frac{Q_{\mathrm{MT}}^{\#1}(\vec{r}\,')}{|\vec{r}_1 - \vec{r}\,'|}, \tag{1.26}$$

where $\vec{r}_1$ is the position of the center of the first sub-system and $Q_{\mathrm{MT}}^{\#1}$ is the "empty" charge density:

$$Q_{\mathrm{MT}}^{\#1}(\vec{r}) = \sum_{j=2}^{P}\left\{\sum_{\beta=1}^{m_j} eZ_{j,\beta}\delta(\vec{r} - \vec{R}_{j,\beta}) - e\sum_{\alpha=1}^{n_j}\varphi_{j,\alpha}^{*}(\vec{r})\varphi_{j,\alpha}(\vec{r})\right\}. \tag{1.27}$$

This is the charge distribution of the entire system except those on sub-system #1. Since the energy level of interest belongs to sub-system #1, it will be known as the "host" of this energy level, while all other sub-systems will be known as the "bath" surrounding the host. The empty potential is the electrostatic potential due to the charges of the bath, found at the empty location of the host when the host has been removed. With this potential assumed to be constant, Eq. (1.25) becomes identical to the HF equation for electrons on the isolated #1 sub-system in Eq. (1.7), except for the constant MT potential. Unaffected by an additive constant in the equation, the wave functions remain the same as for the isolated host. However, HF eigenvalues of the host in the bath are shifted by the additional constant potential to

$$\varepsilon_{1,k} = \varepsilon_{1,k}^{\mathrm{iso.}} - e\bar{V}_{\mathrm{MT}}^{\#1}, \tag{1.28}$$

where $\varepsilon_{1,k}^{\mathrm{iso.}}$ is the energy for the kth orbital of the host in isolation. This equation is valid for all the eigenvalues of the first sub-system, as the empty potential rigidly shifts the HF equations for all electrons the same way. Due to the presence of the bath, it now takes an energy of $\varepsilon_{1,k}$ rather than $\varepsilon_{1,k}^{\mathrm{iso.}}$ to remove an electron from the system and place it at the vacuum level at infinity. The total electronic energy of the entire system can also be worked out as, approximately,

$$E_N^{\mathrm{el}}(\Psi_N) = \sum_{i=1}^{P} E_i^{\mathrm{el}} + \frac{1}{2}\sum_{i=1}^{P} Q_i\bar{V}_{\mathrm{bath}}^{i}, \tag{1.29}$$

where Q_i is the net charge of the ith sub-system and E_i^{el} is the total energy of the ith sub-system in isolation. We should mention that the present selection of a trial wave function of the form Eq. (1.21) is actually unnecessary and is adopted here mainly for an uncluttered derivation of equations. A formal approach starting with a fully anti-symmetrized Slater determinant of N orbitals would have led to approximately the same result, Eq. (1.29), because the exchange interactions between two orbitals on separate sub-systems are small and may be ignored anyways.

When a system consists of spatially separated and essentially independent sub-systems, an equation similar to Eq. (1.29) may also be derived with the KS theory. In DFT, the quantum mechanical interaction between electrons is entirely contained within the $V_{\mathrm{XC}}(\vec{r})$ term. If surrounding subsystems do not contribute to the $V_{\mathrm{XC}}(\vec{r})$ of the KS equations for orbitals on one sub-system, then the quantum mechanical interactions between subsystems are effectively cut off. Making the same argument as above for the HF method, it is straightforward to show that the KS orbitals for a particular sub-system are unchanged from those for the isolated system, and that the eigenvalues to the KS equations are rigidly shifted:

$$\varepsilon_{\mathrm{KS},q,k} = \varepsilon_{\mathrm{KS},q,k}^{\mathrm{host}} - e\bar{V}_{\mathrm{MT}}^{\#q}, \tag{1.30}$$

where $\varepsilon_{\mathrm{KS},q,k}^{\mathrm{host}}$ is an orbital energy for the qth sub-system in isolation and the empty potential energy is defined the same way as Eq. (1.26) above for HF calculations. Again, the orbital energy can be broken into a quantum mechanical part due only to its host and a Coulomb energy part due to the other charged entities of the bath. Since energy eigenvalues are known approximately as the one-electron energy levels, the host–bath concept introduced in this section paints a simple picture of how an energy level varies with its environment, which is the basis for the understanding of all energy level alignment problems in general.

1.4.1 *Host example: Isolated redox ion in solvent*

There is perhaps no better example to demonstrate the concept of host and bath than a discussion of the energy levels of two (charged) redox species in a solution, which, on its own, is a topic of much interest and importance in the field of electrochemistry. The liquid solvent here plays the role of the bath, which is in the shape of a bath! A small ion (red#1) is assumed to reside in a small spherical cavity of radius R_1 in a liquid solvent. We are interested in the HOMO of this donor ion, which is at an energy of $-I_{\mathrm{ion}\#1}^{\mathrm{iso}}$ with respect to the vacuum level when the ion is in isolation. The presence of molecules in the liquid naturally contributes to an empty potential in the cavity that modifies this energy level. As it depends on the instantaneous configuration and structure of the solvent molecules, this empty potential fluctuates as a function of time. In the classical Born model, the solvent is assumed to behave linearly as a polarizable continuum medium with a static dielectric constant of $\varepsilon_{\text{solv}}$.[4] The equilibrium, or time-averaged, energy level of the ion can be calculated under this assumption. At the edge of the cavity, the electric potential would have been $q_{\text{red}\#1}/(4\pi\varepsilon_o R_1)$ without the solvent and is now $q_{\text{red}\#1}/(4\pi\varepsilon_o\varepsilon_{\text{solv}} R_1)$ with the solvent. The difference between the two is the empty potential due to the polarization of the solvent. Additionally, a solution always provides an average MT potential (Volta potential), $\tilde{V}_{\text{MT}}^{\text{sol}}$, for all the molecules in the solution. Adding these two contributions together, the empty potential inside the cavity is

$$V_{q_{\text{red}\#1},\text{MT}}^{\text{equil.}} = -\frac{q_{\text{red}\#1}(1-\varepsilon_{\text{solv}}^{-1})}{4\pi\varepsilon_o R_1} + \tilde{V}_{\text{MT}}^{\text{sol}}. \tag{1.31}$$

The superscript "equil" is a reminder that this empty potential is for a solvent that behaves linearly, which is the most probable solvent configuration when averaged over time. Using Eq. (1.28), we can write the occupied one-electron energy level of the donor ion as

$$E_{q_{\text{red}\#1},\text{HOMO}}^{\text{equil.\_QM}} = -I_{q_{\text{red}\#1}}^{\text{iso}} - e\tilde{V}_{\text{MT}}^{\text{sol}} + \frac{eq_{\text{red}\#1}(1-\varepsilon_{\text{solv}}^{-1})}{4\pi\varepsilon_o R_1}, \tag{1.32a}$$

where the superscript "QM" indicates that this is the energy needed to remove an electron from the ion and put it at infinity in a virtual quantum mechanical process.

A bath made up of polar solvent molecules in this example provides an opportunity to discuss the dependence of the energy needed to remove an electron on the speed of its removal. If an electron is removed or added adiabatically ("ad"), with time for the solvent to react to the change of $+e$ in the net charge inside the cavity, the adjusted empty potential raises the energy level slightly, to

$$\begin{aligned} E_{q_{\text{ion}},\text{HOMO}}^{\text{equil.\_ad}} &= E_{q_{\text{ion}},\text{HOMO}}^{\text{equil.\_QM}} + \frac{e^2}{8\pi\varepsilon_o R_{\text{ion}}}(1-\varepsilon_{\text{solv}}^{-1}) \\ &= -I_{q_{\text{ion}}}^{\text{iso}} - e\tilde{V}_{\text{MT}}^{\text{sol}} + \frac{e(q_{\text{ion}}+e/2)(1-\varepsilon_{\text{solv}}^{-1})}{4\pi\varepsilon_o R_{\text{ion}}}. \end{aligned} \tag{1.32b}$$

To get the above result, we have integrated the work required to remove charge incrementally from the cavity, against a solvent configuration that is always optimized for the remaining charge in the cavity, up to a total removed charge of $-e$. A more realistic time scale for electron transfer processes to take place is the Franck–Condon condition ("FC"), under which the positions and orientations of the solvent molecules do not have time to adjust, but the electron density on the solvent molecules does.[5,6] Under such a circumstance, the relevant energy change is

$$E^{\text{equil.\_FC}}_{q_{\text{ion}},\text{HOMO}} = E^{\text{equil.\_QM}}_{q_{\text{ion}},\text{HOMO}} + \frac{e^2}{8\pi\varepsilon_o R_{\text{ion}}}(1-\varepsilon_\infty^{-1}), \tag{1.32c}$$

where ε_∞ is the high-frequency dielectric constant of the solvent. Note that the second term of Eq. (1.32c) describes the response of all electrons on the solvent to the movement of one electron and is thus independent of the charge of the ion in the cavity. This contribution of the solvent is unavoidable in an ET process.

Using procedures just described, the unoccupied energy level for an acceptor ion (ox#2) in a separate spherical cavity of radius R_2 in the solution can be written as

$$E^{\text{equil.\_ad}}_{q_{\text{ox\#2}},\text{LUMO}} = -A^{\text{iso}}_{q_{\text{ox\#2}}} - e\tilde{V}^{\text{sol}}_{\text{MT}} + \frac{e(q_{\text{ox\#2}} - e/2)(1-\varepsilon_{\text{solv}}^{-1})}{4\pi\varepsilon_o R_2}, \tag{1.33a}$$

for an electron to be added adiabatically. Under FC conditions, the unoccupied level is slightly raised

$$E^{\text{equil.\_FC}}_{q_{\text{ox\#2}},\text{LUMO}} = E^{\text{equil.\_ad}}_{q_{\text{ox\#2}},\text{LUMO}} + \frac{e^2(\varepsilon_\infty^{-1} - \varepsilon_{\text{solv}}^{-1})}{8\pi\varepsilon_o R_2}. \tag{1.33b}$$

A comparison of Eqs. (1.33a) and (1.32b) shows that the adiabatic energy level for the LUMO of the oxidized (acceptor) version of a redox couple is the same as that for the HOMO of the reduced (donor) ion. Therefore, the adiabatic energy level in the equilibrium solvent configuration is the redox potential with respect to the vacuum level. Note that the same cannot be said about the FC energy levels. An implicit assumption so far is that the ion is in a dilute solution with no other ions in its vicinity to contribute significantly to its empty potential.

It takes energy to create the "equilibrium" solvent polarization surrounding an isolated ion in a cavity. Starting from a neutral solvent without average local polarization, a cavity is created, a neutral host is brought into the cavity, and then charged, which induces polarization in the solvent. The total potential energy required to create this host is

$$G^{\text{equil.}}_{q_{\text{red\#1}}} = E_{\text{cavity}} + E^{\text{internal}}_{\text{red\#1}} + q_{\text{red1}}\tilde{V}^{\text{sol}}_{\text{MT}} - \frac{(1-\varepsilon_{\text{solv}}^{-1})}{8\pi\varepsilon_o}\frac{q^2_{\text{red\#1}}}{R_1}, \tag{1.34}$$

where E_{cavity} and $E^{\text{internal}}_{\text{red\#1}}$ are, respectively, the minimum amount of energy required to create a cavity in the solvent and the internal energy of the host. An intriguing aspect of an ion in a solvent is that the energy level of this ion, with respect to vacuum level, can be shifted up or down by thermal fluctuations of the solvent. Even though the time-averaged results are predictable by the Born model, it is of interest to know how frequently we can expect the FC energy level of ion red#1 to fluctuate to a value δE_1 above or below the equilibrium level given in Eq. (1.32c). In other words, what is the free energy ΔG_1 needed to change the energy level by δE_1? This is of great importance to the rates of ET reaction. Obviously, such an energy level shift requires the empty potential in the cavity, for a short period of time, to shift by $\delta E_1/e$. This is caused by a non-optimized solvent configuration that happens to exist at that moment. From Eq. (1.31), the equilibrium empty potential varies with the charge on the host. Therefore, to decrease (increase) the empty potential, we need a solvent arrangement that ordinarily surrounds an ion with a more positive (negative) charge than the actual charge on the host. The energy needed to create a solvent field that is not optimized for the charge in the cavity may be obtained in two steps: (i) Calculate the energy needed from a solvent optimized for a fictitious ion with a charge of $q_{\text{red}\#1} + \delta q_1$. (ii) Remove δq_1 from the ion under the Franck–Condon condition to recover the actual ionic charge (in a non-optimized solvent configuration).

According to Eq. (1.34), the electrostatic energy needed to create a solvent configuration optimized for $q_{\text{red}\#1} + \delta q_1$ is

$$G^{\text{equil.}}_{q_{\text{red}\#1}+\delta q_1} = E_{\text{cavity}} + E^{\text{internal}}_{\text{red}\#1} + q_{\text{red}\#1}\tilde{V}^{\text{sol}}_{\text{MT}} - \frac{(1-\varepsilon^{-1}_{\text{solv}})}{8\pi\varepsilon_o}\frac{(q_{\text{red}\#1}+\delta q_1)^2}{R_1}. \tag{1.35}$$

For step (ii), the removal of charge in the cavity incrementally from $q_{\text{red}\#1} + \delta q_1$ to $q_{\text{red}\#1}$ requires an energy of

$$\int_0^{\delta q_1} dq' \frac{(1-\varepsilon^{-1}_{\text{solv}})(q_{\text{red}\#1}+\delta q_1) - (1-\varepsilon^{-1}_{\infty})q'}{4\pi\varepsilon_o R_1}$$
$$= \frac{(1-\varepsilon^{-1}_{\text{solv}})2q_{\text{red}\#1}\delta q_1 - (2\varepsilon^{-1}_{\text{solv}} - \varepsilon^{-1}_{\infty} - 1)\delta q_1^2}{8\pi\varepsilon_o R_1}. \tag{1.36}$$

The energy level of the ion with the non-optimized solvent configuration is shifted from that for the optimized configuration by an amount equal to the change in the empty potential:

$$\delta E_1 = -\frac{e\delta q_1(\varepsilon^{-1}_{\infty} - \varepsilon^{-1}_{\text{solv}})}{4\pi\varepsilon_o R_1}. \tag{1.37}$$

The total energy required to create such a non-optimized solvent configuration for an ion with a charge of $q_{\text{red}\#1}$ is the sum of Eqs. (1.35) and (1.36).

Making use of Eq. (1.37), the potential energy required to create a host with its energy level shifted by δE_1 from equilibrium, $G_{q_{\text{ion}}}(\delta E_1)$, is

$$G_{q_{\text{red}\#1}}(\delta E_1) = G^{\text{equil.}}_{q_{\text{red}\#1}} + \frac{2\pi\varepsilon_o R_1(\delta E_1)^2}{e^2(\varepsilon_\infty^{-1} - \varepsilon_{\text{solv}}^{-1})}. \tag{1.38}$$

It costs energy to shift the energy level either up or down and, furthermore, the energy cost is quadratic in the amount of energy level change. In other words, moving the energy level by an amount from equilibrium, either up or down, requires the same potential energy! This means that over time, it is equally as probable to observe the energy level at $E^{\text{equil.\_}FC}_{q_{\text{red}\#1},\text{HOMO}} + \delta E_1$ as it is to find it at $E^{\text{equil.\_FC}}_{q_{\text{red}\#1},\text{HOMO}} - \delta E_1$. As expected, the smaller the cavity radius (R_1) and the more polarizable the solvent ($\varepsilon_{\text{solv}}^{-1} \ll 1$), the easier it is to shift the energy level.

1.4.2 *Example: Two redox reactants in solution*

As we are interested in the alignment of energy levels, we need more than one energy level in the solvent on spatially separated hosts. How these energy levels are aligned is important for the dynamic processes of the system. To discuss these issues, the donor and acceptor thus far treated in isolation are now considered together. From Eq. (1.34), the amount of energy needed to create the two reactants in optimized solvent configuration is

$$G^{\text{equil.}}_{\text{reac.}} = G^{\text{equil.}}_{q_{\text{red}\#1}} + G^{\text{equil.}}_{q_{\text{ox}\#2}} + \frac{q_{\text{red}\#1} q_{\text{ox}\#2}}{4\pi\varepsilon_o\varepsilon_{\text{solv}} r_{12}}, \tag{1.39}$$

where r_{12} is the distance between the two ions. We now consider the energy required to create the optimum solvent configuration for the "products", $G^{\text{equil.}}_{\text{prod.}}$, when charges of the ions are $q_{\text{ox}\#1}$ ($= q_{\text{red}\#1} + e$) and $q_{\text{red}\#2} = q_{\text{ox}\#2} - e$, respectively. Of course, this only requires changes in some subscripts in Eqs. (1.32)–(1.39). Writing both energies out in full shows they differ by the difference in the redox potentials of the two species:

$$\Delta G = G^{\text{equil.}}_{\text{prod.}} - G^{\text{equil.}}_{\text{reac.}} = E^{\text{equil.\_ad}}_{q_{\text{ox}\#2},\text{LUMO}} - E^{\text{equil.\_ad}}_{q_{\text{red}\#1},\text{HOMO}}. \tag{1.40}$$

In writing Eq. (1.40), we have assumed that the oxidized forms of the two reactants have the same charge $q_{\text{ox}\#1} = q_{\text{ox}\#2}$ to keep the equations simple without missing important physics of the problem. The energy gain $-\Delta G$ is commonly viewed as the thermodynamic driving force for the reaction to proceed.

Even though the total potential energy required for the ET process is ΔG under equilibrium solvent configuration, this is not the difference in the energy levels of the two reactant ions. The energy levels derived for the two ions in isolation, Eqs. (1.32) and (1.33), need to be modified because the presence of red#1 nearby contributes to the MT potential of ox#2 and *vice versa*. Under optimized solvent configuration, the LUMO level of ox#2 is higher than the HOMO level of red#1 by

$$\Delta E^{\text{equil\_FC}}_{\text{reac\_2}-1} = \Delta G + \lambda_o. \tag{1.41}$$

A "reorganization energy", λ_o, of the form

$$\lambda_o = \frac{e^2(\varepsilon_\infty^{-1} - \varepsilon_{\text{solv}}^{-1})}{8\pi\varepsilon_o}(R_1^{-1} + R_2^{-1} - 2r_{12}^{-1}) \tag{1.42}$$

is required to move an electron between the two cavities at a distance r_{12} apart under FC conditions. This is a positive energy that is always required, regardless of the direction of the electron transfer.

Since the energy level difference is crucial for the rates of possible ET reactions, it is important to understand the free energy (potential energy) required to decrease (or increase) the energy level difference of the reactants by a certain amount. As shown above for the energy level of the isolated host, the energy level difference of the reactants, expressed in Eq. (1.41) under equilibrium, can also be adjusted by spontaneous solvent configuration. Obviously, the most efficient route to change the difference between two levels is to move one level down while shifting the other up. The minimum potential energy required to shift the energy level difference is calculated by first writing down the energy required to create equilibrium solvent configurations for fictitious charges of $q_{\text{red}\#1} + \delta q$ and $q_{\text{ox}\#2} - \delta q$ in the two cavities, before removing δq and $-\delta q$ under the Franck–Condon condition. Using procedures described earlier, the required energy is

$$G^{\delta q}_{\text{reac.}} = G^{\text{equil}}_{\text{reac}} + \lambda_o\left(\frac{\delta q}{e}\right)^2. \tag{1.43}$$

Under the non-optimized solvent configuration created by the use of δq, the difference in FC energy levels changes by

$$\delta\Delta E_{2-1} = \Delta E^{\delta q\_\text{FC}}_{\text{reac\_2}-1} - \Delta E^{\text{equil\_FC}}_{\text{reac\_2}-1} = -2\lambda_o(\delta q/e). \tag{1.44}$$

The additional potential energy required again depends quadratically on the shift in level difference from equilibrium, after plugging Eq. (1.44) into

Eq. (1.43):

$$G_{\text{reac.}}(\delta\Delta E_{2-1}) = G_{\text{reac}}^{\text{equil}} + \frac{(\delta\Delta E_{2-1})^2}{4\lambda_o}. \tag{1.45}$$

This equation shows that the magnitude of λ_o controls how easily the energy level difference can be modified by the solvent configuration. As expected, the larger the dielectric constant of the solvent $\varepsilon_{\text{solv}}^{-1} \ll 1$ and the smaller the cavity size of redox species (R_1, R_2), the easier it is for solvents to change the LUMO level with respect to the HOMO. A schematic demonstration of how empty potentials in spherical cavities in a solvent may be modified by spontaneous solvent configuration is shown in Fig. 1.1. The quadratic dependence of the potential energy, which is a straightforward result from the semiclassical host–bath approach, is important, as it is both an assumption and a foundation of the Marcus–Hush theory of ET.[7]

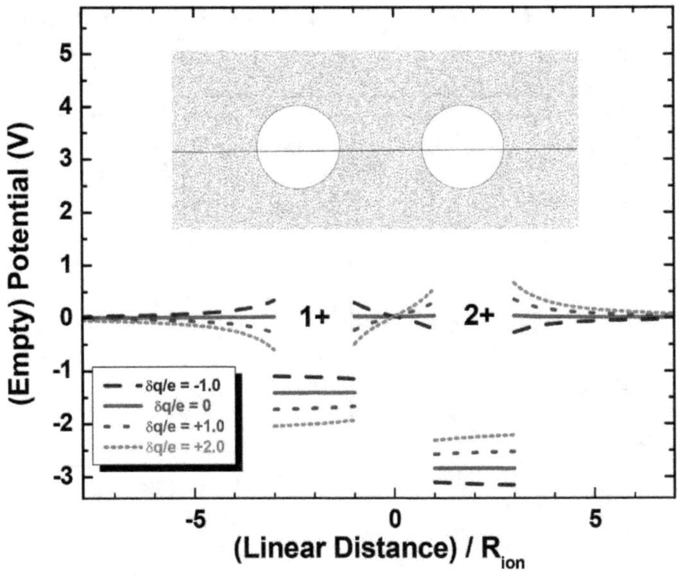

Fig. 1.1. Electric potential profile along a straight line connecting the cavities for a donor (1+) and an acceptor (2+) in a solvent, under different asymmetric shifts (δq). The full potential is shown for the solvent, whereas the empty potential is shown inside the cavities. In this example, the inter-ionic distance r_{12} is set at 4 times the cavity radius, $R_{\text{ion}} = 1\,\text{nm}$. The static and high-frequency dielectric constants of the solvent are assumed to be 80 and 3.5, respectively. These properties translate to a λ_0 of $\sim$0.30 eV, which can be seen to be approximately the shift, in opposite directions, in the empty potential of the cavities per $\delta q/e$. Note that the potential is small in the solvent at equilibrium ($\delta q/e = 0$) and yet the shifts in empty potential can be significant.

With the energies for the reactants known, those for the products can also be written down with only a $1 \leftrightarrow 2$ switch in some subscripts of the equations above. With λ_o remaining unchanged, the extra potential energy required to shift the energy level difference also remains the same:

$$G_{\text{prod.}}(\delta\Delta E_{2-1}) = G_{\text{prod.}}^{\text{equil}} + \frac{(\delta\Delta E_{2-1})^2}{4\lambda_o} = G_{\text{reac.}}^{\text{equil}} + \Delta G + \frac{(\delta\Delta E_{2-1})^2}{4\lambda_o}. \tag{1.46}$$

Since the electron is on ion #2 for the products, the relevant energy level difference is between the LUMO of ion #1 and the HOMO of ion #2:

$$\Delta E_{\text{prod\_1-2}}^{\text{equil\_FC}} = -\Delta G + \lambda_o. \tag{1.47}$$

Prominently featured in the expressions for the energies of both the reactants and the products is the reorganization energy λ_o, which appears as a natural energy "unit" for a redox couple. When both the energy level difference and the potential energy are normalized by this unit, their relationship becomes universal. In Fig. 1.2, the G for reactants, Eq. (1.45),

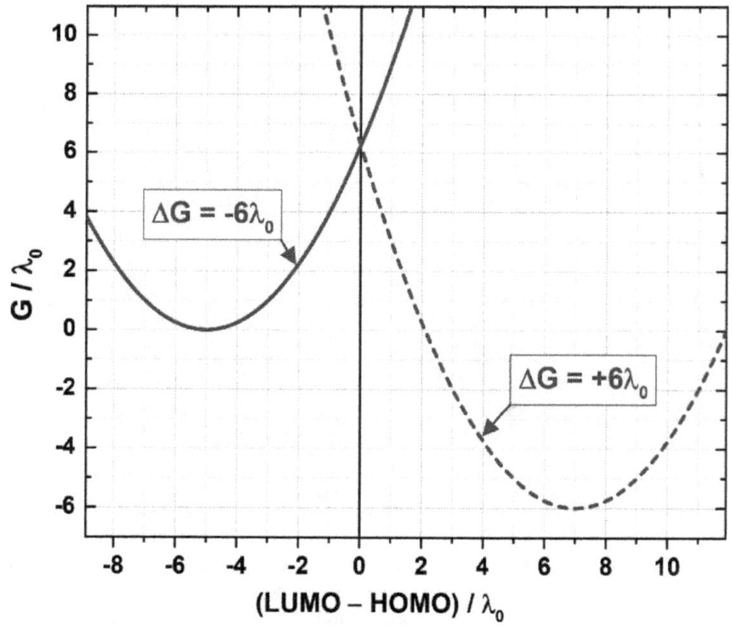

Fig. 1.2. The sum of internal energies of the hosts and the potential energy of the solvent, G, is shown against the difference in FC energy levels for the reactants (solid lines) and the products (broken lines), with $\Delta G = -6\lambda_0$.

is plotted as the solid curve against the energy level difference with an assumed $\Delta G = -6\lambda_o$. The equilibrium solvent configuration appears on the unitless horizontal axis for this exothermic reaction at $-5 = (\Delta G/\lambda_o) + 1$, which means that solvent configuration only needs to move $5\lambda_o$ to make the energy level difference disappear. An ET reaction should be most efficient in the absence of a difference in donor and acceptor levels. For the endothermic reaction in the reverse direction shown as the dashed curve, the equilibrium point is at +7 on the horizontal axis and −6 on the vertical axis, according to Eqs. (1.46) and (1.47). It is harder for the solvent to cancel the energy level difference ($7\lambda_o$) when the reaction is endothermic. Note that the relative position between the equilibrium of the two curves is fixed by ΔG and always gives a kinetic advantage to exothermic over endothermic reactions.

Before finishing the discussion on how energy levels may be shifted, we should mention that the two hosts in the solvent may be close enough to fit inside a single cavity. For a simple geometry, such as a spherical cavity of radius R with two ions positioned r_{12} apart at the center of the cavity ($r_{12} < 2R$), analytic equations for the energy levels and the required free energy can be written down for the assumed charges on the ions. Although the energy levels and the free energies have somewhat different forms and dependences on the interionic distance and radius of the cavity, the main results derived for two cavities in Eqs. (1.41)–(1.45) remain unchanged, except for the replacement of λ_o in these equations with an expression appropriate for the single cavity (1C):

$$\lambda_{1C} = \frac{e^2 r_{12}^2}{4\pi\varepsilon_0 R^3} \frac{(2\varepsilon_{\text{solv}} - 2)}{(2\varepsilon_{\text{solv}} + 1)}. \tag{1.48}$$

Since the overall dependence of potential energy on energy level shift remains unchanged, all inferences that can be drawn from Fig. 1.2 remain valid for a single cavity.

1.5 Summary of Semiclassical Approach

We use a system that consists of several molecules as an example to explain the host–bath concept. In this semiclassical approach, all quantum mechanical interactions are fully kept for the molecules separately, as shown on the top panels of Fig. 1.3, resulting in the energy levels on the right. Between each host and the bath, quantum mechanical interaction is neglected, although classical electrostatic interaction is always present.

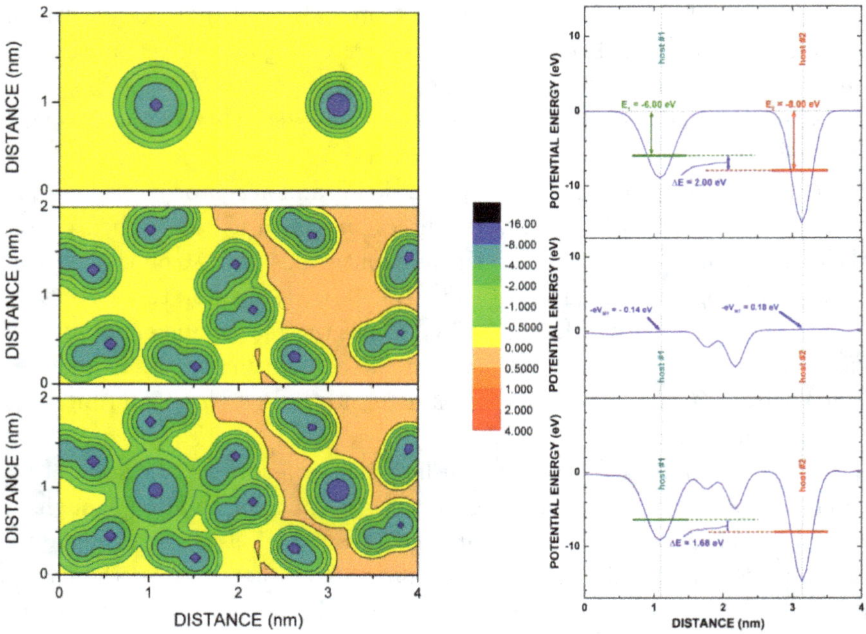

Fig. 1.3. Two-dimensional electric potential energy plots of molecules are displayed on the left, while the potential energy on the horizontal line through the middle of each panel on left is plotted on the right. The top panels show two molecules, called "hosts", approximately in isolation, whose energy levels shown on the right are of interest to us. In the middle panels, various molecules known as the "bath" are arranged. In each empty space between the molecules, the electric potential energy is approximately constant, and these "empty potential energy" ($-eV_{\mathrm{MT}}$) values are marked on the right middle panel. At the bottom, the hosts are brought in with the bath to create a possible configuration encountered in an actual study. The difference in the energy levels of the two hosts in the bath has been modified from that observed in isolation to $(-6.00 - 0.14) - (-8.00 + 0.18) = 1.68\,\mathrm{eV}$ due to shifts from empty potential energies.

Without quantum mechanical interactions, the energy level of each host is rigidly shifted by the electric potential of the bath, called the "empty potential", indicated as $-eV_{\mathrm{MT}}$ at the location of each host in Fig. 1.3. Therefore, the alignment of the energy levels on the hosts is a result of and is controlled by the bath, as shown in the bottom panels of Fig. 1.3. Note that the effect the bath can exert on the alignment of energy levels is entirely electrostatic in nature, which means that this influence has nothing to do with the internal electronic structures, such as energy levels, of the bath molecules but only their charge distributions. The alignment condition of the energy levels on the two hosts can shift from its fixed quantum mechanical contribution (see top panel) by the distribution of charge of the bath molecules. The simple message of this chapter is that energy level alignment is controlled by charge distribution.

References

1. P. Hohenberg and W. Kohn, *Phys. Rev.* **136**, B864–B871 (1964).
2. W. Kohn and L.J. Sham, *Phys. Rev.* **140**, A1133–A1138 (1965).
3. T. Koopmans, *Physica B & C* **1**, 104–113 (1934).
4. M. Born, *Z. Physik* **1**, 45 (1920).
5. J. Franck, *Trans. Faraday Soc.* **21**, 536 (1926).
6. E. Condon, *Phys. Rev.* **27**, 640 (1926).
7. R.A. Marcus, *J. Chem. Phys.* **24**, 979–989 (1956).

Chapter 2

ENERGY BANDS AND BULK REFERENCES OF CRYSTALLINE SOLIDS

The alignment of energy bands at solid interfaces is of primary interest in this book. For small molecules in Chapter 1, the alignment of energy levels has been shown to be conveniently analyzed through the host–bath approach. For crystalline solids, the energy bands are their energy levels. Therefore, the alignment of energy bands at a solid interface can also be understood and analyzed through the same approach. A slight complication for the solid interfaces arises from the need to identify the hosts to represent the two crystals. While the definition of the host is unmistakable for small molecules before, there is some ambiguity and freedom in choosing the host for a solid. The formation of energy bands for a crystalline solid and available choices to select "model solid" as the host of the energy bands are discussed in this chapter. The extra step taken to employ a model solid allows the energy bands of a crystal to be referenced to the vacuum level, similar to the energy levels of an isolated molecule in Chapter 1. With the energy band positions of both crystals of an interface known with respect to the vacuum level, how they are aligned can be analyzed the same way as for the molecules. As is the case for molecules, the bath, which is the charge distribution of the interface after the removal of the two hosts, plays a crucial role in deciding the alignment condition.

2.1 Energy Bands of Crystalline Solids

In principle, all properties of an electronic system can be extracted from its wave function, which may be obtained through the Schrödinger equation. For a finite system such as a molecule, this process is straightforward, as described in Chapter 1. The same approach becomes impractical for solids, as the number of atoms in a solid is essentially infinite. Fortunately, the equilibrium structure of a solid needs to minimize its energy, often resulting in crystalline structures for solids in nature. Periodicity considerably simplifies the task of studying and describing the electronic structure of a solid. Since the alignment of energy levels at solid interfaces is the main topic of this book and the majority of materials of practical importance are crystalline solids, energy levels of crystalline solids naturally become our main concern. In this chapter, we pave the way for later discussions on the alignment of energy levels at solid interfaces by going over the formation of energy bands in crystalline solids and the strategies to identify hosts to represent crystalline solids in interface analysis.

The electronic structure of crystalline solids is a familiar topic in solid-state physics.[1] Therefore, in our discussions, some of the well-known results are stated without elaboration. Starting from the known atomic structure of a solid crystal, three primitive vectors $\vec{a}_1$, $\vec{a}_2$, and $\vec{a}_3$ can be defined such that any translational vector constructed from them, i.e. $\vec{R} = n_1\vec{a}_1 + n_2\vec{a}_2 + n_3\vec{a}_3$, brings an infinite crystal back to itself. The collection of all these lattice vectors forms the (Bravais) lattice of the crystal. A real crystal with its periodic arrangement of atoms may be regarded as the result of adding a unit cell of atom(s) to every lattice site. The three primitive vectors span a parallelepiped with the volume of the unit cell $V_{\text{cell}} = (\vec{a}_1 \times \vec{a}_2) \cdot \vec{a}_3$. A unit cell may also be defined with the full symmetry of its lattice by taking on the form of a "Wigner–Seitz cell" (WSC) for the lattice. A WSC centered about a lattice point is defined by including all points of the crystal that are closer to this lattice point than to any other lattice point. Even though the primitive unit cell is the smallest repeating unit, it is sometimes more convenient to adopt a larger "conventional unit cell" for easier visualization and description. Most notably, the face-centered cubic (f.c.c.) and the body-centered cubic (b.c.c.) lattices are defined by the size of the cube, which has four and two times the volume of the respective primitive unit cell. Another structure commonly found for crystals of common metals and semiconductors is the hexagonal close-packed (h.c.p.) structure, with two inequivalent atoms in a unit cell. Selected unit cells of crystalline solids commonly encountered in band alignment problems for interfaces are drawn in Fig. 2.1.

Due to the periodic arrangement of atoms, some important properties of a crystalline solid, such as the charge density and the electric potential, are also periodic. Fourier transformation can significantly simplify the description of periodic functions. A Fourier transformation of the Bravais lattice would lead to non-vanishing values only for discrete wave vectors, $\vec{K}$.

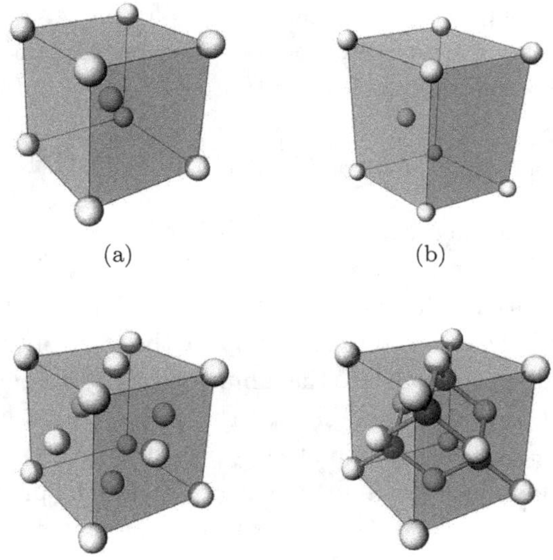

Fig. 2.1. Unit cell models of common crystalline solids. (a) body-centered cubic, (b) hexagonal close-pack, (c) face-centered cubic, and (d) cubic zincblende structure.

These vectors form a three-dimensional periodic lattice in the "reciprocal space" that may be generated by three primitive vectors

$$\vec{b}_1 = 2\pi \frac{\vec{a}_2 \times \vec{a}_3}{(\vec{a}_1 \times \vec{a}_2) \cdot \vec{a}_3}, \quad \text{etc.} \tag{2.1}$$

such that all reciprocal vectors may be reached by $\vec{K} = n_1\vec{b}_1 + n_2\vec{b}_2 + n_3\vec{b}_3$. These properties show that the reciprocal lattice of a Bravais lattice is itself a Bravais lattice. Conventionally, a WSC centered about the origin (Γ) of the reciprocal lattice is known as the first Brillouin zone of this crystal, with a volume of $\Omega = (\vec{b}_1 \times \vec{b}_2) \cdot \vec{b}_3$ $(=8\pi^3/V_{\text{cell}})$.

To calculate the electron levels for a solid crystal, we proceed with the Kohn–Sham approach (Eq. 1.16) from a Hamiltonian written for all the electrons of the crystal:

$$\left\{ -\frac{\hbar^2}{2m}\nabla^2 - eV_{\text{ext}}(\vec{r}) - eV_{XC}(\vec{r}) + \frac{e^2}{8\pi\varepsilon_0}\int d\vec{r'}\,\frac{\rho(\vec{r'})}{|\vec{r} - \vec{r'}|} \right\}$$
$$\times\, \phi_{KS,i}(\vec{r}) = \varepsilon_{KS,i}\phi_{KS,i}(\vec{r}), \tag{2.2}$$

where the Born–Oppenheimer approximation has been applied. In Eq. (2.2), the electric potential energies due to all the positive ions, $-eV_{\text{ext}}(\vec{r})$, and all the electrons are periodic. Although the exact form of the $-eV_{XC}(\vec{r})$ term is not known, it is expected to be periodic because the electron density is periodic. The KS equation may be written as

$$\left\{-\frac{\hbar^2}{2m}\nabla^2 - eV_{\text{eff}}(\vec{r})\right\}\phi_{n\vec{k}}(\vec{r}) = \varepsilon_{n\vec{k}}\phi_{n\vec{k}}(\vec{r}), \tag{2.3}$$

where $V_{\text{eff}}(\vec{r})$ is an effective potential with the same periodicity as the crystal lattice. Since the charge density of the crystal is not available without a calculation, the wave functions here have to be sought self-consistently. We have changed the subscript of the electron state to indicate explicitly that this is a solution to the infinite crystal. Ordinarily, the spatial dependence of this wave function over the entire crystal is needed as a solution. However, when the potential is periodic, Bloch's theorem states that the wave function solution may be expressed as the product of a plane wave and a function, $u_{n\vec{k}}(\vec{r})$, that has the same periodicity as the Bravais lattice of the crystal:

$$\phi_{n\vec{k}}(\vec{r}) = e^{i\vec{k}\cdot\vec{r}}u_{n\vec{k}}(\vec{r}). \tag{2.4}$$

Due to its periodicity, the function $u_{n\vec{k}}(\vec{r})$ only needs to be solved within the volume of a unit cell, which greatly simplifies the problem. A consequence of the periodicity is that the energy and the wave function of each state are now associated with a specific wave vector, written as one of the subscripts. The other subscript n is a band index to distinguish different states with the same wave vector. Since it is not meaningful to distinguish two wave vectors differing by a reciprocal lattice vector, the wave vector for any wave function can be confined to within the first Brillouin zone.

For a finite system, the number of occupied electronic states is fixed and so is the number of equations that need to be solved. For any crystal, that number is infinite. Yet, the task of calculating the electron states may be kept manageable by imposing the Born–von Karman periodic boundary condition (PBC). The full wave functions including their phase factors, Eq. (2.5), are now assumed to repeat themselves in all three dimensions over multiple times (N) the size of the original unit cell. In order for the wave functions not to diverge in any direction, all wave vectors for the PBC must be real. This requirement limits the number of distinct $\vec{k}$ vectors in a Brillouin zone to be N. The accuracy of the calculations depends on the size of the periodic boundaries. Obviously, the larger the N is, the more accurate the calculation is, although the complexity and the time required for the calculation also increase. The calculation of the electronic states individually via Eq. (2.3) is straightforward with appropriate functionals.

The wave function of a particular state is known once $u_{n\vec{k}}(\vec{r})$ is solved within a unit cell. Since this quantum mechanical problem is finite in all dimensions, many discrete eigenvalues and eigenfunctions can be solutions for the same $\vec{k}$ vector. These are distinguished by their band index. The word "band" is used because the dependence of the electron energy on $\vec{k}$ is continuous in reciprocal space. States with different $\vec{k}$'s in a Brillouin zone can be connected into separate energy bands. Energy dispersion of electronic states in the reciprocal space determines transport properties of the crystal, such as the group velocity and the effective mass of the electron (or hole). Each band has the same number of states as there are different $\vec{k}$'s in a Brillouin zone, which is N. Therefore, every filled band contributes two electrons (of different spins) to each unit cell of the crystal. An energy band contains states within a finite range of energy, and the ranges of different bands may overlap on the energy scale. The ground state of a crystal at low temperatures is the result of filling the states, beginning at the bottom, until the total number of electrons of the system is reached. The crystal is metallic if the highest lying band(s) is partially filled. The highest occupied state of a metal is the Fermi level (FL), above which all states are empty and below which all states are filled, at low temperatures.

An insulator/semiconductor has every energy band either completely filled or completely empty. The HOMO of an insulator is the valence band maximum (VBM), the LUMO the conduction band minimum (CBM), and the difference between these two is the fundamental band gap of the insulator or semiconductor, as shown in Fig. 2.2. With the inclusion of impurity states, the electrochemical potential for a doped semiconductor is determined by Fermi–Dirac statistics to be close to the CBM and the VBM, respectively, for n-type and p-type semiconductors. The exact location of the electrochemical potential with respect to the CBM for n-type semiconductor, or the VBM for p-type semiconductor, depends on the doping level, the temperature, and the density of states of the conduction and the valence bands. The density of states of the conduction band varies with energy due to energy dispersion, as shown in Fig. 2.2. However, for the purpose of calculating the electrochemical potential, it can be lumped into a single energy level at the CBM, with an "effective density of states" of

$$N_C = 2\left(\frac{2\pi m^* k_B T}{h^2}\right)^{3/2}, \tag{2.5}$$

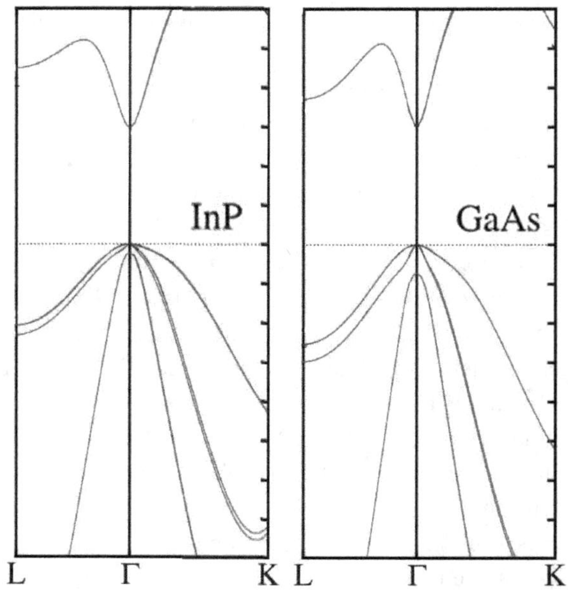

Fig. 2.2. Energy levels along major symmetry axes of the Brillouin zone, as a function of energy, calculated for (a) InP and (b) GaAs semiconductors.

Source: Reprinted with permission from Kim *et al.*[2] Copyright 2010 by the American Physical Society.

where m^* is the effective mass of electrons in the conduction band. The effective density of the valence band, N_V, can be similarly written. Using the effective density of the conduction band, charge neutrality requires the electrochemical potential for bulk semiconductor with a density N_D of n-type dopants to be at an energy of

$$eV_N = k_B T \ln(N_D/N_C), \tag{2.6}$$

with respect to the CBM. For small donor densities, the electrochemical potential is in the band gap, while it moves inside the conduction band ($eV_N < 0$) for a degenerate semiconductor with high doping level. In deriving the above result, the energy level for the donor in the band gap has been assumed to be so close to the CBM that the difference between the two can be ignored. This is a good approximation for semiconductors lightly to moderately doped with "shallow" dopants.

An important aspect of the formation of energy bands from the periodic potential of an infinite solid is the fixed relationship between the energy levels and $-e\bar{V}_{\text{cryst}}$, the average electrostatic potential energy of the solid.

Since the potential is periodic with the lattice structure, it can always be Fourier transformed into components with reciprocal lattice wave vectors. Upon averaging over the entire unit cell, no Fourier components make a non-zero contribution to the average potential energy of the crystal, except that associated with $\vec{K} = 0$. This Fourier component is arbitrary and cannot be defined for an infinite crystal. Nevertheless, the relative position of any specific energy level of the crystal is definitive and fixed with respect to this average electrostatic energy. For an insulator, the difference between the VBM and the average potential energy is a fixed energy known as the internal chemical potential, μ_e^{int}:

$$E_{VBM} = \mu_e^{\text{int}} - e\bar{V}_{\text{cryst}}. \tag{2.7}$$

Being higher than the VBM by the band gap energy, E_g, the CBM is also fixed with respect to the average electrostatic energy of the solid:

$$E_{CBM} = E_g + \mu_e^{\text{int}} - e\bar{V}_{\text{cryst}}. \tag{2.7a}$$

It is through these fixed relations that the band position of a solid may be compared to that of another solid, with the difference in the average electrostatic energies of the two solids providing the connection. Examples of how the VBM position is referenced to the average electrostatic potential energy can be found in Table 2.5 for common zincblende semiconductors. For a metal, the difference between the Fermi level and the average electrostatic potential energy, μ_e^{int}, is also a fixed quantity of the crystal. Through the relationship

$$E_F = \mu_e^{\text{int}} - e\bar{V}_{\text{cryst}}, \tag{2.8}$$

the FL of the metal can be compared with the energy levels of other crystals.

2.2 Host of Energy Bands of Solid Crystal: Model Solids

The alignment of energy levels at a solid interface can take the form of the Schottky barrier height between a metal and a semiconductor or the offset in the valence or conduction band between two semiconductors. In order to perform an analysis for solid interfaces, the hosts to represent solid need to be identified. Since energy bands are derived from the atomic structure of a crystalline solid of infinite size, the host for energy bands is technically the infinite crystal. However, no real crystals are infinite and the crystalline solids forming an interface are definitely finite in size. A charge distribution that can "fit" next to an interface and still adequately represent an infinite

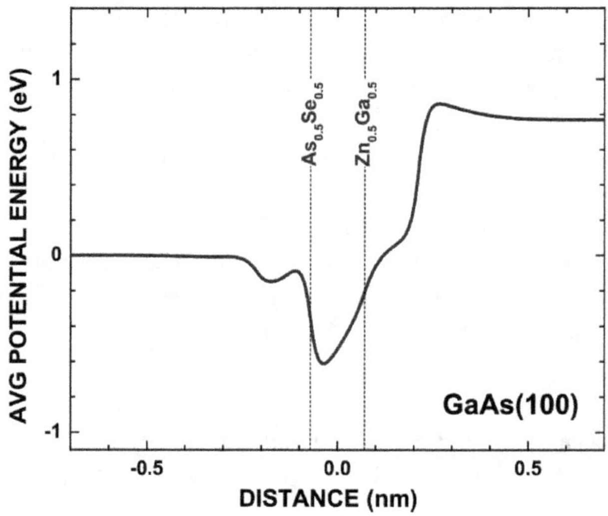

Fig. 2.3. Macroscopically averaged potential energy of a GaAs crystal near the location where one-half monolayer of ZnSe has been embedded. Band edges of the GaAs are also rigidly shifted by ~0.8 eV by this embedded dipole layer.

crystal should serve as the host of the energy bands. Consider the example of a GaAs single crystal artificially embedded with one-half monolayer of ZnSe. The average electrostatic potential energy drawn in Fig. 2.3 is drastically different to the left and right of this dipole layer, emphasizing that it is the energy band at specific location(s) that is relevant. Naturally, the hosts for energy bands have to be local. Here, we describe strategies to choose a representative of a crystalline solid as the local host of the energy bands. There is no one-size-fits-all choice for the host for all crystalline solids. The optimal method to define a host may well depend on the type of chemical bonds that bind the crystal together.

We shall first address a question that often arises at this point of the discussion, which is "why can't we use the infinite crystal itself as the host, especially since the energy bands are properties of an infinite crystal?" For example, it seems that we can assume that at a location of interest next to an interface, we can map the energy bands of the infinite crystal to the local potential. An infinite crystal cannot be used as the host for energy levels in this fashion because the average electrostatic potential of an infinite crystal cannot be defined with respect to the vacuum level. The main problem is that, as the crystal expands indefinitely, the surface contribution to the crystal's average potential energy does not vanish but remains constant. Furthermore, this average potential energy depends critically on the structure assumed for the solid surface, although it does not vary much with the size of the crystal.[3] The critical task of choosing a host for the energy bands of a solid is deciding how the periodic charge distribution of an infinite crystal should be terminated to create a "model solid" with energy bands that can be referenced to the vacuum level. A model solid is not a real crystal but a device of a finite size with the charge distribution of a perfect crystal.[4] Such an object does not exist in nature because its charge distribution, especially near its edges, is not a solution of the Schrödinger equation. Nevertheless, the electrostatic potential of a model solid, as a result of its charge distribution, is invaluable for the analysis of interfaces. In order to search for appropriate model solids, we first examine what properties a terminated periodic charge distribution should possess in order to qualify as a model solid.

Since the internal charge distribution of a model solid is identical to that of the infinite crystal, it may appear that the boundaries on which the charge distribution terminates are arbitrary. This is technically correct, as long as the average potential energy of the terminated crystal is constant inside and may be referenced to the vacuum level outside. But naturally, some model solids are preferred over others because they are more useful in predicting the interface charge distributions and the band alignment. From Chapter 1, we know that what remains from the total charge distribution of an interface, after the hosts have been chosen and removed, is the bath of the band alignment problem. Since the explanation of any band alignment problem depends on how successfully the effect of the bath can be estimated, the true figure of merit for the choice of the model solid is how accurately the effect of the bath can be estimated!

We are now primarily concerned with the mechanics of defining appropriate model solids. Since it is the average potential energy of the model solid that will be used explicitly in the analysis of band alignment, a properly constructed model solid should see its average potential energy converges quickly to a constant value inside and drops off rapidly outside. Essentially, the electric fields due to the charge distribution of a model solid should be confined as much as possible to within the model solid itself. When the electric fields due to charge distributions of the model solids are confined to the model solids, the electric fields due to the bath are naturally also confined to the bath. This makes the electrostatic effect of the bath easier to handle. This requirement on confinement, which also guarantees that the empty potential in the space vacated by the model solid is uniform, restricts the manner a model solid should be terminated. For example, one can wonder whether the method of termination can change for surfaces on different sides of a model solid, or from region to region on the same surface? The answer is clearly no, as non-uniformity in surface termination would lead to long-range inhomogeneity in the electric potential and significant electric fields external to the model solid. Therefore, besides being neutral, all surfaces of the model solid should be electrically equivalent on the scale of the lattice parameters. This suggests that the charge distribution of a model solid should be assembled with identical units of charge distribution. Furthermore, the distribution of charge in each unit, called a model solid cell (MSC), should confine its own electric fields. A model solid so constructed would be electrically equivalent on all surface orientations and confines its own electric fields. A drawing illustrating the construction of a model solid from MSCs is shown in Fig. 2.4.

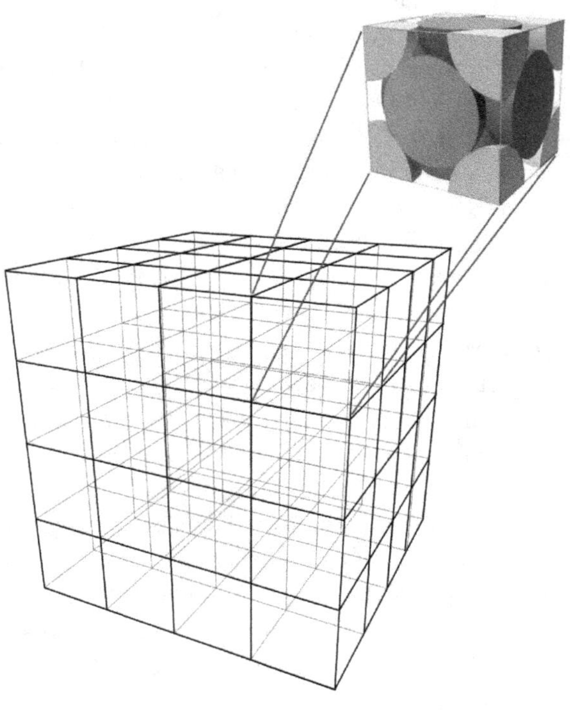

Fig. 2.4. Schematic illustration of how the charge density of a model solid cell may be used to construct a model solid of the crystal.

To identify how an appropriate MSC may be chosen so that it confines electric fields, we start from the relationship between electric potential and charge distribution. From standard textbooks, the electric potential at an exterior location $\vec{r}$ due to a charge distribution of a finite size centered at $\vec{R}_{\text{center}}$ can be expressed in a multipole expansion:

$$V(\vec{r}) = \frac{1}{4\pi\varepsilon_o}\left\{\frac{q_{\text{cell}}}{|\vec{r}-\vec{R}_{\text{center}}|} + \frac{\vec{P}_{\text{cell}}\cdot(\vec{r}-\vec{R}_{\text{cell}})}{|\vec{r}-\vec{R}_{\text{center}}|^3} + \sum_{i,j}\frac{\overleftrightarrow{Q}_{ij}\cdot(\vec{r}-\vec{R}_{\text{center}})_i\cdot(\vec{r}-\vec{R}_{\text{center}})_j}{2|\vec{r}-\vec{R}_{\text{center}}|^5} + \cdots\right\}, \tag{2.9}$$

where q_{cell}, $\vec{P}_{\text{cell}}$, and $\overleftrightarrow{Q}_{ij}$ are the total charge, dipole moment, and quadrupole moment $(i, j = x, y, z)$, respectively, of the charge distribution about the point $\vec{R}_{\text{center}}$. To keep the average potential energy inside a model solid constant, the electric fields due to each MSC should fall off rapidly. It is obvious that all the lower multipole moments of the charge distribution of the cell should vanish for the electric fields to be extremely short ranged. MSCs should be chosen to have maximum symmetry such that the net charge, dipole, and all the quadrupole moments vanish, in which case the potential falls off more rapidly than r^{-5}.[1]

2.2.1 *Average potential of model solids assembled with cells*

As mentioned, the surface makes a constant contribution to the average potential of a crystal independent of its size. Here, the average potential of a model solid is shown to be completely determined by the properties of the MSC used for its construction. We assume that an MSC with the charge distribution of one unit cell of a crystalline solid has been properly chosen. This MSC, with vanishing net charge, dipole, and quadrupole moments, is now used to build a model solid. We are interested in the average electrostatic potential of this model solid with respect to the vacuum level. To derive the average potential energy of a model solid, it is convenient to use the "slab" geometry, which is semi-infinite in a chosen planar direction and finite in the direction perpendicular to the plane. At the end, the derived average potential energy from the slab geometry will be shown to be valid for model solids of any shape. For a slab parallel to a particular (*hkl*) direction, two in-plane lattice vectors, $\vec{a}_1$ and $\vec{a}_2$, are used to generate a single and complete sheet of lattice points at $n_1\vec{a}_1 + n_2\vec{a}_2$, for $n_1 = -N_1, \ldots, -1, 0, +1, \ldots, N_1$ and $n_2 = -N_2, \ldots, -1, 0, +1, \ldots, N_2$, where N_1 and N_2 are large numbers. The two-dimensional unit cell has an area of $a_1\, a_2 \sin\theta$, where θ is the angle between $\vec{a}_1$ and $\vec{a}_2$. Without loss of generality, we shall assume $\vec{a}_1$ to point in the x-direction and that the plane spanned by $\vec{a}_1$ and $\vec{a}_2$ is the x-y plane. A third vector $\vec{a}_3$ is chosen that would translate the lattice points on the sheet at $z = 0$ to those on an adjacent sheet, with an interplanar distance of d_{hkl}. The volume of a unit cell is

$$(\vec{a}_1 \times \vec{a}_2) \cdot \vec{a}_3 = (a_1 a_2 \sin\theta) d_{hkl} = V_{\text{cell}}, \tag{2.10}$$

where d_{hkl} is the z-component of the lattice vector, $\vec{a}_3$. To each lattice point, a cell of charge centered about the lattice point is now inserted, as schematically shown in Fig. 2.5. We first consider the electric potential due to the charge distribution of only one monolayer of cells, with the centers of the cells lying on the $z = 0$ plane. The charge density of the monolayer can be written as a simple sum of all the cells in this layer:

$$\rho_{ml}(x, y, z) = \sum_{n_1=-N_1}^{N_1} \sum_{n_2=-N_2}^{N_2} \rho_{\text{cell}}(x - n_1 a_1 - n_2 a_2 \sin\theta, y - n_2 a_2 \cos\theta, z), \tag{2.11}$$

where $\rho_{\text{cell}}(x, y, z)$ is the charge distribution of the "home" cell, centered at $x = y = z = 0$, and "ml" stands for monolayer. The potential due to

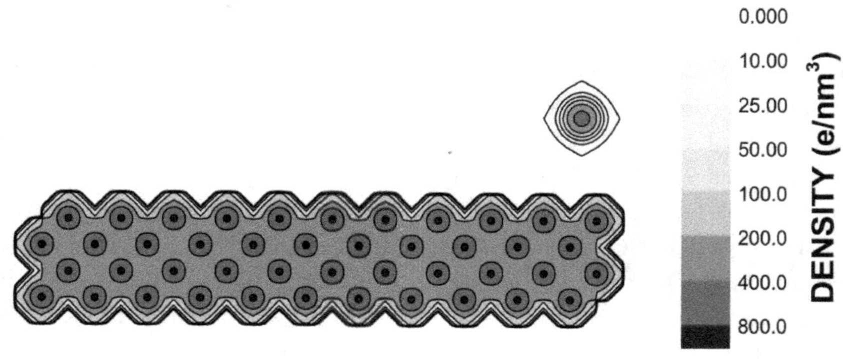

Fig. 2.5. Electron density contours of the cross-section of a two-dimensional semi-infinite slab assembled with four layers of model solid cells. A cell used for the construction is shown in isolation.

this charge distribution of a monolayer of cells, $V_{ml}(x, y, z)$, has a planar average of

$$\bar{V}_{ml}(z) = \frac{1}{a_1 a_2 \sin\theta} \int_{-a_1/2}^{a_1/2} dx \int_{-a_2 \sin\theta/2}^{a_2 \sin\theta/2} dy\, V_{ml}(x, y, z). \tag{2.12}$$

Taking the Laplacian of both sides of Eq. (2.12), one gets

$$\frac{\partial^2}{\partial z^2}\bar{V}_{ml}(z) = -\frac{\bar{\rho}_{ml}(z)}{\varepsilon_0}, \tag{2.13}$$

which now becomes an electrostatic problem in one dimension. Equation (2.13) can be integrated twice to get $\bar{V}_{ml}(z)$ as an integral of $\bar{\rho}_{ml}$, using the boundary condition $\bar{V}_{ml}(-\infty) = 0$ (vacuum level). We leave it as an exercise for the reader to show that, from there, an average potential in the interior of the model solid may be obtained

$$\begin{aligned}\bar{V}_{\text{ModSol}} &= \frac{-1}{2\varepsilon_0 d_{hkl}} \int_{-\infty}^{\infty} dz z^2 \bar{\rho}_{ml}(z) \\ &= \frac{-1}{2\varepsilon_0 V_{\text{cell}}} \int_{-\infty}^{\infty} dx \int_{-\infty}^{\infty} dy \int_{-\infty}^{\infty} dz z^2 \rho_{\text{cell}}(x, y, z).\end{aligned} \tag{2.14}$$

Attention should be paid to the fact that the integrals in Eq. (2.14) are over the volume of one single MSC, ρ_{cell}, and not a layer of cells. The average model solid potential in Eq. (2.14) has an apparent dependence on the direction chosen for the slab because of the z^2 factor in the integration over the cell. However, the requirement that the quadrupole moments of an appropriate MSC vanish allows us to rewrite it in a form independent of $(hkl)^5$:

$$-e\bar{V}_{\text{ModSol}} = \frac{e\Theta_{\text{cell}}}{6\varepsilon_0 V_{\text{cell}}}, \tag{2.15}$$

where Θ_{cell} is the spherapole of the cell, defined as:

$$\Theta_{\text{cell}} = \int d\vec{r} r^2 \rho_{\text{cell}}(\vec{r}). \tag{2.16}$$

Equation (2.15) is the main result that directly connects the charge distribution of a single MSC with the average potential energy, with respect to the vacuum level, of the model solid built with thousands or millions of the same cells. Note that this average potential energy is independent of the shape and surface orientation of the model solid. It should also be pointed out that the charge distribution of an MSC does not need to be confined to the volume of a unit cell for Eq. (2.15) to be valid. As long as an MSC has the same total number of electrons and ions as that in a unit cell and can assemble the entire crystal charge density inside the model solid, it can be used to construct a model solid, as illustrated in the following for charged ions.

2.2.2 *Planar and WSC model solids*

For some analyses, the bulk reference for a crystal solid is needed only for a single interface orientation, in which case the model solid does not need to be finite in all three dimensions but very well can be in the shape of a two-dimensional semi-infinite slab. Such a model solid can be created by partitioning an infinite crystal into two parts along the desired orientation, and the most straightforward way to do this is to make a cut on a flat crystallographic plane. A model solid that ends on flat surfaces is the result of construction with MSCs in the shape of a parallelepiped. An important requirement for the abruptly terminated surface is that after the cut the two halves should be neutral and electrically equivalent, such that average potential energy can be obtained meaningfully for both. Along any crystallographic direction, there is a limited number of locations that can be used to make such a neutral cut. We'll use the average electric potential energies of Ag, GaAs, and SrTiO3 along the (100) direction, shown in Fig. 2.6, to illustrate how these locations might be chosen. Each 1D potential energy plot can be seen to go through minima at the locations of the atomic planes and maxima between planes of atoms. At a maximum, the vanishing slope assures that a cut there would leave two neutral surfaces with identical average potentials. Neutral model solids can also be specified on a plane of atoms. In that case, the nuclear charges on the plane need to be shared in a specific way between the two halves in order to make both neutral and electrically equivalent. Such a model solid makes no accommodations for the presence of either bonds or atoms and is not as chemically astute as model solids to be discussed that attempt to preserve the integrity of atoms and/or bonds.

Model solids with flat surfaces just described may be viewed as having been constructed with parallelepiped-shaped unit cells. A model solid can also be constructed with MSC in the shape of the WSC of the lattice. Even though the shape of a WSC is completely fixed by the Bravais lattice of a crystal, its center may be chosen anywhere in a unit cell. When

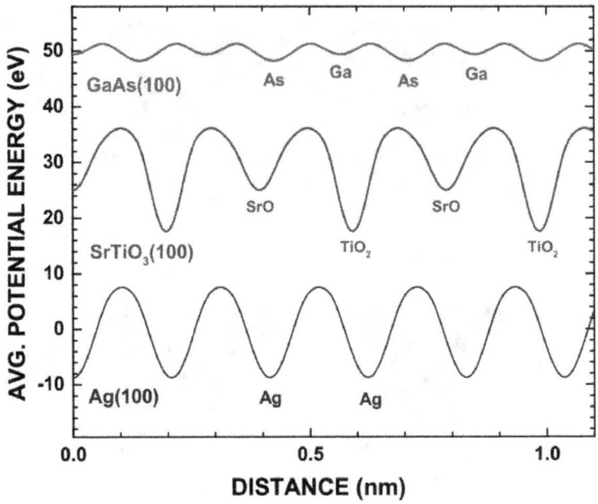

Fig. 2.6. Potential energy, planar averaged in the (100) direction, of three crystalline solids: GaAs, $SrTiO_3$, and Ag.

centered about a point of high symmetry, a WSC can have vanishing dipole and quadrupole moments and would be appropriate for model solid construction. For a monatomic unit cell, a WSC centered about the atom may be used to construct the model solid. With different species of atoms and polarized bonds in a compound solid, the choices of the center of the WSC appropriate for model solid construction are limited. To avoid a dipole moment, WSC center for a zincblende crystal can be located only at either the cation or the anion site. An anion-centered WSC (AWS) has a quarter ($^1/_4$) cation at four of the eight three-fold corners for a total of one full cation–anion pair in the cell. Conversely, a cation-centered WSC (CWS) has four $^1/_4$ anions at its corners, as schematically shown in Fig. 2.7. Both have vanishing dipole and quadrupole moments and are convenient building blocks for the construction of model solids. Some properties of model solids constructed with the AWS and the CWS, computed with the help of Eqs. (2.15) and (2.16), are given in Table 2.1 for common zincblende semiconductors. Model solids for compound materials constructed with WSC do not preserve the wholeness of atoms on their surfaces. Yet, directional bonds are preserved, which makes them useful in the analysis of band offsets, as demonstrated for interfaces between zinc blende semiconductors in Chapter 5.

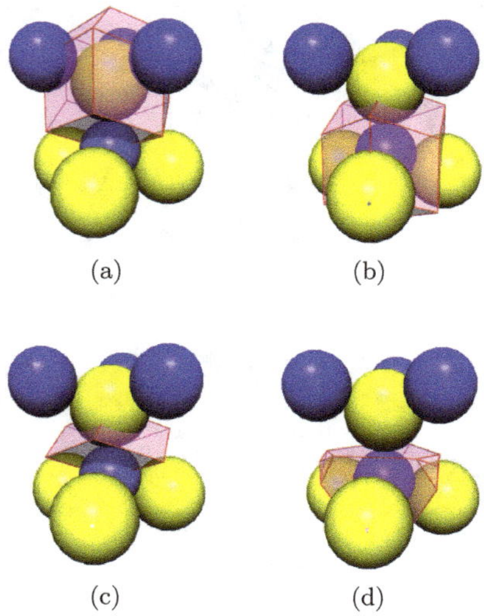

Fig. 2.7. Ball-and-stick models of the zinc-blende lattice, with cations and anions displayed in blue and yellow colors, respectively. (a) Anion-centered Wigner–Seitz (AWS) cell, (b) cation-centered Wigner-Seitz (CWS) cell, (c) quadrant, (d) "neutral polyhedron" on the cation site.

Table 2.1. Valence band minimum (eV) for various neutral model solids with respect to the vacuum level, calculated from the charge density of bulk semiconductors. Cation plane (100) is a model solid terminated in the (100) direction on the plane of cations. The model solids for mid-plane (110) and neutral point (111) end at those two locations.

Semiconductor	CWS	AWS	Cation Plane (100)	Anion Plane (100)	Mid-Pln (110)	Neutral Pt. (111)
AlP	10.034	6.712	8.571	13.063	−0.300	−2.775
GaP	10.517	6.586	8.510	13.408	−0.134	−2.475
AlAs	9.957	6.154	7.982	12.858	−0.365	−2.719
GaAs	10.490	6.083	7.967	13.261	−0.132	−2.364
InP	8.437	6.201	7.877	11.127	−0.683	−2.788
AlSb	10.343	5.187	7.012	13.003	−0.197	−2.475
GaSb	10.694	4.969	6.835	13.246	−0.107	−2.305
InAs	8.324	5.575	7.219	10.891	−0.837	−2.815
InSb	8.818	4.690	6.340	11.225	−0.741	−2.665
ZnS	10.773	16.531	20.142	13.882	−1.591	−3.676
ZnSe	11.368	15.323	18.832	14.411	−1.262	−3.361
CdS	8.501	17.578	21.196	11.329	−2.059	−3.978
ZnTe	11.899	12.966	16.241	14.777	−0.999	−3.156
CdSe	8.998	16.227	19.714	11.763	−1.906	−3.762
CdTe	9.828	14.124	17.407	12.492	−1.627	−3.485
Si	9.385	9.385	11.867	11.867	0.613	−1.968
Ge	9.045	9.045	11.412	11.412	0.607	−1.721

2.2.3 *Model solids assembled with charged ions*

At an interface, the atomic structure often switches abruptly from that of one crystal to that of the other. However, the charge distribution cannot be switched so suddenly. Some charge relaxation is inevitable. Additionally, chemical adjustments may be required for new bonds at the interface. The region of the interface with modified charge distribution is part of the interface specific region (ISR) and acts as the bath for the interface with the model solids, playing the roles of the hosts. Ideally, model solids should produce an initial charge distribution of the interface that resembles the final charge distribution so that the effect of charge relaxation is minimized. To accomplish this, the charge distributions of the two model solids, when stitched together, should possess the main character of the bonds at the interface. From that perspective, it is desirable for MSCs to be defined in keeping with the chemical bonds in the crystal. Since chemical bonds are different for covalent and ionic compounds, the best choice of model solid may also be different! For interfaces between covalent semiconductors, the preservation of the charge density associated with each directional bond has been argued as a strategy to minimize charge rearrangement at the interface. This argument does not necessarily hold for ionic compounds with non-directional bonds. Rather, it may be more advantageous to replicate the net charge for each ion at the interface according to that for each ion in bulk crystal.

To construct a model solid that preserves the individuality of charged ions, one first needs a strategy to define ions/atoms in a compound. Several methods exist, the most popular of which is Bader's atoms in molecules (AIM) theory.[6] Zero density flux boundaries are employed by AIM to partition the charge density of a molecule into basins or atoms.[7] Bader atoms defined for a compound are invariably charged. Being both easy to implement and in wide use in the community of chemistry and materials, AIM has been used for the construction of model solids of ionic compounds. As charge neutrality is an important requirement of model solids, the use of charged Bader atoms as the building blocks requires some planning. The goal is to assemble, with charged Bader atoms, an MSC that has vanishing net charge, dipole moment, and quadrupole moment. We will use two classes of material, zincblende semiconductors and perovskite oxides, as examples to illustrate how AIM-based MSCs may be chosen.

In a binary zincblende compound crystal, one cation and one anion defined by AIM occupy the entire volume of a primitive cell. From the 1:1

stoichiometric ratio, the Bader anion and cation have equal but opposite charges, Q_{cat} and $Q_{\text{ani}}(= -Q_{\text{cat}})$. To construct a model solid for zincblende based on AIM, the MSC chosen should include equal numbers of Bader cations and anions. For the smallest MSC containing one of each, the cation and anion cannot be placed side by side to avoid a dipole moment for the cell. Instead, they have to be arranged symmetrically. In a anion-centered AIM model solid cell (AC-AIM), a full anion is placed at the center and four Bader cations, each with $1/4$ of the charge density, are placed at the tetrahedral corners, an illustration of which is shown in Fig. 2.7. A $1/4$ cation occupies the same volume as a full cation but contains only 25% of the electron density and nuclear charge. All dipole and quadrupole moments of an AC-AIM obviously vanish from symmetry. Due to the quarter ions, the spatial extent of an AC-AIM cell extends well beyond the volume of a primitive cell. This is not a problem as all the charge in this spatially expanded volume comes from one unit cell, and the charge distribution of the crystal can be fully assembled with these cells. A model solid built with AC-AIM cells terminates on all surfaces with a fractional layer of anions, as illustrated in Fig. 2.8. This is a necessary result in order to keep all the surfaces neutral. There is no issue with using such model solids in the analysis of band alignment, as the mating model solid always provides the fractional anions needed to complete a full layer. Alternatively, an AIM model solid for the zincblende semiconductors could terminate on a layer of partial anions, if cation-centered cells (CC-AIM) were used in its construction. By analogy, a CC-AIM has a Bader cation at the center and four $1/4$-anions at the corners.

The average electrostatic potential energy of a model solid constructed with the AC-AIM and the CC-AIM cells can be computed from the spherapole of the MSC. The Bader cation and anion of a crystal are first analyzed in isolation to obtain their spherapoles, Θ_{cat} and Θ_{ani}, about their nuclear positions. Note that these spherapoles measure the distribution of negatively charged electrons about positive ions and therefore are negative in values. From the lattice structure, the distribution of electrons is symmetric and thus each Bader ion should have no net dipole moment or quadrupole moments about its nuclear position. Let us first examine, in general, how spherapoles of various charged parts of a system can be combined to yield a spherapole for the entire charge distribution. We assume that the entire charge distribution of a system can be partitioned into N different parts, each with a volume of Ω_i, a charge of Q_i, no dipole moment, and a spherapole of Θ_i about its center at $\vec{R}_i$. Naturally, the sum of Ω_i's is

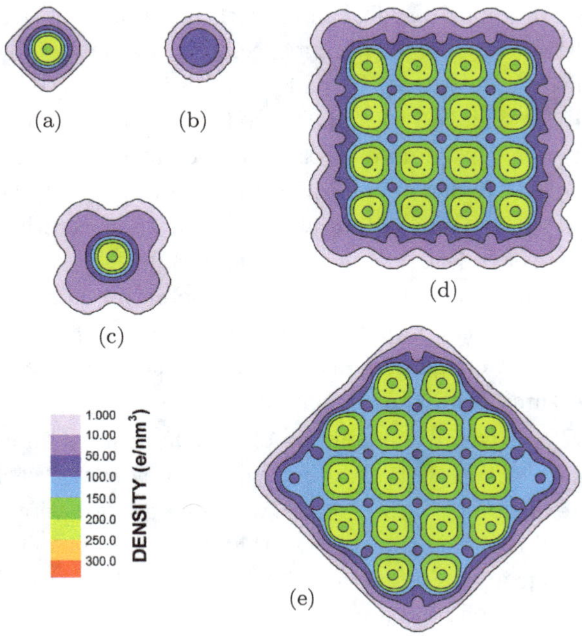

Fig. 2.8. Electron density contours, in [001] view, of structures from Bader analysis of GaAs crystal. (a) Arsenic Bader atom; (b) Ga Bader atom; (c) anion-centered Bader model solid cell (AC-AIM) consisting of an As atom at the center and four 1/4 Ga Bader atoms in tetrahedral directions; (d) model solid of GaAs assembled with As-centered Bader MSC; (e) non-polar model solid of GaAs assembled with Bader atoms.

the volume of the unit cell and the sum of Q_i's vanishes. Using the origin as the center of the cell, the spherapole of the entire unit cell is written as

$$\Theta_{\text{cell}} = \sum_{i=1}^{N} \int_{\Omega_i} d\vec{r}\, r^2 \rho(\vec{r}) = \Theta_{\text{sum}} + \sum_{i=1}^{N} Q_i R_i^2. \tag{2.17}$$

The total spherapole not only has the sum of the individual spherapoles ($\Theta_{\text{sum}} = \sum_{i=1}^{N} \Theta_i$) but also includes the $Q_i R_i^2$ values from all the individual parts. The quantity $Q_i R_i^2$ represents the spherapole contribution of a point charge Q_i placed at the center of the i-th part. In arriving at the above result, use has been made of the dipole moment of each part being zero. A model solid constructed with cells with Θ_{sum} as the total spherapole has an average potential energy of

$$-e\bar{V}_{\text{ModSol}}^{\text{non-polar}} = \frac{e\Theta_{\text{sum}}}{6\varepsilon_0 V_{\text{cell}}}, \tag{2.18}$$

which is that for a model solid ending with non-polar surfaces, e.g. (110) for the zincblende crystals. Returning specifically to the zincblende crystals, the spherapoles for the AC-MSC and the cation-centered (CC) model solids are $\Theta_{\text{cat}} + \Theta_{\text{ani}} + 3a^2Q_{\text{cat}}/16$ and $\Theta_{\text{cat}} + \Theta_{\text{ani}} - 3a^2Q_{\text{cat}}/16$, respectively. These spherapoles can be used to calculate the average potential energies of the AC and CC Bader model solids through Eq. (2.15). The difference between these two average potential energies is the potential drop across one-half monolayer of Bader cations and anions.

The reason for Eq. (2.18) being the potential energy for a non-polar model solid is the implicit requirement of $\sum_{i=1}^{N} Q_i R_i^2 = 0$ for its construction. For a compound in general, this sum can be broken down into separate sums from the cations and anions, $= \sum_{i=1}^{N_{\text{cat}}} Q_{\text{cat},i} R_{\text{cat},i}^2 + \sum_{j=1}^{N_{\text{ani}}} Q_{\text{ani},j} R_{\text{ani},j}^2$, where the sum for cations is positive and that for the anions is negative. When the two individual sums cancel each other, the average squared distance of the cations, weighted by the cation charge, from the assumed cell center is the same as that for the anions. Therefore, this MSC is non-polar about the assumed cell center. The exposed surfaces of such a "generic" model solid are also non-polar, with a surface composition identical to that of the compound crystal, independent of the number of different species of cations or anions in the compound. An example of a non-polar GaAs model solid based on AIM is shown in Fig. 2.8. It can be verified that all surfaces of such a model solid indeed have an average potential energy equivalent to that for the non-polar (110).

In the unit cell of a perovskite oxide, there are two different cations: one at the corner (A site) and the other at the body center position (B site) of the cube, and three oxygen anions at the face center positions. From AIM analysis, they are found to have net charges of Q_A, Q_B, and Q_{Ox}, respectively, and spherapoles of Θ_A, Θ_B, and Θ_{Ox}. Following the zincblende example above, there are two obvious ways to arrange these ions symmetrically in an MSC. Centered about the body-center ion, we can place eight 1/8 corner cations at the eight corners and six 1/2 oxygen atoms on the face center positions of a cubic unit cell. A model solid constructed with these MSCs will terminate on all surfaces with one-half monolayer of AO, e.g. 1/2 monolayer of SrO for the $SrTiO_3$ crystal. From Eq. (2.27), we write the spherapole for the unit cell centered on the B cation as $\Theta_{B-\text{perov}} = \Theta_{\text{sum}} + 3a^2Q_A/4 + 3 \cdot a^2Q_{Ox}/4$, where $\Theta_{\text{sum}} = \Theta_A + \Theta_B + 3\Theta_{Ox}$. Alternatively, we can put the corner cation (cation A) at the center of the cube, eight 1/8 B-site cations at the corner positions, and twelve 1/4 oxygen atoms midway on the edges of the cube. This A-centered model solid will terminate on one-half monolayer of BO_2, e.g. $1/2$ monolayer of TiO_2 for $SrTiO_3$, and has an average potential energy corresponding to its spherapole $\Theta_{A-\text{perov}} = \Theta_{\text{sum}} + 3a^2Q_B/4 + 3 \cdot a^2Q_{Ox}/2$. The average potential energy and energy level position of the generic model solid for selected perovskite oxides are shown in Table 2.8.

Many of the ionic compounds routinely used for technological applications are binary oxides, with a M_aO_b composition. Even though the chemical formula may appear simple, the structure of the unit cell of some binary oxides may be too complicated for the construction of a model solid with a desired surface termination. However, a generic Bader model solid for any binary oxide can always be constructed, with its average potential energy given by substituting the sum of spherapoles of the Bader atoms, $\Theta_{\text{sum}} = a \cdot \Theta_M + b \cdot \Theta_{Ox}$, into Eq. (2.15). The exposed surfaces of such a model solid would have a fractional layer of Bader cations and oxygen atoms in the composition of the binary oxide. In other words, the AIM references listed in Table 2.6 correspond to generic model solids with non-polar surfaces.

2.3 Neutral Polyhedra Model Solids

2.3.1 *Motivations for neutral polyhedron*

In searching for model solids that can approximately reproduce a charge distribution at an interface, we first look back at how electron density depends on atomic positions in general. From the perspective of the DFT, the distribution of electrons in a solid is a quantum mechanical reaction of the electron density to the electric potential from ionic charges. For example, the electron density in the immediate vicinity of an ion core is nearly entirely dependent on that ion core, regardless of whether the ion core is in a bulk crystal or at an interface. This close connection of the charge distribution to the local atomic structure has been described as a "nearsightedness" of the electron density.[8] It thus seems advantageous to begin a simulation of the interface charge distribution by having the volume around each atomic position filled with the electronic density usually associated with that atom in a solid. AIM model solids just described have the desired property that the individual identities of the atoms are maintained. While effective for the explanation of band alignment at some ionic interfaces, AIM model solids are not suited for covalent compounds.

Table 2.2. Lattice constant, experimental work function, NPT model solid work function, and intermediate charge density for common elemental metals.

Metal	Lattice Constants (Å)	Exp. WF (eV)	NPT ModSol WF (eV)	Intermediate Density (e/nm^3)
Cs	6.14	2.14	2.08	10.5
Rb	5.66	2.16	2.04	13.1
Sr	6.09	2.59	2.10	39.9
Na	4.21	2.75	2.09	29.6
Ca	5.59	2.87	2.27	54.1
Li	3.51	2.90	2.40	50.2
Mg	3.21 × 5.21	3.66	2.43	89.8
Hf	3.20 × 5.05	3.90	2.05	174.9
Zr	3.23	4.05	2.19	159.4
In	3.25 × 4.95	4.12	1.64	91.9
Cd	2.98 × 5.62	4.22	2.13	135.6
Pb	4.95	4.25	1.98	87.3
Ta	3.30	4.25	0.51	235.1
Ag	4.15	4.26	2.10	143.2
Al	4.04	4.28	0.33	176.2
Nb	3.30	4.30	0.62	222.4
V	2.98	4.30	0.67	251.3
Ti	2.95 × 4.69	4.33	2.04	179.0
Zn	2.67 × 4.95	4.33	1.67	177.1
Sn	5.83 × 3.18	4.42	1.88	113.2
Cr	2.85	4.50	−0.05	311.0
Fe	2.87	4.50	0.75	255.7
W	3.17	4.55	−0.50	289.9
Mo	3.15	4.60	−0.23	274.2
Cu	3.62	4.65	0.99	216.1
Rh	3.82	4.98	1.18	263.2
Co	2.51 × 4.07	5.00	1.92	265.0
Au	4.15	5.10	2.17	178.9
Pd	3.93	5.12	2.63	197.5
Ni	3.52	5.15	1.03	254.0
Ir	3.87	5.27	0.41	310.0
Pt	3.97	5.65	1.92	248.5

More useful would be a model solid constructed with atoms that are electrically neutral. From detailed analysis of the charge distribution of zincblende semiconductors, a method to define neutral atoms suggested itself.[9] In this section, we describe how these neutral atoms, called neutral polyhedra, may be defined for any crystalline solid and why they are useful for band alignment investigations of the majority of interfaces.

In each primitive unit cell of a zincblende crystal, there are one cation, one anion, and four directional sp^3 bonds between them. The four bonds are between the center ion and four $1/4$ ions at the corners of the WSC, as shown in Fig. 2.7(a) for anion-centered WSC and Fig. 2.7(b) for cation-centered WSC. This suggests a way to partition a unit cell of a zincblende crystal into sub-volumes without breaking the individuality of bonds. The overlapped volume between a cation-centered WSC and a WSC centered on a nearest neighbor anion is known as a "quadrant", as shown in Fig. 2.7(c). There are four quadrants, oriented in tetrahedral directions, in every cation-centered or anion-centered WSC. Each quadrant contains $1/4$ cation and $1/4$ anion at its apexes, occupies $1/4$ the volume of the unit cell, and has two valence electrons. The main character of a single sp^3 bond is clearly contained within each quadrant. Using quadrants, the charge density of a zincblende bulk crystal may be constructed bond-by-bond. However, in order to define atoms in a zincblende semiconductor, these covalent bonds need to be severed. To meet the overall goal of defining atoms in the crystal that are transferable, each quadrant needs to be severed to form the cation and the anion from a point where the cut cation can seamlessly mate with a different anion into a quadrant of a different semiconductor, and vice versa. Therefore, a detailed knowledge of the charge distribution inside a quadrant of each zincblende semiconductor is essential.

A number of technologically important II–VI zincblende semiconductors comprise a group II cation (Be, Mg, Zn, Cd) and a group VI anion (S, Se, Te). The family of III–V zincblende semiconductors contains a cation from group III elements (B, Al, Ga, In) and an anion from group V elements (N, P, As, Sb). Zincblende semiconductors provide a good venue to investigate the transferability of neutral atoms because the same anion (or cation) can be found in several different crystals with identical crystal structure. To determine how much of the local charge distribution around the nucleus of each anion (cation) may be reasonably assigned to that anion (cation), the formation of the total dipole moment in a quadrant is analyzed in detail. The "accumulated dipole" of charge distribution in a single quadrant, defined as

$$P(x) = \int_{-\infty}^{x} dx'(x' - x) \iint dydz\rho(x', y, z), \tag{2.19}$$

is calculated from the cation end for all the III–V semiconductors and shown in Fig. 2.9(a). Curves for different cations have been rigidly shifted for clarity. Nitrides are calculated in the zincblende structure, even though

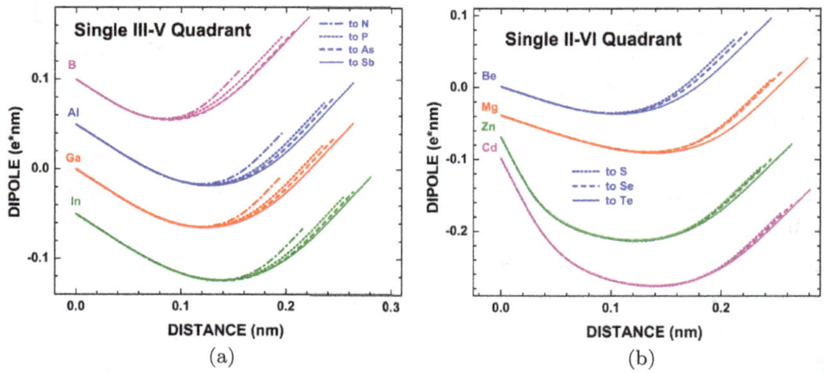

Fig. 2.9. Dipole moment accumulated from the cation end of selected (a) III–V and (b) II–VI Zincblende semiconductor quadrants, calculated at the equilibrium lattice parameter of each semiconductor. For clarity, curves have been vertically shifted.

Source: Reprinted with permission from Tung and Kronik.[9] Copyright 2016 by the American Physical Society.

for some the wurtzite structure is more common in nature. Similar to those displayed in this figure are the accumulated dipoles for II–VI semiconductors shown in Fig. 2.9(b). Starting from the cation end, each of the curves of Fig. 2.9 reaches a minimum before rising toward the anion. The minimum in each curve has zero slope and marks the location where the quadrant is separated into two parts that are each neutral. For example, the number of electrons on the cation side of the quadrant is 0.75 up to the minimum in each of the curves in Fig. 2.9(a) because the positive charge on the $^1/_4$ cation is +0.75 *e*. Quite striking from Figs. 2.9(a) and (b) is the nearly identical charge distribution on the cation end of the quadrant for each cation, independent of the identity of the anion in the quadrant. The same can actually be said about the anion side of the quadrants, when the data of Figs. 2.9(b) are replotted with the minimum of each set at the origin in Fig. 2.10(b). The charge distribution within the charge-neutral polyhedron of each group II cation (or group VI anion) is nearly identical, independent of the semiconductor.[9] In other words, the neutral polyhedra containing either the cation or the anion are transferable among different group II-VI zincblende semiconductors. From Fig. 2.9(a), similar conclusions can be drawn for the III-V semiconductors.

Figures 2.10(a) and (b) bring up another interesting point: all dipole curves for an element end roughly at the same location on the length axis, i.e. the length of the cation section in equilibrium is essentially

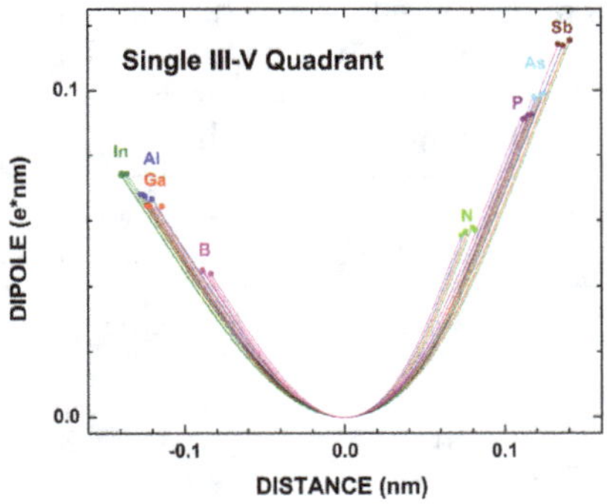

(a)

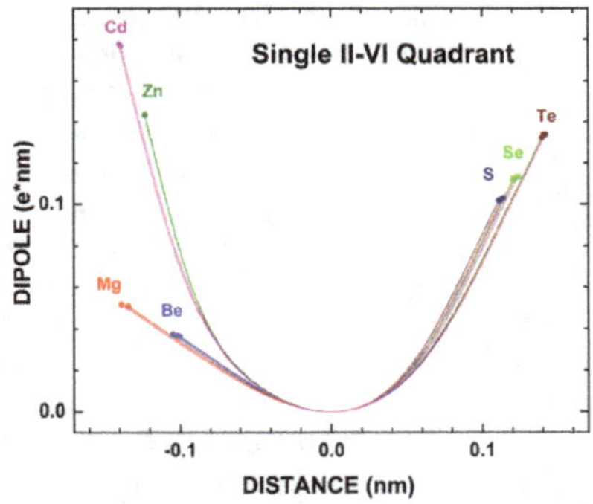

(b)

Fig. 2.10. Accumulated dipole moment curves of ZB semiconductors from Fig. 16, redisplayed with the neutral point for each quadrant shifted to origin. For clarity, the end points of each curve are marked by dots. (a) III–V semiconductors, (b) II–VI semiconductors.

Source: Reprinted with permission from Tung and Kronik.[9] Copyright 2016 by the American Physical Society.

independent of its anion partner, and vice versa! The equilibrium bond length for any of these ZB semiconductors is, therefore, the sum of additive lengths, or "radii", of the two elements. Since the minimum in each dipole distribution curve is the neutral point for that quadrant, the individual radius so determined for an element seems to fit the concept of the element's covalent radius. The covalent radii for all elements determined from the dipole profile of Fig. 2.10 are in reasonable agreement with the reported single bond molecular covalent radii[10,11] and the covalent radii for solids,[12] as shown in Table 2.3. This agreement is remarkable given that the covalent radii are arrived at for zincblende semiconductors on the scientific basis of electrical neutrality, whereas other compilations of covalent radii are determined statistically through empirical fitting of the interatomic distances in thousands of materials.

2.3.2 *The polyhedron shape and energy minimization*

In arriving at the separation point between neutral cation and neutral anion in a quadrant of a zincblende crystal, a planar boundary perpendicular to the line connecting the cation and the anion is used out of convenience of the analysis. In looking for neutral atoms with maximum transferability, we want to know whether the polyhedron shape is necessary or whether a better choice exists. Transferability is dependent on the lack of strong interactions

Table 2.3. The distance from an atom to the neutral point of the quadrant (nm), determined from analysis of zincblende semiconductors.[9] Comparison is made with the covalent radius (C. R., in nm) fitted in single-bond molecules[11] and solids,[10] and from analysis of the Cambridge Structural Database.[13]

Element	Neutral Length (NPT)	Single-Bond C.R.	Solid C.R.	Cambridge Database C.R.
Al	0.125	0.126	0.129	0.121
Ga	0.122	0.124	0.128	0.122
In	0.138	0.142	0.146	0.142
Be	0.102	0.102	0.106	0.096
Mg	0.135	0.139	0.141	0.141
Zn	0.123	0.118	0.130	0.122
Cd	0.139	0.136	0.148	0.144
Si	0.118	0.116	0.118	0.111
Ge	0.123	0.121	0.123	0.120
P	0.115	0.111	0.108	0.107
As	0.122	0.121	0.117	0.119
Sb	0.140	0.140	0.136	0.139
S	0.113	0.103	0.104	0.105
Se	0.123	0.116	0.115	0.120
Te	0.141	0.136	0.134	0.138

between the neutral atoms, which is achievable if the electric fields due to each neutral atom are confined to within its own volume. Therefore, neutral atoms with the maximum transferability are the ones that confine electric fields maximally. The more tightly neutral atoms confine the electric fields, the more rapidly the electric fields drop off outside the model solid, and the higher (less negative) the average potential energy of the model solid. We now show that when neutral atoms are in the shape of polyhedra, the model solid has the highest average potential energy among all assembled model solids for the same crystal. Therefore, neutral polyhedra are maximally transferable. This only requires showing that the sum of the spherapoles is least negative for neutral polyhedra because of the relationship of Eq. (2.15).

Assume that inside a bound, overall neutral, charge distribution, $\rho(x, y, z)$, there are two atomic centers. We place ion core A, with a charge of Q_A, at the origin $r_A = (0, 0, 0)$, and ion core B with Q_B, on the positive z axis at $r_B = (0, 0, d_0)$. We seek a boundary, defined by $Z_I(x, y)$, to partition the total charge distribution into two neutral parts in a way that minimizes the sum of the absolute values of the two spherapoles. The constraint that part A is neutral is

$$\int dx \int dy \int_{-\infty}^{Z_I(x,y)} \rho(x, y, z) dz = Q_A, \tag{2.20}$$

with a similar equation for part B. The two spherapoles are given by

$$\Theta_A = \int dx \int dy \int_{-\infty}^{Z_I(x,y)} \rho(x,y,z)[x^2 + y^2 + z^2]dz, \tag{2.21a}$$

$$\Theta_B = \int dx \int dy \int_{Z_I(x,y)}^{+\infty} \rho(x,y,z)[x^2 + y^2 + (z - d_0)^2]dz. \tag{2.21b}$$

The boundary $Z_I(x, y)$ is sought by finding the quantity $\Theta_A(Z_I) + \Theta_B(Z_I)$ stationary, subject to the constraint $Q_A(Z_I) = const$. Using Lagrange's multiplier method, this leads to

$$\int dx \int dy \rho(x,y,Z_I(x,y))[2d_0 Z_I(x,y) + d_0^2 - \lambda]\delta Z_I(x,y) = 0, \tag{2.22}$$

where λ is the unidentified multiplier. The solution $2d_0 Z_I(x, y) + d_0^2 = \lambda$ is clearly independent of x and y, i.e. it describes a plane $Z_I(x, y) = const$. Therefore, the boundary between two centers that defines neutral cells with a minimized sum of the absolute values of spherapoles is a plane perpendicular to the line connecting these two atomic centers.[14] Applying this argument repeatedly, one can conclude that, when the average potential energy of a model solid is maximized, all boundaries of all neutral atoms are such planes, i.e. all neutral atoms are in the shape of polyhedra.

When charge distribution is removed from real crystal surfaces to construct a model solid, the potential energy of the crystal increases. The NPT model solid may be called "irreducible", in the sense that there is no more surface dipole that can be removed to make the average potential energy any higher. Therefore, there are many advantages of the NPT model solids in modeling band alignment. The maximized potential energy of the model solid indicates that each neutral polyhedron confines its electric fields and has good transferability. With their sizes set by their covalent radii, the neutral polyhedra on opposing surfaces, when stitched together at interfaces, are already at optimal bond lengths from each other. Atomic relaxations may be expected to be small. The minimized total energy of the charge distribution of each NPT model solid further justifies the expectation that the effect due to charge relaxation is also small. For these reasons, the NPT should be useful to analyze interfaces of all kinds, not just between zincblende semiconductors or between covalent materials.

2.3.3 *Algorithm to define neutral polyhedra*

For some crystalline solids, the method to define neutral polyhedra is obvious. NPT model solids for metals with the b.c.c. and f.c.c. structures in Table 2.2 are assembled with WS cells. The definition of neutral atoms is also self-suggestive for elemental materials with multiple atoms in a unit cell, such as metals with the h.c.p. structure and semiconductors with the diamond structure. Elemental semiconductors, C, Si, Ge, and α-Sn, have the diamond lattice, with a two-atom basis for the unit cell. The two atoms in a unit cell are electrically equivalent yet topologically inequivalent. Using the same idea behind the WSC for the monoatomic unit cell, one can define atoms for the diamond lattice by partitioning the crystal into "proximity cells". A proximity cell about an atom includes the volume closer to this atom than to any other atoms. For the diamond lattice, each proximity cell is neutral, has one-half the volume of the unit cell, and has vanishing lower-order multipole moments. There are two types of proximity cells corresponding to the two positions of the atom in the unit cell, shown in Figs. 2.11(a) and 2.11(c), which are related through inversion. The MSC for the diamond lattice is now made up of two proximity cells, one of each kind, with a total spherapole being the sum of the spherapoles for the proximity cells. Neutral polyhedra of the cation and the anion of a zincblende compound semiconductor can be defined by the plane that leads to neutrality in each quadrant in Fig. 2.10. One of each of the neutral polyhedra of the anion and the cation together makes the MSC for the zincblende crystal. The fact that adding the two polyhedra of a zincblende crystal side-by-side creates a shape for the MSC

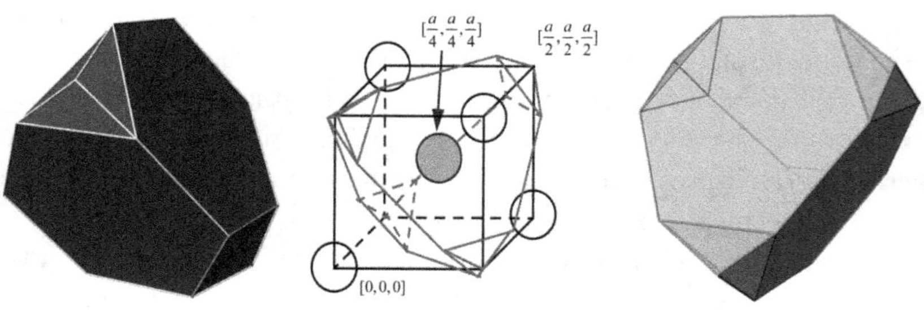

(a) (b) (c)

Fig. 2.11. "Proximity cells" for atoms in the diamond and zinc-blende lattices. (a) and (c) Proximity cells for anions and cations, respectively, of the zinc-blende lattice. A schematic of 1/8 of the conventional unit cell is drawn in (b) to indicate the orientation of the proximity cells.

that is not highly symmetric does not affect the vanishing of all of its lower multipole moments. Since the cation and the anion polyhedra are individually neutral, polyhedra of either kind can be added or removed from the surface of a model solid of the zincblende crystal without affecting its average potential energy. The independence of the average potential energy on the orientation and structure of the surface is a general property of all model solids constructed with neutral polyhedra. The properties of NPT model solids of some zincblende semiconductors are listed in Table 2.7.

For crystals with small unit cells, the definition of neutral polyhedra can be self-evident as just described. However, the proper way to define neutral polyhedra for crystalline solids with large unit cells and more complicated structures is not always obvious. Keep in mind that the definition of the neutral polyhedra for a crystal has to lead to a model solid with maximized average potential energy. Therefore, it is not enough to define a symmetric volume around each atom that is neutral and in the shape of a polyhedron. The sum of all the spherapoles of a unit cell also has to be least negative! To identify possible constraints among different polyhedra, we first assume that the partitions to all the neutral volumes have been properly set such that the total spherapole is already at an extreme. This assumption then allows the deduction of common rules/constraints among optimized neutral polyhedra. This is illustrated with Fig. 2.12, which depicts boundaries

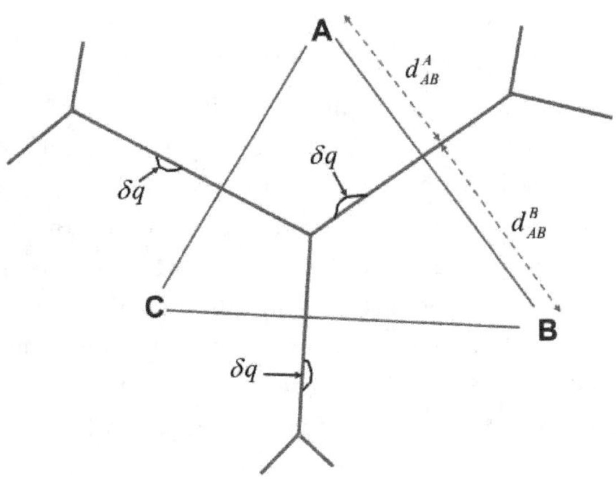

Fig. 2.12. Schematic drawing of the space around three atoms A, B, and C in the unit cell of a crystal where partitions to define optimized neutral polyhedron about every atom of the unit cell have already been set. The solid lines drawn perpendicular to the vectors connecting the atoms represent boundaries of the neutral polyhedra, which are planes in reality. Three small volumes containing identical charges are drawn. The vertical distance from atom A to the boundary separating atoms A and B, d^A_{AB}, is indicated.

separating atoms A, B, and C. Since it is already least negative, the total spherapole is stationary against any incremental changes in the boundaries that preserve the neutrality condition. We focus on the three small volumes that contain the same infinitesimal amount of charge δq on the three boundaries in Fig. 2.12. Whether the three small volumes on the boundaries belonged to the respective atoms in their clockwise direction or the atoms in the counterclockwise direction, the neutrality conditions of all the atoms would be satisfied. The fact that the sum of the spherapoles should also remain stationary now gives the identity

$$[(d_{AB}^{A})^2 - (d_{AB}^{B})^2] = [(d_{CA}^{A})^2 - (d_{CA}^{C})^2] + [(d_{BC}^{C})^2 - (d_{BC}^{B})^2], \qquad (2.23)$$

where, as drawn in the figure, d_{AB}^{A} is the perpendicular distance from atom A to the boundary separating A and B, etc.[15] Note that the lateral distances along the boundary may be ignored altogether since they always contribute the same amount to the total spherapole, regardless of which atom each small volume belongs to. Equation (2.23), which may also be derived from geometry, identifies an important relationship between the locations of different boundaries. This constraint is all that is needed to define the neutral polyhedra that minimize the overall spherapole of a crystal! Since the sum of the individual spherapoles has to remain unchanged against any possible incremental changes in the boundary locations that satisfy the neutrality requirement, the diagram in Fig. 2.12 can be redrawn for any three or more connected atoms of the unit cell and an equation similar to Eq. (2.23) can be written down for that diagram. Therefore, the NPT boundary between any two adjacent atoms in a unit cell is linked to that between any other atomic pair. This coupling significantly simplifies the search process and assures that the neutral polyhedra defined through Eq. (2.23) combine to minimize the absolute value of the overall spherapole. When the crystal contains multiple atoms of a certain species in its unit cell, the search process is usually further simplified. For example, rutile TiO_2 has two titanium and four oxygen atoms in its unit cell. The question may arise as to whether each Ti or O atom should be partitioned independently. In this particular case, all the Ti atoms are on equivalent sites and all the O atoms are also equivalent. Thus, it is only necessary to define two neutral polyhedra, one each for Ti and O, to calculate the average potential energy of the NPT model solid. In the cubic In_2O_3 unit cell, there are 32 In atoms occupying two different Wyckoff positions and 48 O atoms at equivalent positions. In this case, all oxygen atoms occupying equivalent positions may be treated as identical. Yet, the In atoms occupying inequivalent sites should be treated as if they were different chemical species altogether in the search for neutral polyhedra. Thus, to obtain the NPT reference for In_2O_3, one needs to define neutral polyhedron for one O atom and then separately for two inequivalent In atoms. Generally speaking, the number of neutral polyhedra needed to be analyzed for the computation of the NPT reference of a crystal is equal to the number of inequivalent elements/sites in its unit cell. From the charge density output of a bulk electronic structure calculation, it is straightforward to iteratively define the neutral polyhedra and compute the average electrostatic potential of the crystal through Eq. (2.15). A Fortran program to define neutral polyhedra from charge density output of any unit cell is available from the author upon request.

2.3.4 *Intermediate charge density*

To analyze the band alignment at an interface, two model solids are initially placed face-to-face to represent the bulk crystals, and, by doing so, they also create an unrelaxed version of the charge distribution at the interface. From the known properties of the model solids, the band alignment condition is known at this stage. The final equilibrium charge distribution of the interface is reached only after the charge and the lattice are fully relaxed, driven by energy minimization. In order to accurately predict or explain the actual band alignment condition, the effect of charge relaxation needs to be estimated. Among processes associated with charge relaxation at the interface, one can be identified with predictable result. A quick comparison of the stacked frozen densities with typical equilibrium distribution shows one clear difference. There is usually a finite difference in the charge density on the surfaces of the two model solids that is not allowed for any charge distribution in equilibrium! Any discontinuity in density has to smooth out to reach equilibrium, leading to a modification of the initial band alignment condition. In order to more accurately account for this effect, the initial density difference should be known. The average charge density on the face of a model solid is readily available when the neutral polyhedra or the Bader atoms are defined. However, that may not be the most appropriate representative of the density that needs to be smoothed out. As the density equilibration process spreads out over a considerable distance, a more appropriate quantity to use for each crystal is the average density over a more extended region between ion cores. There is arbitrariness in how such an "intermediate density" may be defined. With both NPT and AIM analyses of the bulk charge distribution available, a convenient way to define this is to use the average density in the volume between the NPT and the Bader boundaries. For example, an As Bader atom is analyzed to have a volume of 28.6 $Å^3$ and a surplus of 0.681 electron in GaAs, while an As neutral polyhedron has a volume of 22.4 $Å^3$. These lead to an intermediate density of 0.681/(28.6-22.4) = $0.110\, e/Å^3$ for GaAs. Note that exactly the same intermediate density would have been calculated if the Bader and NPT properties of a Ga atom were used instead! For a perovskite SrTiO3 compound with more than one cation in the unit cell, one can use the volume and charge data for the only anion of the system, oxygen, to find the intermediate electron density as 0.177 e/$Å^3$ for STO.

2.3.5 *Transferability of atomic/ionic density*

The main purpose of model solids is to be able to explain or predict the band alignment condition at interfaces well. This ability more or less rests with the charge distribution of the actual interface not to deviate far from that created by the model solids for the interface. Discussions so far have suggested that the more transferable the atoms/ions used to build the model solid are, the less likely will such atoms/ions relax at an interface. Therefore, it is desirable to learn which of the two main model solids, NPT and AIM, has atoms/ions that are more transferable? Due to the difference in chemical bonds, the answer may be different for covalent and ionic solids. Here, the transferability or interchangeability of atoms/ions defined by these methods is studied among different crystals that contain the same atoms/ions. To meaningfully compare the transferability of a partitioned charge density of an atom, the local chemical environment of the atom should remain similar. This limits the study to materials from the same family, i.e. with the same crystal structure. Among large classes of materials with the same lattice structure, semiconductors with the zincblende structure are chosen to represent covalent compounds and perovskite oxides for ionic compounds. Since the spherapole of an atom or ion is the property that directly affects the average potential energy, it is used here to assess the transferability of an atom or ion among different compounds containing the same atom. In Table 2.4, the spherapoles of the neutral polyhedron and the Bader atom for each cation and anion are compiled for zincblende semiconductors. It shows that the spherapoles of either cations or anions are approximately transferable among different zincblende compounds, as already explicitly demonstrated in Figs. 2.8 and 2.9. In comparison, the Bader atoms in Table 2.4 are clearly not interchangeable among different

Table 2.4. Spherapoles ($e\text{Å}^2$) of neutral polyhedra (NPT) and Bader atoms of cations and anions of the zincblende semiconductors. Spherapoles of a particular element are listed in two columns, with the identity of the accompanying element in the semiconductor indicated by rows.

III–V Cation	Al NPT	Al Bader	Ga NPT	Ga Bader	In NPT	In Bader
With P	4.90	0.87	4.59	3.00	5.87	3.91
With As	5.06	1.06	4.71	3.35	6.02	4.25
With Sb	5.22	1.51	4.82	4.20	6.11	5.04
III–V Anion	P NPT	P Bader	As NPT	As Bader	Sb NPT	Sb Bader
With Al	5.76	12.52	6.48	13.42	8.59	15.92
With Ga	5.87	8.49	6.59	9.08	8.74	10.66
With In	5.99	8.78	6.74	9.46	8.88	11.21
II–VI Cation	Mg NPT	Mg Bader	Zn NPT	Zn Bader	Cd NPT	Cd Bader
With S	3.95	0.39	6.76	4.83	9.68	7.46
With Se	4.04	0.44	6.81	5.08	9.76	7.76
With Te	4.44	0.55	6.93	5.54	9.86	8.28
II–VI Anion	S NPT	S Bader	Se NPT	Se Bader	Te NPT	Te Bader
With Mg	6.01	11.35	7.15	13.13	9.80	17.35
With Zn	6.04	8.94	7.19	10.12	9.74	12.72
With Cd	6.12	8.95	7.30	10.17	9.87	12.77

Source: Reproduced from Tung *et al.*[15] open access.

zincblende semiconductors. From the data for perovskite oxides in Table 2.5, both the NPT and the AIM seem to define individual atoms that are approximately transferable among different oxides, although AIM should be given a slight edge as the scatter in Bader spherapole data is smaller than NPT. Based entirely on density transferability, NPT seems to be the preferred technique for all interfaces, except for those between some ionic compounds that may be slightly better handled by AIM.

2.4 Library of Model Solid Properties

In the development of advanced technologies, a wide range of materials have been employed and explored in various capacities for device applications. Therefore, the band alignment condition at the interfaces of a wide variety of solid materials is important to know and understand. To explain and model the properties of interfaces of scientific interest and technological importance, a database of the properties of model solids constructed for various crystalline materials is required. Tables 2.6, 2.7 and 2.8 list the properties of different classes of materials commonly encountered in

Table 2.5. Spherapoles (eÅ^2) of neutral polyhedra (NPT) and Bader atoms of the cations of II–IV-Ox perovskite oxides. Spherapoles of a particular element are listed in two columns, with the identity of the accompanying cation in the oxide indicated by rows. Conducting compounds are marked by asterisks. For example, the spherapoles of the Ca neutral polyhedron and the Ca Bader atom in $CaTiO_3$ are 9.08 eÅ^2 and 4.26 eÅ^2, respectively.

A-Site Cation	Ca NPT	Ca Bader	Sr NPT	Sr Bader	Pb NPT	Pb Bader	Ba NPT	Ba Bader
With Ti	9.08	4.26	11.08	5.95	13.84	9.07	14.27	8.74
With Sn	9.69	4.29	11.67	6.01			14.86	8.96
With Hf	9.81	4.29	11.74	6.06				
With Zr	9.92	4.31	11.87	6.05	14.46	9.46	14.97	9.05
With Ru*			11.24	5.98			14.33	8.91
With Nb*			11.27	6.03			14.68	8.95
With V*			10.75	5.98			13.99	8.86
With Ir*							14.52	8.85
B-Site Cation	**Ti NPT**	**Ti Bader**	**Sn NPT**	**Sn Bader**	**Hf NPT**	**Hf Bader**	**Zr NPT**	**Zr Bader**
With Ca	7.75	4.01	10.42	5.71	10.35	5.28	10.91	5.70
With Sr	7.88	4.00	10.52	5.63	10.41	5.27	11.01	5.67
With Pb	7.98	3.94					11.04	5.85
With Ba	8.05	3.96	10.67	5.78			11.11	5.70

Source: Reproduced from Tung *et al.*[15] open access.

Table 2.6. The crystal structure and lattice parameters of materials for various applications. The calculated VBM positions and the averaged surface electron densities of the NPT model solids are listed. A bulk reference for the VBM of each material is also calculated from NPT. The VBM positions for AIM were calculated using a simple sum of the spherapoles of the Bader atoms.

Material	Crystal Struc.	Lattice Param. (Å)	NPT (eV)	Density (Å^{-3})	Bulk (eV)	AIM (eV)
Al2O3	Rhomb	a=4.82, c=13.12	−4.47	0.315	−5.66	−3.73
AlN	Hex	a=3.13, c=5.02	−3.43	0.284	−4.50	−3.84
Bi2O3	Monoc	4.22×3.67×5.58	−4.07	0.175	−4.73	
BiFeO3	Rhomb	a=5.59, c=13.91	−3.81	0.230	−4.68	−2.16
CdS	Wurtz	a=4.17, c=6.78	−4.81	0.087	−5.14	−4.85
CH3NH3PbBr3	Orthorh	8.07×8.78×12.08	−5.02	0.113	−5.45	
CH3NH3SnI3	Orthorh	8.51×9.15×13.16	−4.39	0.099	−4.77	

(*Continued*)

Table 2.6. (*Continued*)

Material	Crystal Struc.	Lattice Param. (Å)	NPT (eV)	Density (Å^{-3})	Bulk (eV)	AIM (eV)
CIGS	Tetra	a=5.7, c=11.4	−3.47	0.098	−3.84	−3.58
CsPbBr3	Cubic	a=5.949	−4.76	0.054	−4.96	
CsPbI3	Cubic	a=6.275	−4.53	0.026	−4.63	
CsSnI3	Cubic	a=6.286	−3.88	0.047	−4.05	
Cu2O	Cubic	a=4.25	−3.06	0.218	−3.88	
Cu2S	Cubic	a=5.55	−1.52	0.152	−2.09	
Cu2S	Monoc	9.11×5.93×5.63	−4.72	0.097	−5.09	
Cu2Se	Monoc	13.81×12.63×15.65	−2.10	0.170	−2.74	
Cu2Te 224	Cubic	a=5.65	−5.22	0.082	−5.53	
Cu2Te	Hex	a=4.3, c=8.52	−2.50	0.129	−2.98	
Cu2Te	Monoc	7.693×3.803×7.248	−1.76	0.156	−2.35	
Cu2Te_225	Cubic	a=6.15	−1.63	0.129	−2.12	
CuGaSe2	Tetra	a=5.6, c=11.2	−3.36	0.101	−3.74	−4.92
CuInSe2	Tetra	a=5.81, c=11.62	−3.57	0.090	−3.91	−5.00
Fe2O3 α	Rhomb	a=5.09, c=13.77	−4.14	0.256	−5.10	
Ga2O3 β	Monoc	12.28×3.05×5.82	−5.01	0.245	−5.93	
GaN	Hex	a=3.24, c=5.29	−3.65	0.236	−4.54	−3.31
Gd2O3	Monoc	14.14×3.56×8.76	−4.82	0.219	−5.64	
HfO2	Monoc	5.09×5.15×5.3	−4.31	0.281	−5.37	−2.14
HfSiO4	Tetra	a=6.55, c=5.96	−5.01	0.303	−6.15	−3.56
HgCdTe	Tetra	a=4.58, c=6.48	−3.84	0.070	−4.10	
In2O3	Cubic	a=10.18	−5.28	0.201	−6.03	−3.91
InGaAs	Tetra	a=4.14, c=5.87	−3.05	0.093	−3.40	
InN	Hex	a=3.55, c=5.73	−3.43	0.176	−4.09	−2.74
InSe β	Hex	a=4.04, c=1.743	−3.75	0.081	−4.06	
Lu2O3	Cubic	a=10.26	−4.93	0.220	−5.75	
MgO	Cubic	a=4.194	−4.12	0.296	−5.23	−3.55
RuO2	Tetra	a=4.48, c=3.11	−2.94	0.269	−3.95	−1.76
Si3N4 α	Trig	a=7.75, c=5.62	−3.02	0.293	−4.12	
SiC 4H	Hex	a=3.08, c=10.07	−1.61	0.293	−2.72	−2.56
SiC 6H	Hex	a=3.08, c=15.1	−1.57	0.300	−2.70	
SiC β	Cubic	a=4.35	−1.68	0.274	−2.71	−2.16
SiO2 α	Trig	a=4.92, c=5.43	−6.72	0.250	−7.66	−6.30
SnO2	Tetra	a=4.76, c=3.21	−5.05	0.254	−6.00	−3.24
Ta2O5	Tetra	a=3.88, c=26.27	−5.72	0.210	−6.51	−4.19
TiO2_ana	Tetra	a=3.78, c=9.62	−5.35	0.259	−6.33	−4.08
TiO2_rut	Tetra	a=4.6, c=2.96	−4.51	0.281	−5.57	−3.10
ZnO	Hex	a=3.24, c=5.22	−4.84	0.207	−5.62	−4.35
ZrO2	Monoc	5.15×5.23×5.33	−4.52	0.268	−5.53	−2.36
ZrSiO4	Tetra	a=6.62, c=6.00	−5.15	0.298	−6.27	

Source: Reproduced from Tung *et al.*[15] open access.

Table 2.7. Properties of model solids constructed for semiconductors with the zincblende/diamond lattice structure, computed with the hybrid HSE functional. The average potential energy and the VBM position for each NPT model solid are given, as is the VBM position for the non-polar Bader model solid. From the intermediate density and the VBM position of the NPT model solid, a bulk reference is established for each semiconductor.

Semiconductor	Lattice Const.	VBM-Avg. Pot. (eV)	Non-polar AIM VBM (eV)	Intm. Density (e/A^3)	NPT Avg. Pot. (eV)	NPT VBM (eV)	Bulk Ref. VBM (eV)
AlP	5.45	4.55	−5.43	0.13	−7.93	−3.39	−3.89
GaP	5.45	4.76	−3.81	0.11	−7.76	−3.03	−3.45
ZnS	5.45	5.05	−5.21	0.07	−9.56	−4.49	−4.76
Si	5.45	5.67	−3.17	0.12	−8.16	−2.51	−2.97
AlAs	**5.66**	**4.40**	**−5.23**	**0.12**	**−7.66**	**−3.28**	**−3.73**
GaAs	**5.66**	**4.57**	**−3.70**	**0.10**	**−7.49**	**−2.95**	**−3.33**
Ge	**5.66**	**5.42**	**−2.86**	**0.10**	**−7.65**	**−2.26**	**−2.65**
ZnSe	5.66	5.20	−4.92	0.08	−9.34	−4.12	−4.42
BeTe	5.66	4.75	−7.03	0.12	−8.10	−3.35	−3.81
InP	5.87	3.72	−3.84	0.08	−7.04	−3.36	−3.66
InGaAs	5.87	4.12	−3.12	0.10	−7.16	−3.05	−3.42
InAlAs	5.87	4.06	−3.69	0.11	−7.25	−3.18	−3.61
CdS	5.87	4.71	−5.07	0.05	−9.44	−4.71	−4.92
AlSb	6.09	4.42	−4.89	0.11	−7.36	−2.95	−3.36
GaSb	6.09	4.55	−3.39	0.08	−7.22	−2.70	−2.99
InAs	6.09	3.54	−3.78	0.07	−6.78	−3.28	−3.55
CdSe	6.09	4.70	−4.88	0.05	−9.13	−4.42	−4.60
ZnTe	6.09	5.20	−4.55	0.09	−8.92	−3.70	−4.04
InSb	6.48	3.59	−3.61	0.06	−6.62	−3.05	−3.28
CdTe	6.48	4.70	−4.63	0.06	−8.76	−4.04	−4.27
HgTe	6.48	5.65	−4.18	0.08	−9.36	−3.71	−4.01

Source: Reproduced from Tung *et al.*[15] open access.

Table 2.8. Charge distribution in bulk perovskite oxide analyzed by the method of neural polyhedra theory. The potential energy, face density, and VBM are the average potential energy, the charge density at the surface, and the valence band maximum position calculated with HSE functional, respectively, of model solids constructed from neutral polyhedra of bulk oxides. The average charge density in the volume bound by neutral polyhedra and Bader atoms is listed as intermediate density.

Oxide	Latt. Const. (A)	Interm. Dens. (e/A^3)	Avg. Pot. Eng. NPT (eV)	NPT VBM (eV)	Generic AIM VBM (eV)	Bulk Ref. VBM (eV)
LaAlO3	3.80	0.22	−14.17	−3.79	−1.66	−4.64
SrTiO3	3.94	0.18	−12.83	−4.38	−2.62	−5.06
LaGaO3	3.94	0.19	−13.23	−4.41	−2.15	−5.12
NdGaO3	3.94	0.18	−12.54	−4.34	−2.17	−5.02
NaTaO3	3.97	0.15	− 9.20	−4.94	−3.61	−5.51
PbTiO3	3.97	0.19	−13.96	−4.05	−2.33	−4.77
BaTiO3	4.02	0.18	−13.65	−4.60	−2.68	−5.29
NaNbO3	4.02	0.15	−10.29	−5.10	−3.76	−5.66
KTaO3	4.02	0.18	−10.03	−4.97	−3.47	−5.64
KNbO3	4.07	0.17	−11.04	−5.13	−3.64	−5.77
CaSnO3	4.07	0.16	−12.20	−4.78	−2.92	−5.37
SrSnO3	4.12	0.16	−12.70	−4.85	−2.86	−5.46
CaHfO3	4.12	0.15	−11.83	−4.65	−2.73	−5.20
CaZrO3	4.15	0.15	−11.86	−4.73	−2.80	−5.29
SrHfO3	4.15	0.16	−12.41	−4.65	−2.60	−5.25
BaSnO3	4.18	0.17	−13.50	−4.99	−2.89	−5.63
SrZrO3	4.18	0.15	−12.41	−4.74	−2.68	−5.32
PbZrO3	4.18	0.17	−13.53	−4.84	−2.88	−5.49
NdInO3	4.18	0.15	−11.39	−4.38	−2.12	−4.93
LaInO3	4.22	0.14	−11.79	−4.52	−2.17	−5.07
BaZrO3	4.22	0.17	−13.34	−4.80	−2.63	−5.42
SrVO3	3.84	0.21	−13.53	−0.98	0.71	−1.78
CaTiO3	3.88	0.17	−12.34	−4.23	−2.62	−4.89
YTiO3	3.88	0.20	−13.86	0.97	3.00	0.23
KTiO3	3.88	0.18	−11.12	−5.18	−4.11	−5.86
LaVO3	3.88	0.22	−15.38	0.33	2.58	−0.48
BaVO3	3.94	0.20	−14.24	−1.57	0.25	−2.33
LaRhO3	3.97	0.19	−14.76	−2.28	0.00	−3.01
SrRuO3	3.97	0.18	−14.00	−2.47	−0.71	−3.15
SrNbO3	4.02	0.18	−13.14	0.90	2.84	0.22
BaRuO3	4.02	0.19	−14.96	−2.68	−0.80	−3.41
BaIrO3	4.07	0.19	−13.80	−3.04	−1.13	−3.74
BaNbO3	4.15	0.18	−13.60	−0.31	1.78	−0.98

Source: Reproduced from Tung *et al.*[15] open access.

scientific and device investigations. The charge distribution and electronic structure of these materials were calculated by DFT with the hybrid HSE functional and analyzed by the NPT and AIM techniques.

2.5 Summary

How the energy bands of two solids are aligned at a solid interface is the focus of this book. In order to analyze any energy level alignment situation, the hosts for the energy levels of the two materials should be identified. From the analysis of the charge distribution of an infinite crystal, this chapter describes how a "model solid" can be constructed to serve as the host of the energy bands for a crystalline material. A model solid is an artificial object that has the charge distribution identical to that of an infinite crystal but terminates in all directions on surfaces that are neutral. The energy bands of a model solid can be referenced to the vacuum level outside, in the same way the energy level of an isolated molecule can. And just like for isolated molecules before, the relative position of energy bands of different crystals is established through the empty potential at the site from which the model solid has been removed. This process is graphically demonstrated in Fig. 2.13. The energy band position of a crystal at location #1 is of interest and can be obtained by first removing the charge density of a model solid of that crystal from that location to reveal the empty potential $\bar{V}_{\mathrm{MT}}(1)$ in the vacated hole, as drawn at the bottom of the figure. The energy band position of the model solid with respect to the vacuum level is then studied in isolation, as schematically shown in the upper left part of the figure. Now, imagine putting the model solid back in the hole it came from, where the potential is $\bar{V}_{\mathrm{MT}}(1)$. This empty potential rigidly shifts the energy band position of the model solid #1 from its position with respect to vacuum, as shown in the upper middle of the figure. In the actual example drawn in Fig. 2.13, repeating this process allows the energy band positions for crystal #2 to be obtained and compared with those of crystal #1. Note that with the use of model solids, the alignment of energy levels at a solid interface is analyzed in exactly the same way as that described earlier for molecules. This chapter describes different ways a model solid can be defined for the same solid and points out the need for the model solid to be chosen to fit the chemistry of the solid. Two model solids, NPT and AIM, are described in detail, with their properties listed in tables. The establishment of such a scale of the band edge positions for all the crystalline solids commonly studied for their band alignment conditions is

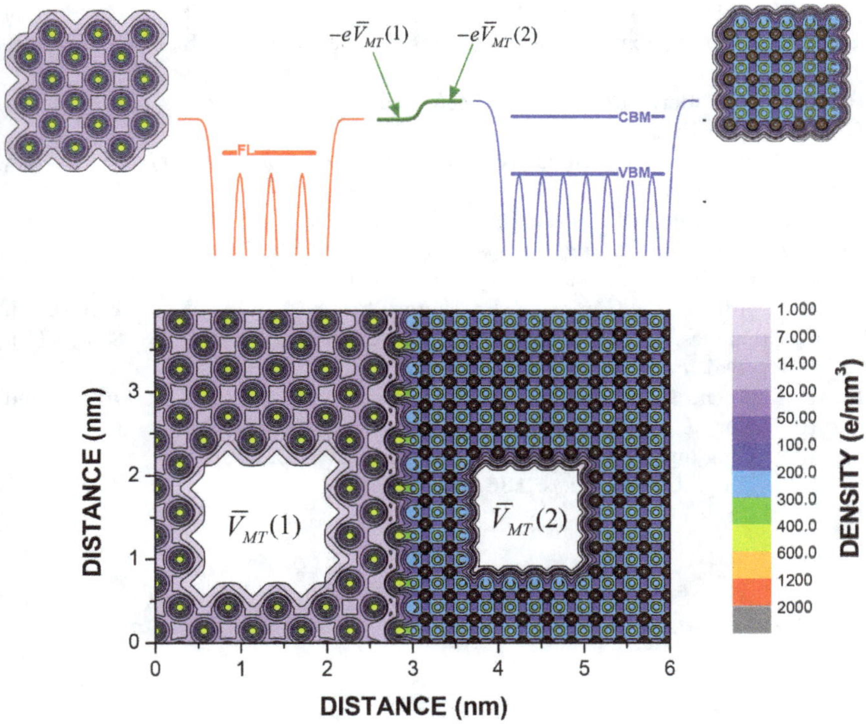

Fig. 2.13. Schematic drawing of the space around three atoms A, B, and C in the unit cell of a crystal where partitions to define optimized neutral polyhedron about every atom of the unit cell have already been set. The solid lines drawn perpendicular to the vectors connecting the atoms represent boundaries of the neutral polyhedra, which are planes in reality. Three small volumes containing identical charges are drawn. The vertical distance from atom A to the boundary separating atoms A and B, d_{AB}^{A}, is indicated.

a prerequisite to the explanation and prediction of interface band alignment by interface charge modeling.

References

1. N.W. Ashcroft and N.D. Mermin, *Solid State Physics*. 1976, New York: Holt, Rinehart and Winston.
2. Y.S. Kim, M. Marsman, G. Kresse, F. Tran, and P. Blaha, *Phys. Rev. B* **82**, 205212 (2010).
3. L. Kleinman, *Phys. Rev.* B **24**, 7412 (1981).
4. L.F. Mattheiss, *Phys. Rev.* **134**, A970 (1964).

5. C. Pisani, R. Dovesi, and C. Roetti, *Hartree-Fock Ab Initio Treatment of Crystalline Systems*, in *Lecture Notes in Chemistry*, V. **48**. 1988, Springer-Verlag: Berlin.
6. R.F.W. Bader, *Atoms in Molecules: A Quantum Theory*. 1994, Oxford: Clarendon Press.
7. G. Henkelman, A. Arnaldsson, and H. Jonsson, *Computa. Mater. Sci.* **36**, 354–360 (2006).
8. W. Kohn, *Phys. Rev. Lett.* **76**, 3168–3171 (1996).
9. R.T. Tung and L. Kronik, *Phys. Rev. B* **94**, 075310 (2016).
10. P. Pyykkö and M. Atsumi, *Chem. Eur. J.* **15**, 186–197 (2009).
11. B. Cordero, V. Gómez, A.E. Platero-Prats, M. Revés, J. Echeverría, E. Cremades, F. Barragán, and S. Alvarez, *Dalton Trans.* 2832–2838 (2008).
12. P. Pyykkö, *Phys. Rev. B* **85**, 024115 (2012).
13. F.H. Allen, O. Kennard, D.G. Watson, L. Brammer, A.G. Orpen, and R. Taylor, *J. Chem. Soc. Perkin Trans.* **2**, S1–S19 (1987).
14. R.T. Tung and L. Kronik, *Phys. Rev.* B **99**, 115302 (2019).
15. R.T. Tung, *J. Appl. Phys.* **136**, 165301 (2024).

Chapter 3

SOLID SURFACES

All solids are finite in size and have to terminate on all sides. Exposed surfaces of a solid crystal have distinctive structures and properties that may vary significantly with how the surfaces are prepared. Surface preparation is an art form and an important field of study in its own right. When adequately prepared, surfaces of crystalline solids can accommodate various chemical and physical processes important for applications. Numerous long-standing scientific journals exist that are dedicated entirely to the science and applications of surfaces.[1] Examples of how surface scientists have been able to control and manipulate the structures and properties of the surfaces for applications have been astonishing to witness. The richness and maturity of surface science notwithstanding, our own discussion of the surfaces will cover only a small part: the determination and explanation of the energy band position with respect to the vacuum level at a crystal surface. Properties of the surface that are related to the energy level alignment problem will be examined. How the electronic states are formed at clean or adsorbate-covered surfaces, what determines the placement of the energy bands of the crystal with respect to the vacuum level, and how the surface potentials and dipoles may be modeled, etc. will be the focus of our discussion. A surface may be viewed as the interface between a "crystal" and "vacuum". As such, the alignment of the crystal bands with respect to the vacuum at a surface perhaps offers the simplest example to examine the general problem of interface band alignment. Since vacuum has no structure and is obviously "simple", one may suspect that the energy level alignment condition for the surface, such as the WF of a metal, is easier to analyze and more accurate to predict than that at interfaces between two solids. This is not the case, however. The replacement of material beyond a plane of a crystal with vacuum to create a surface represents a more severe disruption to the bonds and the charge distribution of the crystal than the replacement with a different solid across the interface. If anything, this makes the explanation of band alignment at a surface harder than that at an interface. In any event, we shall use the metal surface as the first example available to discuss how the magnitude of the band alignment, WF in this case, may be explained by modeling the charge distribution of the surface, albeit only approximately.

3.1 Surface Structure

3.1.1 *Surface crystallography*

The simplest type of surface to think of is a crystalline solid terminated on a crystallographic plane in vacuum without contamination and without lattice reconstruction. Such an ideal surface can be identified simply by the Miller indices of the exposed lattice planes. For a compound crystal, a designation in addition to the crystallographic orientation may be needed to identify the chemical identity of atoms on the topmost layer. Conventionally, the surface plane is assumed to be the x–y plane on Cartesian coordinates. For an unreconstructed surface, the periodicities of the original crystal in the x–y surface plane are preserved, although the periodicity in the surface normal z-direction is lost. Two vectors lying in the surface (x–y) plane, $\vec{a}_1$ and $\vec{a}_2$, may be chosen as the primitive vectors to carry out all possible translational symmetry operations, i.e. all points on the two-dimensional Bravais lattice can be reached by a displacement of $\vec{R}_{\|} = n_1\vec{a}_1 + n_2\vec{a}_2$, where n_1, n_2 are integers, from the origin. The area of the two-dimensional "unit cell" is $A_{2D\_cell} = |\vec{a}_1 \times \vec{a}_2|$. Analysis of point symmetry operations that leave the two-dimensional Bravais lattice unchanged identifies five distinct surface point groups: square, hexagonal, rectangular, centered rectangular, and oblique, as drawn in Fig. 3.1.

Consistent with the 2D periodicity for the atoms, the electric potential and charge distribution for a surface are expected to have the same periodicity, making the description of surfaces most conveniently expressed in the language of Fourier transformation. Through procedures similar to those used for three-dimensional crystals, a 2D reciprocal lattice may be generated with the reciprocal vectors of

$$\vec{b}_1 = 2\pi \frac{\vec{a}_2 \times \hat{z}}{|\vec{a}_1 \times \vec{a}_2|} \quad \text{and} \quad \vec{b}_2 = 2\pi \frac{\hat{z} \times \vec{a}_1}{|\vec{a}_1 \times \vec{a}_2|}. \tag{3.1}$$

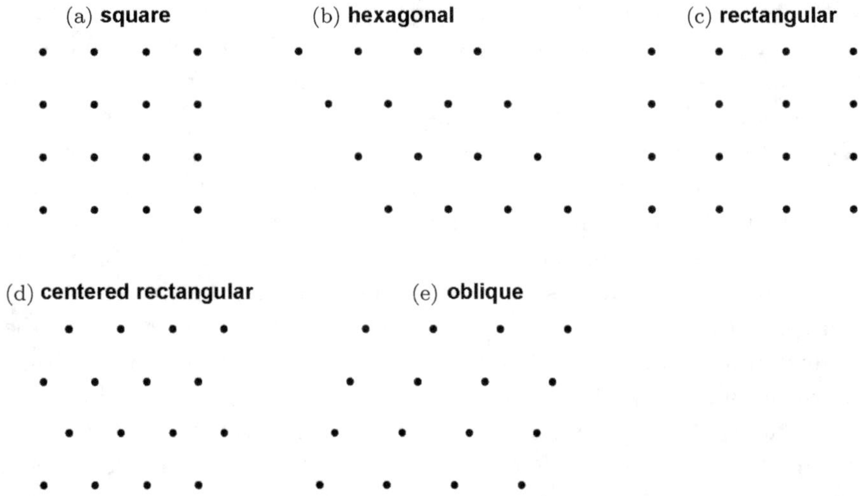

Fig. 3.1. Two-dimensional Bravais lattices. (a) Square, (b) hexagonal, (c) rectangular, (d) centered rectangular, and (e) oblique.

All the points of the reciprocal lattice are reachable from the Γ point by translation with these two reciprocal lattice vectors $\vec{K}_{\parallel} = c_1\vec{b}_1 + c_2\vec{b}_2$ for integers c_1 and c_2. It is easy to verify that $\vec{a}_i \cdot \vec{b}_j = 2\pi\delta_{ij}$, and therefore,

$$e^{i\vec{K}_{\parallel}\cdot\vec{R}_{\parallel}} = 1 \tag{3.2}$$

for all lattice vectors in real space, $\vec{R}_{\parallel}$, and all lattice vectors in reciprocal space, $\vec{K}_{\parallel}$. The area of a 2D Brillouin zone is $|\vec{b}_1 \times \vec{b}_2| 4\pi^2/A_{2D\_cell}$. As before, it is not meaningful to distinguish two vectors differing by an integral number of the reciprocal lattice vector, $\vec{K}_{\parallel}$. Therefore, distinct reciprocal space vectors, $\vec{k}_{\parallel}$'s, may be limited to the first 2D Brillouin zone.

3.1.2 *Surface relaxation, reconstruction, defects, and surface preparation*

The 2D surface unit cell just described, based on the truncated structures of a crystal solid, provides the initial configuration from which the actual equilibrium surface structure evolves. Given enough time and thermal activation to reach equilibrium, the surface relaxes into a structure that may be significantly different. Some degree of surface relaxation of the first two planes is generally unavoidable. The expression "surface relaxation" refers to a rigid shift, in directions perpendicular and/or parallel to the interface, in the positions of all equivalent atoms in a layer. Such a shift does not change the lateral periodicity of the surface nor the size of the unit cell. More commonly, the interplanar distance between the first and the second atomic planes decreases from the bulk distance, leading to a "contraction" of the top surface plane. At many surfaces, atoms occupying equivalent sites of the crystal lattice relax with different lateral and/or perpendicular shifts. A common example of this is the dimerization of atoms at the surface of a crystal to form bonds and lower the surface energy. Such lateral movements cause the surface to "reconstruct" and necessitate a redefinition of the two-dimensional unit cell. Frequently, the new cell vectors are parallel to the original $\vec{a}_1$ and $\vec{a}_2$ but are multiplied by factors of m_1 and m_2, respectively, in which case the surface is known as a $m_1 \times m_2$ reconstruction. The surface unit cell is now $(\mathrm{m}_1 * \mathrm{m}_2)$ times larger, while the 2D Brillouin zone is $(\mathrm{m}_1 * \mathrm{m}_2)^{-1}$ times its original size. Some reconstructed unit cells have primitive vectors that are rotated from the original vectors, in which case the angle of rotation is also specified in the designation. In addition to the various reconstructions found on clean crystal surfaces, investigations of adsorbate-covered surfaces have helped build a large library on the reconstructed and unreconstructed structures of semiconductor and metal surfaces with different coverages of adsorbates. The structures of some of the most famous reconstructed surfaces of clean semiconductors are briefly described in this section. For researchers working in the general area of semiconductor interfaces, familiarity with these surfaces is almost mandatory! Discussions so far seem to suggest that a reconstructed unit cell always has integral multiple times the size of the original 2D unit cell. This is only correct for the vast majority of the known surfaces, as a few extreme exceptions are known where the atomic arrangement of the surface layer(s) is incommensurate with the crystal lattice!

To clearly reveal the relationship between surface properties and atomic structure, impurity-free surfaces with precise orientation and uniform atomic structure need to be prepared. Advances in material fabrication have made available single-crystal surfaces with any orientation of a wide selection of materials. Getting rid of chemical contamination and achieving high structural quality of the surface still requires much experience and hard work in a controlled environment. Processes involving etching, oxidation, ion sputtering, annealing, and overgrowth may be required to achieve pristine surfaces *in situ* in an ultrahigh vacuum environment. Our discussions of surface periodicity may have conjured up the image of a uniform and perfect structure for the surface under study. This is a gross idealization of the structures commonly found on real surfaces. Despite the utmost care, impurities and contaminations are always of concern for the surfaces. Even when a surface can be prepared to be atomically clean, there are structural imperfections ranging from steps and point defects to facets and macroscopic roughness. While surface defects are unavoidable, they can be managed by careful preparation to densities low enough not to significantly affect either the spectroscopic investigation of the surface or the utilization of the surface for the fabrication of technologically important device structures.

3.1.3 *Surface characterization*

Surface characterization techniques are briefly discussed here, as they are routinely used by researchers in many fields.[2] Auger electron spectroscopy (AES) is a standard vacuum accessory to provide information on the chemical composition of the surface under study. High-energy electrons, ~3–5 keV, are used to excite atoms near the surface region and create core holes. An energy-conserving Auger process may then be triggered by a core hole through the deexcitation of a high-lying electron to fill the core hole and the simultaneous promotion of a second electron on this atom to an energy level high enough to escape the surface and be detected. In the spectrum of electrons emitted from the surface, those that have gone through the Auger process have discrete energies reflecting the differences in atomic levels, thus allowing chemical identification of atoms near the surface. The magnitudes of the Auger signals from different chemical elements can then give semi-quantitative information on the chemical composition of the surface region. AES is an essential and convenient technique to monitor the cleanliness of the surface.

The structural order and periodicity of the surface may be conveniently studied by electron diffraction techniques. Electrons, ~20–150 eV, employed in the low energy electron diffraction (LEED) technique have small penetration depth into the surface and thus are sensitive to the structural order in only one or two topmost atomic layers of the surface. Examples of LEED patterns of clean and adsorbate-covered silicon surfaces are shown in Fig. 3.2. LEED is a convenient and robust technique to get a quick glance of the condition of the crystal surface. The overall quality of the LEED pattern is indicative of the quality of the surface, e.g. a low background intensity indicates little diffuse scattering and excellent quality of surface order. The intensities of LEED beams carry additional information about

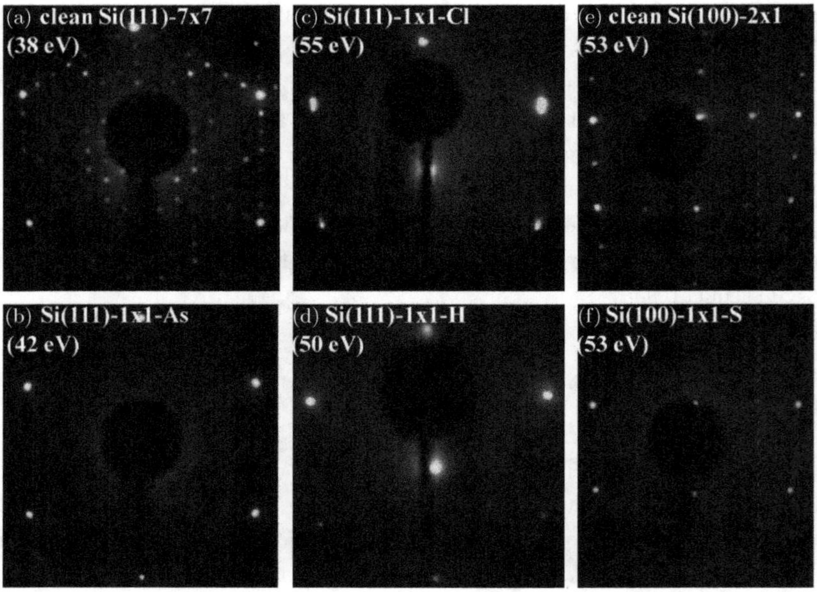

Fig. 3.2. Reverse-view low-energy electron diffraction (LEED) patterns of clean and adsorbate-covered silicon surfaces. (a) Clean Si(111) 7×7, (b) Si(111) 1×1-As, (c) Si(111) 1×1-Cl, (d) Si(111) 1×1-H, (e) clean Si(100) 2×1, and (f) Si(100) 1×1-S. *Source*: Reprinted from Li *et al.*[26] with permission from Elsevier.

the spatial distribution of electron density near the surface. Before advanced scanning techniques were invented, the dependence of the individual beam intensity on the electron energy was the primary technique researchers relied on to determine the fine structure of the surface. Besides structural order in the arrangement of surface atoms, diffracted LEED beams can also reveal morphological details of the surfaces, through the use of additional *in situ* electron optics. Magnified real-space images of the surface can form with a selected reflected beam in a low-energy electron microscope (LEEM). Various surface dynamic processes, including step movements during crystal growth and phase transitions between unreconstructed and reconstructed structures with temperature cycling, have been revealed by continuous real-time LEEM imaging.[3]

A technologically important process at the surface, for which real-time feedback is essential, is the epitaxial growth of material in vacuum or in a reduced-pressure environment. Its excellent surface sensitivity notwithstanding, a beam of low-energy electrons is difficult to implement inside a vacuum apparatus to monitor the crystal growth. The technique of reflective high-energy electron reflection (RHEED) uses a glancing incident angle and thus can be implemented without interfering with the growth process in vacuum. The small incident and reflection angle, ~1–4°, significantly reduces the penetration depth of the ~15–40 keV electron beam to less than ~5 nm to give this technique enough surface sensitivity. During crystal growth, RHEED can monitor surface periodicity in the direction perpendicular to the incident beam in real time. More than revealing just the static periodicity of the surface during growth, its intensity can provide valuable information on the status of active crystal growth.[4] RHEED intensity is at a maximum when the surface is perfectly smooth with few adatoms and drops off when the densities of steps and islands increase,[5] as schematically illustrated in Fig. 3.3. Since it is in synch with the growth of each monolayer, the intensity of any reflected beam serves as a reliable layer-by-layer thickness monitor of the growth. Based on the RHEED intensity, one can terminate the growth at a specific point of the growth cycle for improved surface smoothness or perform "growth interruption" that allows the surface to heal before growing a different material. For certain quantum semiconductor devices requiring thickness control at the atomic level, such an *in situ* capability is crucial. For various

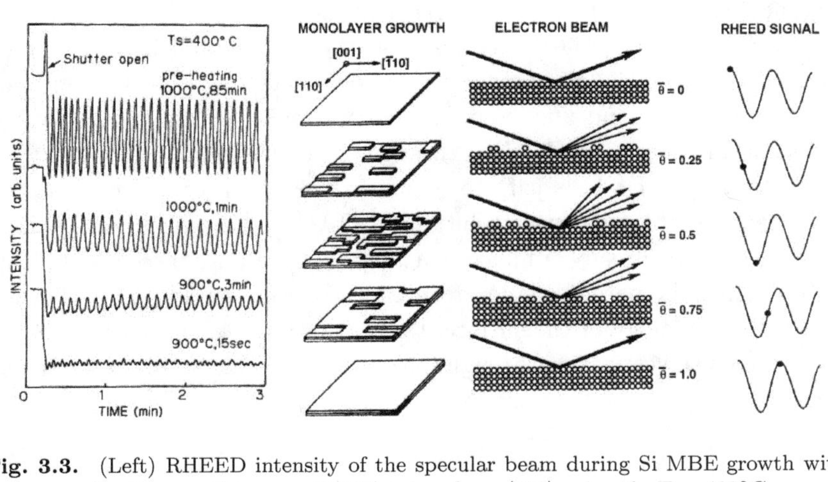

Fig. 3.3. (Left) RHEED intensity of the specular beam during Si MBE growth with various preheating conditions on Si(100) taken from ⟨110⟩ azimuth. T = 400°C.

Source: Reprinted from Sakamoto *et al.*[4] with the permission of AIP Publishing. (Right) Real-time representation of the formation of a single complete layer and corresponding variation in RHEED intensity. Reprinted from Neave *et al.*[5] with permission from Springer Nature.

advantages it offers, a glancing high energy electron gun has become a required attachment of modern growth facilities.

Ion beams have been used in various capacities in the processing, fabrication, and characterization of thin film structures and devices. The scattering of energetic ions off stationary nuclei in solids has well-established cross-sections that depend on the masses of the ion and the nucleus, the ion energy, and the scattering angle. Ion scattering thus carries information that may be used for chemical analysis. Rutherford backscattering spectrometry employing MeV ions and medium energy ion spectrometry (MEIS) using 100–200 keV ions are convenient techniques to obtain composition depth profiles of thin film material. These techniques are not surface sensitive *per se*. When combined with ion channeling, however, valuable information of the surface structure can still be obtained.[6] Low-energy ion scattering has high surface sensitivity, but data interpretation is complicated due to multiple scattering. In addition, the detectability of light elements is poor with ion beam techniques.

By far the most powerful and widely used techniques to provide details of the surface atomic structure in real space are the scanning tunneling microcopy (STM) and its related scanning techniques.[7] Using piezoelectric drives to control the three-dimensional movement of a fine tip, STM can map the height of the surface at cryogenic temperatures with sub-angstrom accuracy, as the STM image of a reconstructed Si(111) 7×7 in Fig. 3.4 shows. Alternatively, the forces between the surface and a tip mounted on a cantilever of the microscope can be detected through the magnitude of the cantilever bend and used to generate high-resolution maps of the surface at ambient temperatures, through the operation of an atomic force microscope (AFM). It is safe to assume that the design, the operation principle, and the amazing resolution of STM, AFM, and other scanning microscopies are already familiar to most readers because these instruments are ubiquitous in materials laboratories in the world. In addition to showing detailed morphology of the surface, the STM can also be used to conduct spectroscopic investigation through the voltage dependence of the tunneling current.

3.1.4 *Structures of selected semiconductor surfaces*

Common elemental and compound semiconductors have simple crystal structures and small unit cells. Yet the surfaces of these semiconductors are anything but simple and some have presented challenges for scientists to

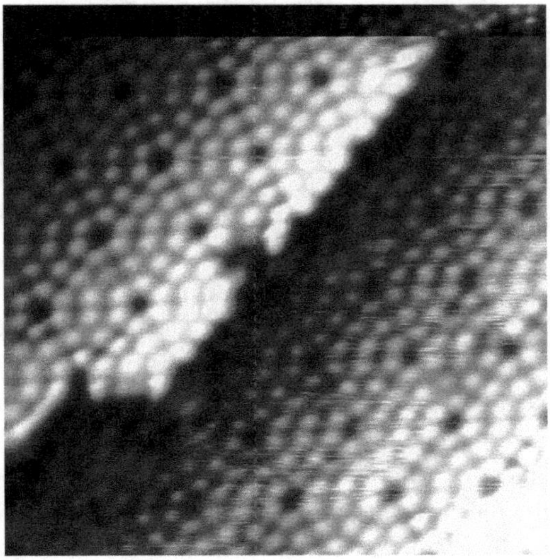

Fig. 3.4. Scanning tunneling microscopic image of 7×7 Si(111) surface. Courtesy and property Prof. V. LaBella.

identify their exact atomic structures. When a diamond/zincblende crystal is terminated at a surface, its network of directional bonds is interrupted and all atoms on the surface plane are left with dangling orbital(s). Clearly, the surface can and will lower its energy by rearranging the positions of its atoms such that the number of broken bonds can be reduced. Based on the minimization of the total energy, a number of *general principles* that seem to govern the formation of semiconductor surface structures have been identified.[8] How these principles, stated here without explanation, actually apply will become clear when actual structures of semiconductor surfaces are described.

1. Atoms on the surface tend to rearrange so as to either "saturate" the surface atoms' dangling bonds through hybridization or to convert them into non-bonding electronic states.
2. In many cases, and in all quasi-one-dimensional ones, surfaces can lower their energies by atomic relaxations leading to semiconducting surface as opposed to metallic surface.
3. The surface structure observed will be the lowest free-energy structure kinetically accessible under the preparation conditions.
4. Surfaces tend to be autocompensated.

5. For a given surface stoichiometry, the atomic geometry is determined primarily by a rehybridization-induced lowering of the surface state bands associated with either surface bonds or (filled) anion dangling bond states.

(a) Si(100): Silicon single crystal with (100) orientation is the most important substrate for the development and fabrication of advanced CMOS devices, even though the reason for this has more to do with the high quality of silicon oxide interfaces fabricated on this surface rather than the structure of the clean Si(100). When a Si crystal is abruptly terminated on a (100) plane, each surface atom has two broken bonds. On a flat terrace, broken bonds on adjacent Si atoms can form σ bonds by tilting toward each other and allowing the orbitals that face each other to overlap. These lateral movements pair the surface atoms up into "dimers" and significantly lower the energy, in accordance with principle #1 stated above. Indeed, dimerization of the surface atoms is the main feature of the clean Si(100) surface, although there are other finer details of the surface structure that vary with surface preparation and temperature. The formation of dimers involves lateral shifts along either the $[011]$ or the $[01\bar{1}]$ direction. Dimers usually form rows in the direction perpendicular to the dimer direction, with the direction flipping by 90° across every step of single atomic height. Even though dimers often appear to be symmetrical at moderate temperature in STM, i.e. with the two Si atoms at the same height, detailed studies have shown that dimers can lower their energy by adopting a buckled asymmetric structure. The tilting of the dimers makes the two Si atoms in a dimer inequivalent, which reduces the total energy according to principle #2 above, by opening up a gap in the π band formed by the (untilted) dangling orbitals on Si atoms. Due to oscillations in the dimers, buckling is not detectable by STM at room temperature. Only at cryogenic temperatures or in the vicinity of structural defects can the buckled dimers be stabilized and clearly imaged by STM. The small gap opened up in the density of states by the buckling of the dimers lowers the surface energy and makes the surface insulating, which otherwise is metallic for symmetric dimers.

(b) Si(111): As the natural cleavage plane of the diamond crystal structure, atomically clean Si(111) surfaces can be obtained by cleaving in vacuum. Cleavage invariably exposes full Si bilayers on both cleaved surfaces, with one dangling bond per surface atom. Obviously, such a surface will lower its energy by lattice relaxation to accommodate the formation of bonds among surface Si atoms. Atoms in the first four layers (two bilayers) of the surface rearrange their positions spontaneously to form a reconstructed (2×1) structure consisting of π-bonded zig-zag chains along the $\langle 01\bar{1} \rangle$ directions, a schematic drawing of which is shown in Fig. 3.5.[9] Chemically etched Si(111) surfaces are hydrogen-terminated and do not reconstruct. However, upon desorption of hydrogen atoms at moderate temperatures, these surfaces also relax to the (2×1) reconstructed structure, which may be understood in terms of the principles stated above. The top-most Si bilayer is bent such that atoms on consecutive rows of the top three layers are almost co-planar. All electrons at the surface are paired in σ bonds except for lone pairs on the top-most Si atoms in filled π bonds. The gap in occupied and unoccupied π bands, due to lattice distortion, lowers the energy and makes the (2×1) surface insulating.

Upon annealing to above ~600°C, clean Si(111) surfaces exhibit unreconstructed 1×1 periodicity regardless of the starting surface structure. Upon cooling to below ~600°C, the 7×7 reconstructed structure becomes the stable structure of the clean Si(111) surface. The transformation from the 1×1 to the 7×7 structure during cooling down has been revealed quite spectacularly by LEEM to be influenced by the presence and the orientation of surface steps.[3] The Si(111) (7×7) is arguably the most famous surface reconstruction of all, and its detailed structure presented a difficult challenge for researchers to identify for many decades. The overall

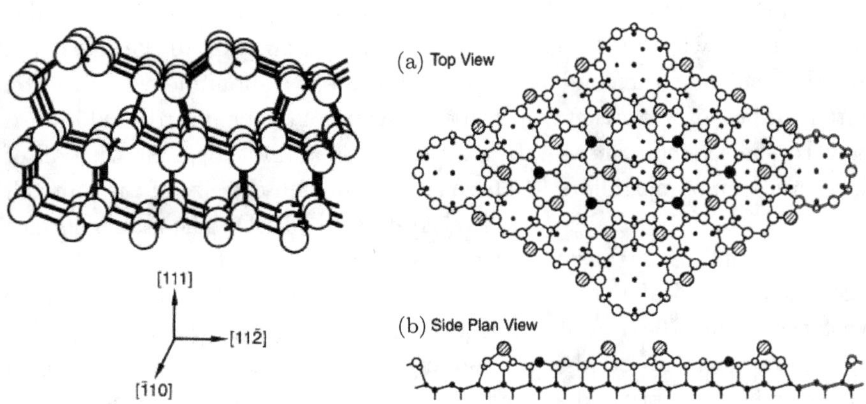

Fig. 3.5. Ball and stick model of clean Si(111) surfaces. (left) The reconstructed 2×1. *Source*: Reprinted from Haneman *et al.*[9] with permission from Elsevier. (right) The DAS model of the Si(111) 7×7. Reprinted from Takayanagi *et al.*[10] with permission from AIP Publishing.

structure of this reconstruction, which was solved in a high-energy electron diffraction investigation and repeatedly confirmed by STM, belongs to the family of $(2m + 1) \times (2m + 1)$ dimer-adatoms-stacking-fault (DAS) models.[10] There are 9 dimers, 12 adatoms, and a stacking fault in a 2D unit cell the Si(111) 7×7, as shown on the right side of Fig. 3.5. Significant rearrangement of positions of atoms in the first two bilayers of Si and the addition of atoms manage to reduce the number of broken bonds per unit cell from 49 in the unreconstructed structure to 19 for the (7×7). From a planar surface, this structure requires adatoms to build and therefore can only be reached after thermal activation. Calculations show that the (7×7) has an energy that is more favorable than the (2×1) but not by much. The closeness in energy is likely the reason that Si(111) (7×7) is the stable structure at ambient temperatures, while the seemingly similar Ge(111) displays a $c(2 \times 8)$ reconstruction. It should be mentioned that the optimum $(2m + 1) \times (2m + 1)$ DAS structure a surface adopts depends on the compressive stress of the surface. Strained Ge or Si_xGe_{1-x} layers grown on Si(111) have been shown to adopt the (5×5) DAS structure.

(c) GaAs(110): The natural cleavage plane for compound semiconductors with the zincblende crystal structure is the non-polar (110). As a result, atomically clean (110) surfaces of compound semiconductors can be prepared with relative ease by cleaving in vacuum, without the need for elaborate *in situ* cleaning and growth capabilities. The unrelaxed zincblende (110) surface contains equal numbers of cations and anions in zigzag chains in a unit cell, with one broken bond per surface atom. For all the III–V and II–VI zincblende semiconductors studied, there is a spontaneous relaxation into a stable surface structure, with the cations of the chains rotated inward toward the semiconductor until each cation is almost coplanar with all of its nearest anion neighbors.[11] There is no change in the size of the unit cell and the structure remains (1×1) according to LEED. The stable surface structure contains a transfer of electrons from the cation toward the filling of the lone pair orbital on the anion. As the cation is pushed to almost in plane with the anions, the orbitals on the cation rehybridize to have mainly planar sp^2 character in its bonds with the anions, while leaving the out-of-plane dangling bond empty and above the CBM, in agreement with principles #1 and #5. As a result of lattice relaxation, the anion dangling bond on the original undistorted surface is filled and lowers its energy to below the VBM of the bulk semiconductor, as prescribed by principle #2. The transfer of an electron from the surface cation to the surface anion preserves surface

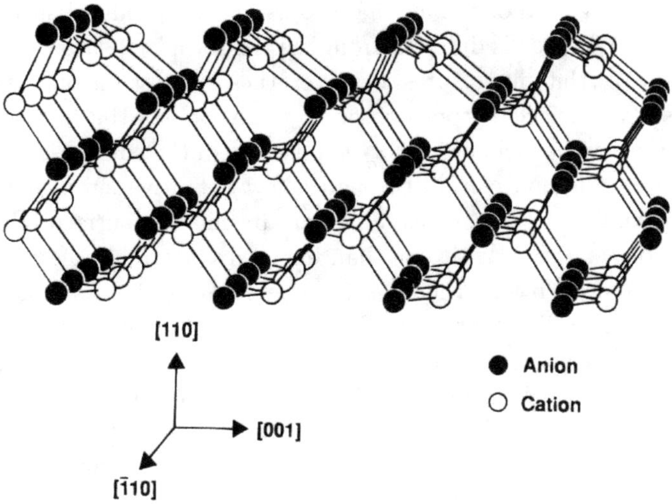

Fig. 3.6. Ball-and-stick model of the relaxed non-polar (110) cleavage faces of zincblende structure binary compound semiconductors.

Source: Reprinted with permission from Duke[11] American Vacuum Society.

neutrality and satisfies principle #4. With no surface states in the band gap, the (110) surfaces of zincblende semiconductors are stable and inert to certain chemical reactions. It should be mentioned that the stable surface structure of the zincblende (110) is independent of how such a surface is fabricated. Cleaving, sputtering, annealing, crystal overgrowth, etc. lead to the same stable structure, as long as the surface is clean and stoichiometric.

(d) GaAs(100): The GaAs(100) and InP(100) surfaces, upon which elaborate structures have been grown and utilized to fabricate advanced optoelectronic devices, are technologically important. The (100) crystallographic planes of the zincblende crystal alternate between the layer of cations and the layer of anions. In principle, the planar (100) surface may end on a full plane of either cations or anions. However, application of the principles on surface structure would rule out such structures and, instead, favor other structures that may depend on the prevailing composition of the surface, in order to reduce the number of dangling bonds and lower the energy. High-quality GaAs(100) surfaces are usually prepared in vacuum by the growth of a buffer layer. At growth temperature, the sticking coefficient of incoming source molecules depends sensitively on the composition of the surface during growth, resulting in the formation of self-limited structures with either the cation or the anion on top. The As-stabilized surface has a structure (2 × 4) that is different from the Ga-stabilized surface structure, c(8×2), at growth temperatures. However, these are not the only structures the GaAs(100) surface has been observed to display. As the surface is made more As-rich, or Ga-rich, or as the temperature of the surface is varied, the surface reconstruction can go through a few definitive surface structures besides a mixture of structures. Since an arsenic-rich surface is expected to have more exposed As atoms than exposed Ga atoms, the main motif of several As-rich reconstructed surface models is the As-dimer, which explains the overall "2 by" periodicity of these surfaces. Out of several (2 × 4) structures that have been observed experimentally to occupy parts of the surface, the structure shown in Fig. 3.7 is known as the $\beta 2(2 \times 4)$.[12] A close examination of this structure, which has the most favorable energy according to theoretical calculations, shows that the surface structure satisfies all the principles stated before. When a GaAs(100) surface receives arsenic in significant excess, an overall c(4 × 4) periodicity is observed. On top of a full As-terminated planar surface, adatom dimers of As-As and/or Ga-As are arranged in a c(4 × 4) pattern. Ga-stabilized GaAs(100) surfaces can display a number of different reconstructions, the most common of which are the c(8 × 2) and the (6 × 6) under moderately Ga-rich surface

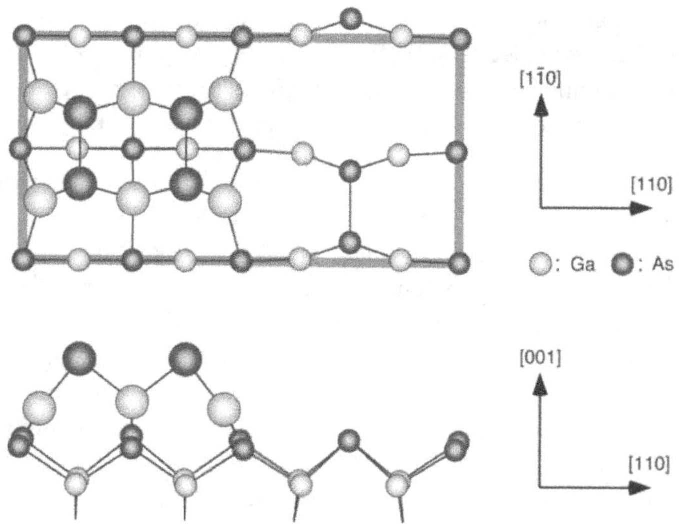

Fig. 3.7. Ball-and-stick model of the As-stabilized (2 × 4) reconstruction of GaAs(100) surface.

Source: Reprinted from Ohtake,[12] with permission from Elsevier.

preparation, and the (4×6) under extremely Ga-rich conditions.[12] The main features of all these Ga-stabilized surfaces are sub-surface Ga-Ga dimers and surface As occupying faulted positions. All observed structures are in agreement with the principles stated above.

3.2 Electronic States at Crystal Surfaces

When a solid crystal is terminated at the surface, bonds that hold the crystal together are broken. Surface structures just described illustrate the drastic changes in the atomic arrangement that can take place in response to the broken bonds, to lower the total energy of the surface. We examine how the electronic states of the crystal are modified to satisfy the boundary condition with vacuum. For simplicity, the surface is assumed to have a uniform atomic structure, even though some forms of non-uniformity may be present on real surfaces. Again, our discussion describes the very last step in a real electronic structure calculation, i.e. when the electronic states calculated actually self-consistently reproduce the charge distribution input for the calculation. Surface electronic states are solved, as usual, through the Schrodinger equation pertaining to the structure of the surface. Within either the Hartree–Fock approximation or the Kohn–Sham approach of the DFT, an equation such as Eq. (2.3) is used to solve for the electronic state. For a crystal surface, we assume that the crystal occupies the space $z \leq 0$, with the last layer of atoms positioned at $z = 0$. It is expected that the first few atomic layers on the surface may be at relaxed positions and the charge distribution in the near-surface region is different from that for the bulk crystal. A depth ($z = z_{\text{Cry-Surf}} < 0$) inside the crystal is identified such that the charge distribution beyond this point is indistinguishable from that in the bulk. This location is the boundary that separates a bulk crystal region from a "surface specific region" (SSR), in which the charge distribution differs from that in the bulk. Even though the electric potential and the charge distribution are no longer periodic in the direction perpendicular to the surface, Bloch theorem still holds for directions parallel to the surface. Each electronic state for the surface is still expressed as a Bloch wave, but with a wave vector, $\vec{k}_{\parallel}$, that is confined to the x-y plane.

To see how an electronic state for the surface may be obtained, we first analyze the vacuum side of the SSR ($z > z_{\text{Cry-Surf}}$). The potential for this region may be written in a Fourier series

$$V(\vec{r}_{\parallel}, z) = \sum_{\vec{K}_{\parallel}} U_{\vec{K}_{\parallel}}(z)\, e^{i\vec{K}_{\parallel}\cdot\vec{r}_{\parallel}}, \tag{3.3}$$

where $\vec{K}_\parallel$'s are the 2D reciprocal lattice vectors. The Fourier coefficients, $U_{\vec{K}_\parallel}(z)$'s, are known for all $\vec{K}$'s and all z's, from the previous iteration in the calculation. A wave function with a specific $\vec{k}_\parallel$ can be written as

$$\varphi_{\vec{k}_\parallel}(\vec{r}_\parallel, z) = e^{i\vec{k}_\parallel \cdot \vec{r}_\parallel} \sum_{\vec{K}_\parallel} w^{\text{surf}}_{\vec{k}_\parallel, \vec{K}_\parallel}(z)\, e^{i\vec{K}_\parallel \cdot \vec{r}_\parallel}, \quad z > z_{\text{Cry-Surf}}, \tag{3.4}$$

where band and spin indices have been suppressed. Equations (3.3) and (3.4) can be plugged into the Kohn–Sham equation, Eq. (2.3), to solve for the $w^{\text{surf}}_{\vec{k}_\parallel, \vec{K}_\parallel}$ coefficients:

$$\left[\frac{\hbar^2}{2m}\left(\left|\vec{k}_\parallel + \vec{K}_\parallel\right|^2 - \frac{\partial^2}{\partial z^2}\right) - \varepsilon_{\vec{k}_\parallel}\right] w^{\text{surf}}_{\vec{k}_\parallel, \vec{K}_\parallel}(z) = e \sum_{\vec{K}'_\parallel} U_{\vec{K}_\parallel - \vec{K}'_\parallel}(z) w^{\text{surf}}_{\vec{k}_\parallel, \vec{K}'_\parallel}(z). \tag{3.5}$$

In theory, the sums in Eqs. (3.3)–(3.5) are over infinite number of 2D reciprocal lattice vectors. However, finer spatial details are unimportant, especially when pseudopotentials are used, justifying the termination of these summations at some appropriate cut-off energy. In reality, the number of terms in these sums should not exceed the number of mesh points used in the real space (N) to avoid overspecification. Since Eq. (3.5) represents coupled second-order differential equations, the solution for the SSR can always be written down if a total of $2N$ independent boundary values are supplied. The requirement that all wavefunctions vanish as z approaches infinity provided N boundary conditions, with the other N coming from the interface with the bulk crystal at $z = z_{\text{Cry-Surf}}$ and the requirement that the wavefunction is normalized.

In the crystal region, $z \leq z_{\text{Cry-Surf}}$, the potential is periodic in three dimensions and is identical to that in a bulk crystal. It is clear that any bulk Bloch function $\psi^{Bloch}_{\vec{k}_j}$ with the same energy and $\vec{k}_\parallel$ would satisfy the Schrödinger equation in this region. A general form for the full wave function on the bulk side of the surface is, therefore,

$$\varphi_{\vec{k}_\parallel, \varepsilon_i}(\vec{r}) = \sum_{\vec{k}_j}^{\vec{k}_\parallel, \varepsilon_i} b^M_{\vec{k}_j}\, \psi^{Bloch}_{\vec{k}_j}(\vec{r}), \quad z \leq z_{\text{Cry-Surf}}, \tag{3.6}$$

where the sum is over all the Bloch states with the same energy and $\vec{k}_\parallel$ ($\vec{k}_j = \vec{k}_\parallel + \vec{k}_{\perp,j}$). It is worth pondering that, for a particular energy and $\vec{k}_{\parallel}$, how many Bloch states there are. Since each energy band of the bulk crystal continuously varies, between band extremes, with $\vec{k}$ in all three dimensions in the repeated-zone scheme, it also varies continuously along any straight line in the Brillouin zone. For each fixed $\vec{k}_{\parallel}$, there are two (local) energy extremes for each band between which each energy is found at two different $k_{\perp}$'s: one with a positive magnitude and one negative. In bulk crystals, these states represent, independently, electrons with velocity component in the +z or −z directions. When the crystal is terminated, these two states become coupled, as one becomes the surface reflection of the other. In equilibrium, the flux of electrons toward the surface at any energy must be equal to that leaving the surface, which requires that the absolute value of the amplitude of these two components be identical. This leaves the relative phase difference between the incoming and the outgoing waves as the only adjustable parameter for such a stationary state, besides the normalization requirement. Since the incoming and outgoing states form a complete set of solutions at this $\vec{k}_{\parallel}$ and energy for the crystal part of the surface, a properly chosen combination of these two states can always satisfy the boundary values at $z = z_{\text{Cry-Surf}}$ for the wavefunction to decay toward vacuum. In other words, such a terminated version of the original bulk states is always allowed at the surface. It should be mentioned that, under certain conditions, bulk states could exhibit "surface resonance", as the amplitude of the wavefunction for the surface region could far exceed that for the crystal region, before decaying toward the vacuum.

At energies and $\vec{k}_{\parallel}$'s where no bulk states are available, there are no contributions of the Bloch kind to the sum of Eq. (3.6). It is still possible for states originating from the surface region to decay toward the crystal region, but only when certain conditions are met. These are true "surface states", as they are peaked at the surface region and decay in both directions away from the surface. Since there are no Bloch states at these energies and $\vec{k}_{\parallel}$'s, solutions to the Kohn–Sham equation in the crystal region, $z \leq z_{\text{Cry-Surf}}$, may still be found that do not satisfy the Bloch condition. Conventionally, this can be accomplished by assuming the wavefunctions to still have an apparent Bloch form, yet with the perpendicular wave vector acquiring an imaginary part, $k_{\perp} = p + iq$. This leads to "evanescent waves" which are bulk wavefunctions that decay with an exponential factor of e^{qz} toward the crystal.[13] Even though the wave vector is assumed to be complex, the eigen energy for such a state still has to be real. Naturally, an evanescent wave in the gap of bulk bands at a certain $\vec{k}_{\parallel}$ can form when it matches the wavefunction with the same energy and $\vec{k}_{\parallel}$ in the surface region that decays

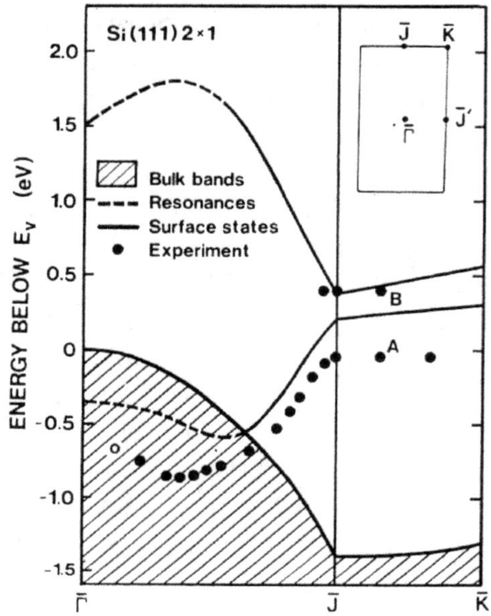

Fig. 3.8. Initial-state energy dispersions for the bonding (A) and antibonding (B) surface states measured experimentally from cleaved Si(111) 2 × 1 surface, compared with the corresponding calculated bands for the π-bonded chain model. Note that the energy is shown with respect to the VBM of silicon. The antibonding surface state (B) is close to and therefore is proposed to pin, the surface Fermi level of this n-type Si at ∼0.40 eV above the VBM.

Source: Reprinted with permission from Martensson *et al.*[14] Copyright 1985 by the American Physical Society.

toward the vacuum. This is not a condition that is easily met. Surface states are observed at the surfaces of some metals and semiconductors, although they may be weakened, or "quenched", by impurities, adsorbates, and other structural defects of the surface. Experimentally measured surface states at cleaved Si(111) 2×1 surface, shown in Fig. 3.8, corroborate earlier discussions of this surface structure.[14]

3.3 Metal Surfaces

3.3.1 *Work function*

The termination of Bloch states at a surface and the formation of electronic states specific to the surface lead to modifications of the charge density in the surface region from that of the bulk crystal. How the bulk energy levels a short depth below the surface are positioned with respect to the vacuum level outside is an important scientific and technological issue that has been subject to experimental and theoretical investigations. At a metal surface, the electrochemical potential is constant in equilibrium and is positioned such that the metal is neutral in the absence of applied electric fields. We shall follow convention and use the term "electrochemical potential" interchangeably with "Fermi level" (FL) for a metal. The "work function", ϕ_M, of the metal surface represents the least amount of energy required to remove an electron from the FL and place it at rest in vacuum. Since the FL is always at a fixed energy μ_M away from the average electrostatic potential energy according to Eq. (2.8), the work function can be written as

$$\phi_M = -E_F = e\bar{V}_M^{\text{surf}} - \mu_M, \tag{3.7}$$

where $-e\bar{V}_M^{\text{surf}}$ is the average potential energy inside the metal surface with respect to the external vacuum level. The common factors controlling the alignment of energy levels at interfaces, in general, are illustrated by this equation. Here, a metal surface may be viewed as the interface between a metal crystal and vacuum, and the characteristic energy levels for the metal and vacuum are, respectively, the FL and the vacuum level. As shown by Eq. (3.7), this difference may be decomposed into an interface term, $-e\bar{V}_M^{\text{surf}}$, that depends electrostatically on the charge distribution and a "bulk" term, μ_M, that contains the many-body effect but does not depend on the surface. There are ordinarily two bulk terms in such an analysis, one for each side of the interface. Only one is present for the surface because the bulk term for vacuum vanishes identically. In the present discussion, with uniform surface structure implicitly assumed, the vacuum level is spatially uniform. For real metal surfaces, the WF varies with the orientation and structure of the surface and experimental measurements of the WF are frequently conducted on surfaces of polycrystalline metals. Structure non-uniformity could lead to non-uniform electric fields just outside the metal surface and make the definition of the vacuum level ambiguous. It is thus important to note that the location where the vacuum level is defined for Eq. (3.7) is not at infinity but rather at a small macroscopic distance outside the surface.

3.3.2 *Work function measurement*

Photoelectron spectroscopy is a technique often used to measure the FL energy of a metal surface with respect to the vacuum level. In such a measurement, a monochromatic light is used to promote electrons in occupied levels near the metal surface to higher unoccupied levels. Some photo-excited electrons have enough energy to escape the metal surface and are energy-analyzed to yield a spectrum of approximately the distribution of occupied electronic states in the metal. Electrons with maximum kinetic energy in the spectrum have come from the highest occupied energy level of the metal, namely, the FL. The energy of the photon subtracted by the maximum kinetic energy of the photoelectron is the work function of the metal surface. Instead of the maximum kinetic energy of the photoemission, the cut-off of the low-energy photoelectron tail in the spectrum can also be used to determine the work function of a metal surface. One notes, however, that on a microscopic scale, the work function of the metal may be non-uniform due to lateral variations in the structure, condition, and orientation of the surface. When the WF is determined by the low-energy limit of the photoemission spectroscopy, the measured value could reflect a weighted average due to the non-uniform distribution of potential energy exterior to the surface.

In laboratories of more conventional settings, Kelvin probe is the standard equipment for WF measurements. Such a probe is a movable metal plate or ring with a precisely known WF that can be placed at a small distance away and parallel to the metal sample under investigation. When the probe and the surface are electrically connected, the difference between the WFs of the surface and the probe leads to a contact potential difference, or Volta potential $\Delta\psi$, in the vacuum gap between the two surfaces. The presence of an electric field in the gap induces charges on the two surfaces: negative on the metal surface and positive charge on the probe in the example drawn in Fig. 3.9(a). This situation is similar to that inside a parallel plate capacitor, except that for the capacitor the charge is induced by applied potential difference, whereas the charge on the Kelvin probe is induced by the contact potential difference. Since $\Delta\psi$ is fixed between two metal surfaces, any change in the width of the vacuum gap leads to a change in the electric field and the amount of induced surface charge on the Kelvin probe. When the Kelvin probe is set up to oscillate back and forth in the direction perpendicular to the surface, an a.c. current flows to change the charge on opposing surfaces. A d.c. bias can be applied between the two surfaces to null the contact potential difference and the current, as schematically shown in Fig. 3.9(b). The bias that nulls the a.c. current is the contact potential difference between the unbiased surfaces. The WF of the surface under investigation can be obtained, with the help

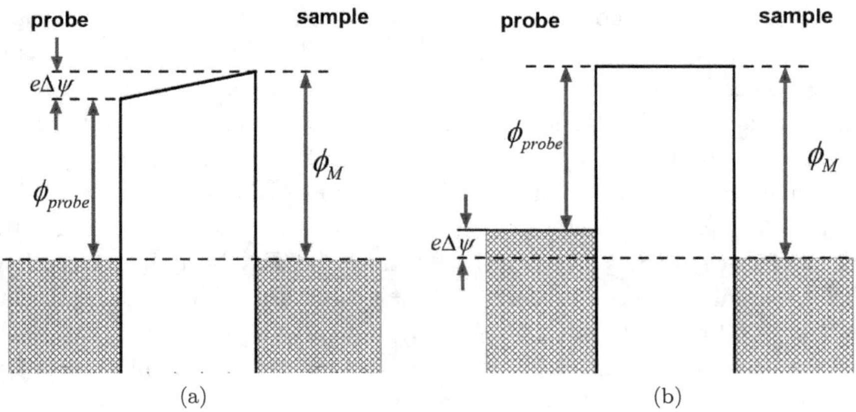

Fig. 3.9. Energy diagram at a small gap between the connected surfaces of a probe and a metal sample. (a) Without applied voltage. (b) With an applied voltage that eliminates the contact potential difference between the two surfaces.

of the known WF of the probe:

$$\phi_M = \phi_{probe} + e\Delta\psi. \tag{3.8}$$

Since the WF is measured with respect to the WF of the probe, investigation of the changes in WF caused by surface treatments should be conducted without modifying the Kelvin probe. For example, a study on how the WF of a metal surface is affected by the adsorption of gas molecules should protect the probe from also receiving gas adsorption! In case of WF non-uniformity, the conventional Kelvin probe obviously gives a measurement of the average of the distribution in contact potential differences between the two surfaces. Along with the birth of the scanning probe microscope, a technique that provides spatially resolved WF maps of metal surfaces has also developed. A scanning Kelvin probe (SKP) uses the conducting tip assembly of an STM as a microprobe. With the oscillatory z-motions provided by the piezoelectric drive, which also provides the scanning mechanism in the two horizontal directions, the additional bias voltage that nulls the a.c. electrical current is recorded as a function of the lateral coordinates. When corrected by the work function of the STM tip, a high-resolution two-dimensional map of the work function of the metal surface is provided.

A phenomenon that helped establish the negative sign of the electron's charge centuries ago and enabled the emitters in vacuum tube transistors can also be used to measure the WF. The thermionic emission of electrons from a heated metal surface in vacuum can be used to measure the

WF, although it is not used for that purpose today because of the high temperature required and the modifications the high temperature may bring to the surfaces. The theory behind thermionic emission is still treated in detail here because the same mechanism is mainly responsible for carrier transport across solid interfaces to be discussed in later chapters. To treat thermionic emission, the conduction band of a metal is assumed to be in the shape of an ellipsoid centered about the Γ point and is filled according to Fermi–Dirac statistics based on the Fermi level at E_F above the conduction band minimum. An energy barrier of $E_F + \phi_M$ above the bottom is present in the $z \geq 0$ region to represent the vacuum. This is the energy that must be surpassed by the kinetic energy associated with the z-component of an electron's motion in order for it to escape.

The current density contribution is the product of the density of such electronic states, $d\vec{k} \cdot 2/(2\pi)^3$ counting both spins, the electronic charge, $-e$, the z-component of its velocity, $\hbar k_z/m_z$, and the probability that these states are occupied. Including only electrons with enough energy and traveling in the +z direction, the total current density in the metal approaching the barrier is

$$J = \int_{k_{\min}}^{\infty} dk_z \frac{2}{8\pi^3} \frac{-e\hbar k_z}{m_z} e^{\frac{-\hbar^2 k_z^2 + E_F}{2m_z k_B T}} \int_{-\infty}^{\infty} dk_x e^{\frac{-\hbar^2 k_x^2}{2m_x k_B T}} \int_{-\infty}^{\infty} dk_y e^{\frac{-\hbar^2 k_y^2}{2m_y k_B T}}, \tag{3.9}$$

where we have used the Maxwell–Boltzmann statistics which is valid when ϕ_M far exceeds $k_B T$. In Eq. (3.9), $k_{\min}(=\sqrt{2m_z(E_F + \phi_M)}/\hbar)$ is the minimum k_z required to overcome the barrier. Only occupied states lying above the plane $k_z = k_{\min}$ in the reciprocal space contribute to the thermionic emission of current. Ignoring quantum mechanical reflections, thermionically emitted current, after carrying out the integrations in Eq. (3.9), is

$$I = -A\frac{4\pi m_{\text{eff}} e k_B^2}{h^3} T^2 \exp\frac{-\phi_M}{k_B T} = -AA_o T^2 \exp\frac{-\phi_M}{k_B T}, \tag{3.10}$$

where A is the area of the metal surface and A_o is known as the Richardson constant. When the mass of the free electron is inserted, A_o has a value of $1.202 \times 10^6\ \text{A} \cdot \text{m}^{-2} \cdot \text{K}^{-2}$. Experimentally measured thermionic emission currents have the correct functional dependence of Eq. (3.10) but are invariably of a smaller magnitude than predicted, likely due to quantum mechanical reflection. This disagreement prompted the replacement of the A_o term with a material-specific "effective Richardson constant", A_{eff}, in

Eq. (3.10) for the description of the thermionic emission process. One notes that, even with uncertainties in the value of A_{eff} and the area A, the WF can still be accurately measured through the dependence of the current on temperature, in a "Richardson plot" of the quantity $\log(I/T^2)$ against $1/T$. One also keeps in mind that the WF measured by thermionic emission is at a very high temperature, which may be different from the WF at ambient temperatures.

Similar to the detection of electrons emitted from light irradiation, the WF measured by thermionic emission may be affected by the inhomogeneity of surface potential when a bias is used to accelerate the electrons into the detector. If an electric field E is used, one should also correct for the so-called "Schottky effect" in the determination of the WF, which is a barrier lowering due to image force. The field-free work function is slightly larger than that extracted directly through the thermionic emission equation:

$$\phi_M = k_B T \ln\{A A_{\text{eff}} T^2/I\} + e\sqrt{eE/(4\pi\varepsilon_o)}. \tag{3.11}$$

When the electric field is large, tunneling through the energy barrier can dominate. In cases of extreme field strength, electrons can be extracted by field emissions from cold metal tips, as in a field emission microscope or in field emission electron guns used in electron microscopes.

3.3.3 *Experimental WF's: Correlations and image-charge models*

The WF is an important property of any solid material and is part of routine characterizations whenever novel compounds and structures are fabricated. There is thus a huge collection of WF data on metals, alloys, and other compounds. Systematic trends in the experimentally measured WF of metals are well known. For example, metals with large electronegativities typically have larger WF. The dependence of the WF on the orientation of the metallic single crystals is usually weak and the majority of the experimental WFs reported in the literature are from polycrystalline surfaces and/or are averages over different surface orientations. WFs measured from the surfaces of elemental metals and compiled in Table 2.2 by Michaelson[15] have been the set of experimental data that many theories and models of the WF aim to explain. Most of the elemental metals crystallize in the f.c.c., b.c.c., and h.c.p. structures. As the WF is the minimum energy required to remove an electron from, or the maximum energy released to add an electron to, the metal crystal, the comparison of the WF to the energies to do so when a metal is still in its atomic form is inevitable. The ionization potential, I, and the electron affinity, A, of a metal atom in isolation are, respectively, the energies needed to remove an electron from and add an electron to an atom. As more metal atoms bond together, the atomic states broaden into molecular orbitals and eventually to energy bands, which closes the gap between I and A and allows a single WF to emerge for the metal. It thus seems that the WF is correlated with, and may even be explained by, the I and A of the metal atom. This correlation is best illustrated by the approximate empirical relationship,

$$\phi_M \approx (I + A)/2 \approx I/2, \tag{3.12}$$

found for many metals. To explore this correlation, the evolution of the I and A for a metal cluster of increasing size has been studied experimentally and calculated theoretically. With an assumed functional for the total energy of a metal cluster, the observed dependence of the I and A on cluster size can be accounted for, although this does not lead to an explanation of the WF of all metals.

Many theoretical calculations have been conducted to explain the magnitudes of the experimental results of Table 2.2. Early numerical calculations were not very accurate as the jellium model of free electrons that ignored the atomic structure of the metal surface was employed. Developments in *ab initio* methods have allowed detailed and routine investigations of the dependence of the WF on various aspects of the metal surface, such as the atomic structure, the orientation, and adsorbates. Even though the experimental WFs can be approximately reproduced by numerical calculations, there is still a desire to quantitatively explain these values from simple pictures based on scientific principles. As suggested by the form of Eq. (3.7), the WF of a metal is completely determined once the charge distribution of its surface is known. Therefore, modeling the surface charge distribution, if possible, is a sure-footed way to explain the WF. The availability of a density-based approach notwithstanding, the WFs are more frequently modeled through the work against the image force that is required to remove an electron from a metal surface. In the immediate vicinity of an electron in an infinite crystal, the instantaneous density of electrons with the same spin is slightly deficient, forming what is called the exchange–correlation hole. The total electron density removed to create the exchange–correlation hole amounts to exactly one electronic charge. Inside a solid, the exchange–correlation hole is generally centered about the electron. As the electron exits the surface, its hole remains inside the surface and flattens in shape. The spatial separation of the electron from its exchange–correlation hole, which can be recognized as the image charge of the electron in classical electromagnetism, leads to an attractive force. This force for an electron at a distance x in front of a flat metal surface is the Coulomb force between that electron and a point charge of $+e$ positioned at a distance of $-x$ inside the metal. The total work required to remove an electron from a position x to infinity in this scenario is

$$W = \frac{1}{4\pi\varepsilon_o}\int_x^\infty dx\frac{e^2}{(2x)^2} = \frac{e^2}{16\pi\varepsilon_o x}, \tag{3.13}$$

assuming that the exact location of a sharp boundary of the metallic surface can be identified. At any real metal surface, with discrete atomic positions and rapid variation of the electronic density in all three dimensions, it is not possible to define a boundary at the surface that is either sharp or flat. Nevertheless, one can imagine that the interaction of a real surface is projectable to that of a sharp mirror plane, in which case a single parameter, x, would allow the WF to be estimated.

Within the image charge concept, the task of predicting the WF for any metal becomes proposing and justifying an x, the only unknown in Eq. (3.13), for that metal. An electron at the Fermi level in a metal can reach a classical turning point when all its kinetic energy has been converted to potential energy. Since the momentum of an electron traveling normal to the surface changes from $\sqrt{2m^*\varepsilon_F}$ to zero, Brodie argued, based on Heisenberg's uncertainty principle, that the turning point is at a distance of $\delta x \sim \hbar/\sqrt{2m^*\varepsilon_F}$ outside the mirror plane.[16] By inserting this distance into Eq. (3.13), the WF for selected metals could be approximately produced. Alternatively, x was proposed, from plasma physics, to be $\sqrt{2\varepsilon_o\varepsilon_F/(ne^2)}$, where n is the density of free electrons in the metal.[17] By inserting this into Eq. (3.13), Halas and Durakiewicz were able to quantitatively explain the experimental WFs of many metals, as shown in Fig. 3.10, although for alkali metals an artificial correction factor needed to be inserted. Based entirely on arguments, the image-force concept is quite successful in explaining the WF. However, it has not been possible to verify the mirror plane assumption underpinning the entire approach.

3.3.4 *Metal surface density compared with atomic density*

As mentioned, the charge distribution at the metal surface is a verifiable approach to estimate the WF. At the surface, the atoms in the outermost

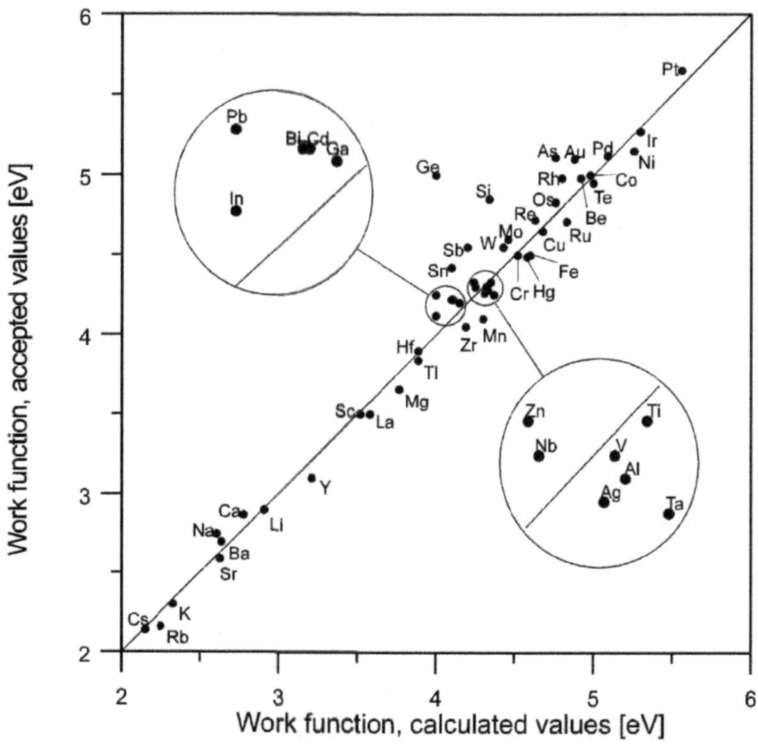

Fig. 3.10. Measured WF values vs. calculated by the metallic plasma model. *Source*: Reprinted with permission from Halas and Durakiewicz.[17] Copyright 1998 by IOP Publishing.

plane have only neighbors within the same plane and the plane further inward to bond with. The loss of all neighbors on the vacuum side of the space seems to return the surface atom to a partially bonded state and invite the comparison of the charge density of the metal surface with that of a plane of metal atoms in an isolated state. For some metals, such a comparison reveals a nearly perfect agreement, as shown in Fig. 3.11(a) for the Pb(111) surface. This shows that the WF for such metals may be accurately modeled by assuming the charge density to have the equilibrium bulk density up to the surface plane and to have the density of unbonded atoms beyond the surface plane. Indeed, WFs so predicted agree well with experiment for these metals. However, only some of the metals have surface charge densities similar to that of isolated atoms. It is more common to find that the charge distribution of the outmost layer of the metal

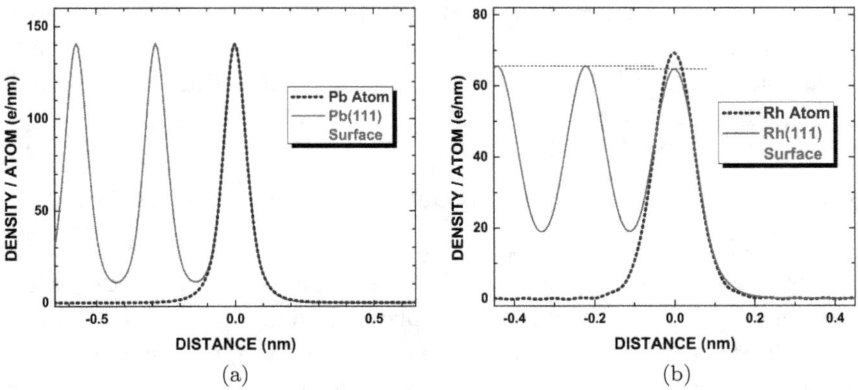

Fig. 3.11. Charge densities integrated over the area of a 2D primitive unit cell are plotted as solid line against the perpendicular distance for (a) Pb(111) and (b) Rh(111). Zero distance on the horizontal axis marks the location of the surface atomic plane. In comparison, the 2D-integrated charge density of an isolated atom is also shown as dotted line.

surface is significantly altered from that of isolated atoms, as shown in Fig. 3.11(b) for the Rh(111) surface. The (111) orientation is chosen in these demonstrations for clarity because it is the f.c.c. direction with the longest interplanar distance to allow layers to be well separated and distinguished in plots such as Fig. 3.11. Another important aspect of charge distribution at the surface, illustrated by the thin dashed lines in Fig. 3.11(b), is the slight drop off in the peak density per plane for the outermost layer of atoms. This drop-off in density toward to surface indicates that the charge relaxation at the surface may involve more than just the first layer of atoms. The significant difference between Figs. 3.11(a) and (b) highlights the important role played by the electronic configuration of the metal atom and likely prevents atomic-density-based models, including those making use of the image force, to predict the WF for all the metals. Notwithstanding, a simple superposition of the atomic densities placed at all the lattice sites of a terminated solid, called "Mattheiss-construct", has been used to simulate the charge distribution of the surface.[18] This amounts to the construction of a model solid using the atomic density as the density of the MSC. Since such a model solid is assembled with identical MSCs, its average potential energy and WF are computed straightforwardly with Eqs. (2.8) and (2.20). Unfortunately, the WFs predicted by this method far exceed, by ~80%, the experimental result. This suggests that the charge distribution at the surface is generally more compact than the density associated with an isolated atom because of bond formation in the crystal. Atomic densities, deliberately shrunk in size, have been shown to better predict the WF, although there is no basis for any artificial modification of the atomic densities in simulating the surface charge density.

3.3.5 *Metal work function modeled by charge distribution*

The WF of a metal surface can be viewed as the difference between the energy levels of two materials: the FL of the metal and the lowest energy state of vacuum. In attempting to explain the magnitude of the WF, we get our first opportunity to try out the density-based approach described in the first two chapters. Therefore, the essential steps of charge distribution modeling are illustrated through the discussion on metal surfaces. This exercise serves to verify the general validity of charge distribution as the foundation of band alignment, although the success with which WFs may be so modeled is only modest. The Ag(100) of Fig. 3.12 is used as an

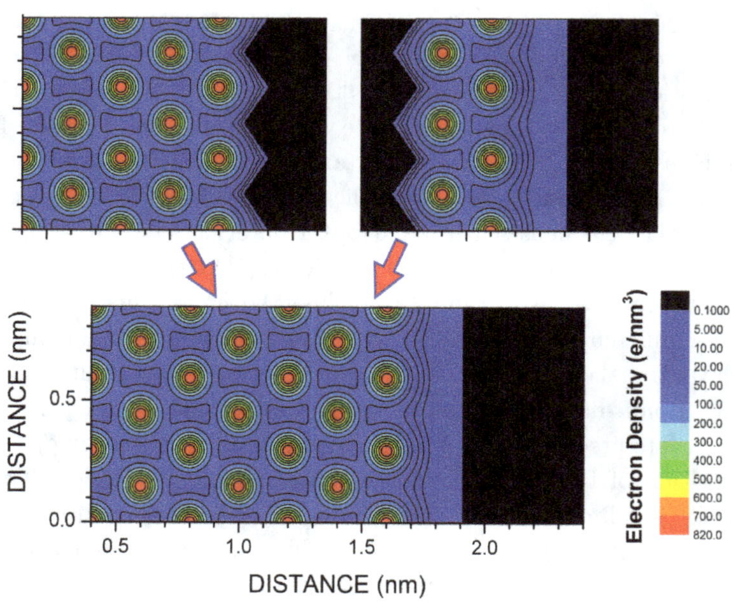

Fig. 3.12. The density of electrons, averaged over the direction perpendicular to the page, numerically calculated for a Ag(100) free surface, shown at the bottom, can be analyzed by partitioning into an NPT model solid of the Ag crystal (upper left), which acts as the host of the Ag energy bands, and a sliver of charge (upper right), known as the surface dipole, which acts as the bath in this analysis.

example to illustrate how its WF may be modeled. Even though the charge distribution is unknown for the entire surface, it is known to be identical to that of an infinite crystal in the interior region. We therefore replace this part of the charge density with that of a model solid of the Ag crystal, as shown in the upper left of Fig. 3.12, as the host of the Ag energy bands. The other host in this assumed "interface problem" is vacuum, which has no charge density. Shown in the upper right of Fig. 3.12 is the SSR that plays the role of the bath in this analysis. Since the surface of the metal and that of the model solid are both electrically neutral without an applied potential, the SSR is also electrically neutral.

The potential energy of each neutral part can be individually calculated and then stitched together to recreate the potential energy of the entire surface. On its own, the model solid of a metal has a work function of ϕ_{ModSol}, which is a known property that may be found in Table 2.6. Separately, the bath contributes a potential energy difference of D_{surf}, which can be added to produce the WF of the metal:

$$\phi_M = \phi_{\text{ModSol}} + D_{\text{surf}}. \tag{3.14}$$

The exact nature of the D_{surf} term, also known as the "surface dipole", becomes clear by noting the following. A surface terminated on the bulk density of the model solid has ϕ_{ModSol} as its WF. The D_{surf} term in Eq. (3.14) emerges when the charge distribution and atomic positions of the surface are allowed to relax. Therefore, the effect of the bath in the upper right of Fig. 3.12 is entirely from charge relaxation of the model solid surface. One should note that the definition of surface dipole is not unique. For example, a different model solid employed in Fig. 3.12 would have led to a different surface dipole being required. For the present analysis, the lack of uniqueness does not present a problem but rather offers a choice. One simply chooses from different model solids the one that "works best", which here means that the effect of the bath is easiest to estimate. In the following, we look for ways to estimate the surface dipole of a metal.

To model the charge distribution at the surface of elemental metals, there is little argument that model solids generated by the NPT method should be preferred. The similarity between surface charge distributions and those of the isolated atoms shown in Fig. 3.11 suggests that model solids assembled with partitioned densities centered about the atomic positions may have charge distributions that resemble real surfaces. The WFs computed for NPT model solids of elemental metals are plotted in Fig. 3.13 against the experimental work function.[19] This is a plot of the

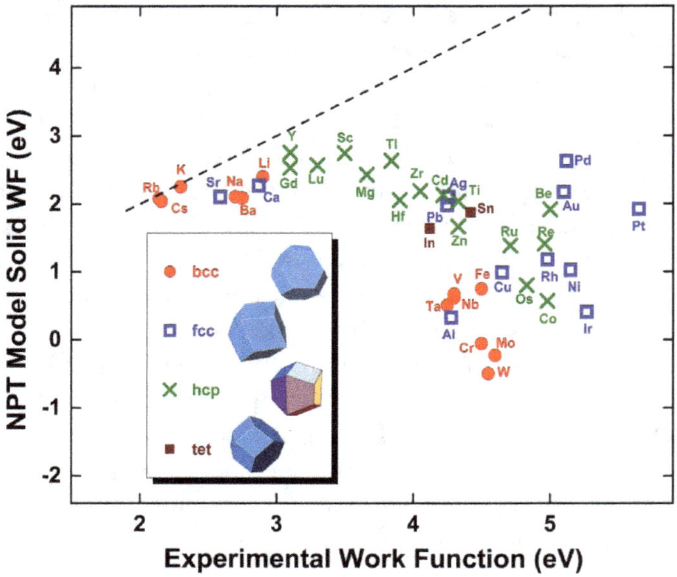

Fig. 3.13. Work function of NPT solids as a function of the experimental work function for various elemental metals. Inset: shape of the neutral polyhedron for different crystal structures. Note that for a monoatomic primitive unit cell, the neutral polyhedron reduces to the Wigner–Seitz cell. The vertical difference between each point and the dashed line, which marks a slope of 1, is the surface dipole for that metal.

Source: Reprinted with permission from R. T. Tung and L. Kronik.[19] Copyright 2021 by the American Physical Society.

first term on the RHS of Eq. (3.13) against the LHS. Surprisingly, there is no clear relationship between these two quantities and the data are quite scattered. The WF of the model solid does not vary linearly, or even monotonically, with the experimental WF. If anything, there is a reverse trend, in that the ϕ_{ModSol}'s of alkali and alkaline earth metals are generally larger than those of transition metals with large electronegativities. For the latter, metals with similar experimental WFs could vary in their ϕ_{ModSol}'s by more than 3 eV! A few metals even have negative NPT-WF, which does not violate any physical laws because model solids are a computational tool, and not real physical objects. Our main interest here is in the estimation of the surface relaxation effect. For each point of Fig. 3.13, the difference between its x- and y-coordinates is the surface dipole for that metal. Equivalently, it is the vertical (or horizontal) difference between each point and the dashed line.

Our ability to explain the WFs of elemental metals from charge distribution rests with a strategy to reasonably model the D_{surf} term. From a comparison of the upper left panel of Fig. 3.12 with the bottom panel, this term covers charge relaxation from an abrupted terminated charge distribution at the uneven surface of a model solid to a completely smooth and uniform charge density decaying into the vacuum. The smoothing out of the density discontinuity is responsible for the surface dipole. From that perspective, the average charge densities on the surfaces of the metal model solids, listed in Table 2.2, are certainly one of the factors that determine the magnitude of D_{surf}. However, the charge relaxation process at the surface of the metal is quite extensive, and the electrons needed from the model solid to create the electron tail toward the vacuum may come from not just a shallow region near its surface but also mobile electrons in the metal. In other words, the density of free electrons of the metal may be responsible for the surface dipole under certain circumstances. A comparison shows little difference between the intermediate density on the face of the NPT model solid and the free electron density, for the majority of elemental metals. However, the two densities differ by a significant fraction for some metals with certain electronic configurations. The other important factor that should influence the magnitude of D_{surf} is the topographical unevenness of the model solid's surface. When charge distribution originally bound by inclined facets is allowed to relax, the total dipole formed naturally depends on the faceted nature of the original surface. Due to different surface terrains of the model solids, more than one method may be required to model the WFs for all the metals. Fortunately, the NPT model solids for all elemental metals with the same crystal structure are identical. Geometrical unevenness can thus affect the surface dipole differently only for crystals with different lattice structures.

Due to the possible effect of crystal structure, the surface dipoles are analyzed for each crystal structure separately. The surface dipoles of all elemental metals are found to depend approximately the same way on the charge density, if the intermediate density is used for metals with the cubic structures and the free electron density is used for all non-cubic metals, as illustrated in Fig. 3.15. A linear fit gives

$$D_{\text{surf}} \approx 0.0175\rho \ \text{eV} \cdot \text{nm}^3, \tag{3.15}$$

where ρ is the characteristic electron density of the metal. Quantum mechanically calculated surface dipoles for jellium with abruptly terminated

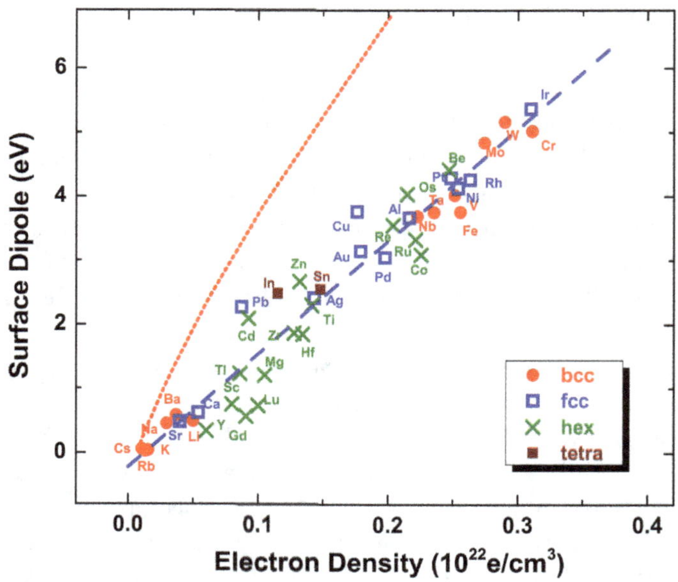

Fig. 3.14. Surface dipole of elemental, defined as the difference between the experimental work function of the metal and that of the corresponding NPT model solid, as a function of the average charge density of the metal. The density on the surface of the NPT model solid is used for metals with the b.c.c. and f.c.c. structures, while the free electron density is used for the non-cubic metals with hexagonal and tetragonal structures. The straight dashed line is a linear fit to the data. A dotted line shows the surface dipole calculated from jellium of uniform electron density.[20]

uniform positive background, also shown in Fig. 3.14,[20] are about a factor of two too large. The very different geometries and definitions of the surface dipole for jellium and real metal crystals should be noted. The linear dependence of the surface dipole on the charge density reflects its origin in the discontinuity of charge density at the surface of the NPT model solid. This forms a density-based foundation to model the WF. As shown in Fig. 3.15, the experimental WFs of metals can be adequately modeled with the use of two bulk properties of the metal: ϕ_{ModSol} and charge density. Due to its dependence on the orientation and the structure of the surface, the WF is clearly not a bulk property even though it seems derivable from bulk properties of the metal. This apparent predictability of the WF is attributable to a consistent distribution in the tails of the charge density toward vacuum.

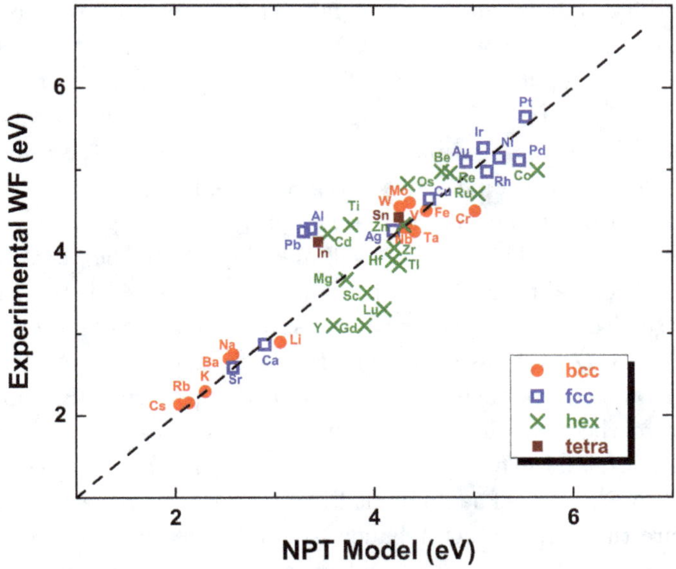

Fig. 3.15. The experimentally measured WF of elemental metals, compiled by Michaelson, is shown against the WF estimated through the use of NPT model solid and a surface dipole proportional to the average electron density of the bulk metal. In these computations, the density on the surface of the NPT model solid is used for metals with the b.c.c. and f.c.c. structures, while the free electron density is used for the non-cubic metals with hexagonal and tetragonal structures.

3.4 Semiconductor Surfaces

3.4.1 *Surface band diagrams*

A fundamental gap separates the completely filled valence band from the empty conduction band of a bulk semiconductor. At the surface of a semiconductor, electronic states specific to the surface are formed in the SSR. The rules governing the formation of electronic states at an insulator surface are not any different from those described for metal surfaces. In the semiconductor band gap, the formation of electronic states is limited to the kind that decays toward both the vacuum and the bulk. These intrinsic surface states vary with the atomic structure of the surface and are allowed only at specific energies and $k_{\|}$'s, as shown in Fig. 3.8. At the surface of a real semiconductor, various structures, defects, and contaminations, etc. contribute to the distribution of electronic states in the SSR and also add to the density of states in the band gap. The FL location with respect to the band gap of a semiconductor surface is a topic of scientific interest and technological importance. Since the FL has to be at a position to make the surface overall neutral without an applied potential, its placement depends almost entirely on the distribution of electronic states in the SSR. At low temperatures, all electronic states in the SSR are filled from the bottom up to a particular point in energy and are empty above this point, to maintain electrical neutrality. This energy level, at ϕ_{CNL} above the VBM, is called the "charge neutrality level" (CNL) of surface states and is the FL for an undoped semiconductor. The density of surface states per energy per area near the CNL, D_{SS}, is assumed to be approximately constant.

When the semiconductor is n-type doped, the surface FL position is still close to the CNL. To keep the interior of the n-type semiconductor neutral, the FL is at an energy of eV_N below the CBM of the semiconductor, according to Eq. (2.6). The different FL positions with respect to the band gap require the semiconductor bands to bend in a space charge region of width W created by exposed dopant atoms, as shown in Fig. 3.17. In the absence of applied potential, the charge of the exposed dopants in the space charge region is offset by an equal and opposite charge residing in the SSR, which shifts E_f slightly above ϕ_{CNL}:

$$E_f = \phi_{CNL} + N_D W / D_{ss}, \tag{3.16}$$

to keep the surface neutral overall. Due to the fixed difference between the CBM and $-e\bar{V}_{SC}$, band bending in the space charge region is governed by Poisson's equation. Rather than dealing with Fermi–Dirac statistics of electron occupation, we adopt the depletion approximation and define the space charge region by a sharp planar boundary, inside which all dopants are fully exposed. With a constant charge density ($+eN_D$) in the depletion region, the spatial variation of the CBM is solved in one dimension and sketched in Fig. 3.17. With respect to the FL, the position of the CBM, ε_{CBM}, varies from the surface, $z = 0$, to the edge of the space charge region, $z = W$, according to

$$\varepsilon_{CBM}(z) = eV_N + (E_G - E_f - eV_N)\left(1 - \frac{z}{W}\right)^2. \tag{3.17}$$

The depletion width W is given by

$$W = \sqrt{\frac{2\varepsilon_S(E_G - E_f - eV_N)}{e^2 N_D}}, \tag{3.18}$$

where ε_S is the electric permittivity of the semiconductor. Equations similar to Eqs. (3.16–3.18) can be written for the surface of a *p*-type semiconductor, with the FL lying slightly below the CNL because the space charge region is negatively charged.

At a homogeneous semiconductor surface, the minimum energy required to remove an electron from the VBM of the semiconductor is the ionization energy I_{SC}. The electron affinity A_{SC} is the energy released when an electron from the vacuum level is placed at the CBM of the semiconductor, as shown in Fig. 3.16. These surface quantities may be analyzed in the same way as for a metal surface. Without kinetic energy, the vacuum level is just the potential energy for the vacuum and therefore the surface quantities for a semiconductor have only two contributions: a semiconductor bulk term and a surface electrostatic potential energy term.

$$I_{SC} = e\bar{V}_{SC}^{\text{surf}} - \mu_{SC}^{(i)}, \tag{3.19}$$

$$A_{SC} = e\bar{V}_{SC}^{\text{surf}} - \mu_{SC}^{(i)} - E_g, \tag{3.19a}$$

where $-e\bar{V}_{SC}^{\text{surf}}$ is the average potential energy of the semiconductor at the surface with respect to the vacuum level. The ionization potential of a

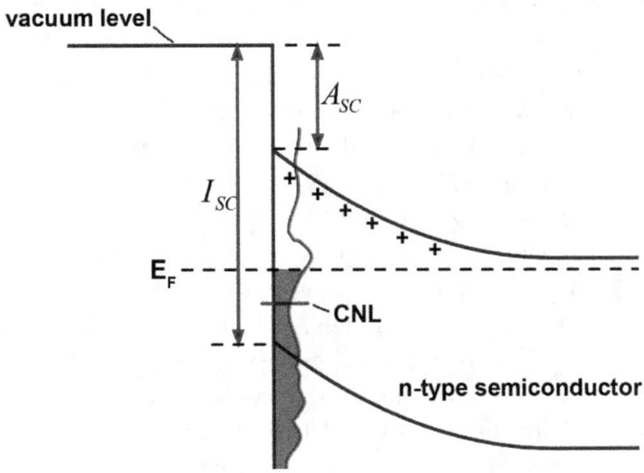

Fig. 3.16. Band diagram of the surface of an *n*-type semiconductor without applied external potential. The CBM and VBM of the semiconductor are drawn as solid lines. The distribution of electronic states at the surface is indicated by the wriggly line and is marked by a charge neutrality level (CNL). When the FL coincides with the CNL, the surface region is neutral by itself. When the semiconductor is doped, the electrons depleted from the space charge region accumulate at the surface to slightly shift the FL from the CNL.

semiconductor surface can be experimentally measured by photoemission spectroscopy. In an inverse photoemission spectroscopy (IPS) experiment, a beam of electrons with well-defined momentum is sent toward a semiconductor surface, and the energies of emitted photons are measured.[21] This allows the surface electron affinity to be measured because electrons relax to the empty conduction band of the semiconductor.

3.4.2 *Fermi level of clean semiconductor surfaces*

An important aspect of any semiconductor surface is whether it is insulating or metallic. Similar to the definition for a solid crystal, an insulating surface is one marked by an energy gap separating occupied states from empty states. The broken bonds on a terminated semiconductor crystal ordinarily add states and leave the surface metallic. However, as discussed, the various ways the semiconductor surfaces relax and reconstruct are driven by energy minimization. The saturation of dangling bonds and the emptying of the non-bonding orbitals during the reconstruction and relaxation processes tend to create or widen an energy gap. Zincblende (110) surfaces offer an excellent example of the effect of surface relaxation on electronic states. Whether cleaved in vacuum or repaired otherwise without impurities, the energy band associated with the lone pair on the anion moves below the VBM and the band due to the empty non-bonding orbital on the cation is raised above the CBM on this surface. These shifts leave no electronic states with energy within the fundamental band gap at the surface. For that reason, zincblende (110) surfaces have been a convenient venue for scientific investigations into surface states introduced by subsequent surface treatments. An energy gap also opens up for some other semiconductor surfaces with lattice relaxation, such as the cleaved Si(111) 2×1 and the Si(100) c4×2 with asymmetric dimers, although the surface bands are not pushed all the way out of the band gap. The band gap in surface states extends to $\sim$0.45 eV above the VBM of bulk Si for the cleaved Si(111) 2×1 surface, where the FL is found by experiment to be approximately located, as shown in Fig. 3.8. For these insulating surfaces just described the requirement of charge neutrality for the surface region would be satisfied if the FL was placed anywhere inside the bandgap. The actual FL position is then decided by the bulk doping type of the semiconductor. Some semiconductor surfaces, such as the Si(111) 7×7, are metallic even with nearly perfect reconstructed structures. The FL is consistently observed to be positioned at $\sim$0.7 eV above the VBM for the 7×7, likely at the partially filled dangling bond level of the adatoms. A uniform and perfect atomic structure is rarely the case for actual semiconductor surfaces. Impurities, roughness, and other structural defects are expected to be present and may add electronic states in the band gap. These may be of specific nature, e.g. acceptor-like or donor-like, that are peaked at specific energies, or they may introduce a broad distribution of states. On the other side of the coin, appropriately chosen adsorbates may significantly reduce the density of dangle bonds and surface states, such as in As-terminated Si(111) 1×1 or S-terminated Si(100).[22]

3.4.3 *Fermi level pinning at semiconductor surfaces*

Electronic states in the band gap at a semiconductor surface were of much interest to scientists and engineers decades ago. Much of the interest stemmed from a popular idea at the time that these states could survive metal deposition and were responsible for the formation of the SBH at MS interfaces. That idea turns out to be incorrect, as explained at length later in the book. On the other hand, it is likely that surface states associated with certain local (defect) structures may survive the over-deposition of an insulator layer, in which case we should learn how to eliminate these states from the surface and not let them become a nuisance at the eventual semiconductor/insulator interface. It is thus important to know the characteristic energies of surface states associated with common structural defects. Common point defects of the bulk semiconductor, including vacancies, anti-site defects, and particular structures that are only allowed at the surface, contribute to electronic states in the gap at a semiconductor surface. A defect state deep in the band gap is almost always at the same energy in the gap, regardless of the chemical environment of the semiconductor. Deep levels show up as peaks in the surface states density that can significantly limit the movement of the FL in response to applied potential. Therefore, structures responsible for peaks in surface state density should be identified, not only by their energetic positions but also by their electrical activities. Some defects act only as donors and can be either neutral or positively charged (but not negatively charged). There are also deep levels that behave as acceptors. Generally speaking, donors are not limited to the upper half of the band gap, nor are acceptors to the lower half. Then, there are amphoteric defects that can switch between positive and negative charge states. When there are peaks in the surface states spectrum, the FL can be confined close to the location of a peak.

This near constancy in the FL position at a semiconductor surface, known as Fermi-level pinning (FLP), is a common experimental observation for routinely processed semiconductor surfaces. FLP indicates a high density of surface states, likely associated with a specific defect at the energy.

The most well-known example of surface FLP refers to the experimental observation of a nearly constant FL position of the cleaved zincblende semiconductor surface, upon different treatments. Without states in the band gap, the FL position on the cleaved clean GaAs(110) is near the CBM for n-type and near the VBM for p-type GaAs. When a small amount of group III metal is deposited, these FL positions quickly change and saturate at close to 0.75 eV above the VBM for n-type and about 0.5 eV above the VBM for p-type GaAs, independent of the metal used, as shown in Fig. 3.17.[23] Exposure of the surface to oxygen leads to saturated FL at

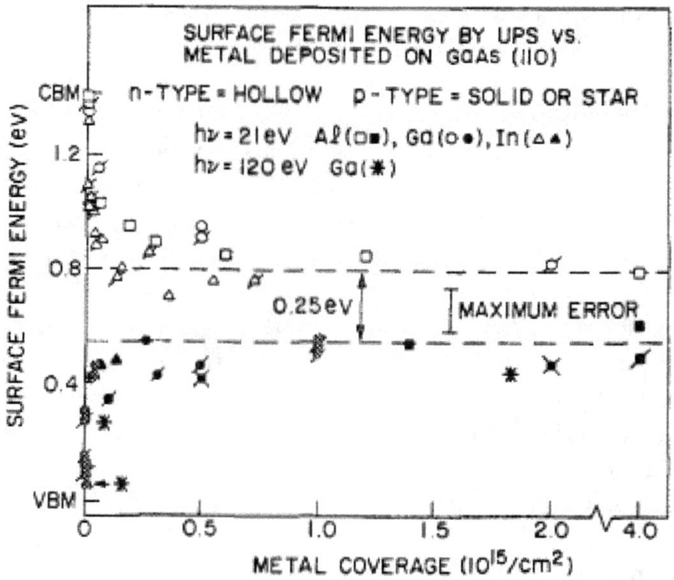

Fig. 3.17. Variation of band bending as a function of metal coverage for Al, Ga, and In on n- and p-type GaAs(110). Note the difference in pinning position on n- and p-type samples; the acceptor levels on n-type GaAs are 0.25 eV above the donor levels seen on p-type GaAs. Error bars for absolute energy position are $+-0.1$ eV and probably smaller for relative energy position. The p-type samples were non-degenerate, and hence the VBM is slightly below E_{FS} on the unpinned p-type surface. The data were obtained from valence band data with $h\nu = 21$ eV except for the starred points which were obtained near maximum surface sensitivity ($h\nu = 120$ eV, probing depth 3–6 Å).

Source: Reprinted from Skeath[23] with permission from AIP Publishing.

similar positions. These changes may be explained by the simultaneous additions of an acceptor level at the 0.5 eV position and a donor level at the 0.75 eV position to the originally empty band gap of GaAs(110), upon metal deposition. It is striking that very significant shifts in FL position take place at a coverage of less than ~0.2 monolayer of deposited material on GaAs(110). While on the topic of defects at surfaces, it should be pointed out that GaAs crystals often contain significant densities of deep levels that also pin the FL of bulk crystals. Similarity with the bulk pinning levels suggests that the energy levels that pin the surface FL may originate from point defects commonly found in the GaAs bulk crystal. The structural defects responsible for these FL positions were speculated to be the As_{Ga} antisite because the As antisite in GaAs has two levels close to those found for the FL on n-type and p-type surfaces. However, the pinned FL position on p-type GaAs surface cannot be entirely attributed to the As antisite defect, as it behaves as a double donor rather than amphoteric. Recently, theoretical calculations suggested that the amphoteric transitions of a structure containing As-As dimer are in agreement with the observed FL pinning effect.[24] By avoiding these structures and employing Ga-rich (4×6) GaAs(100) surface, a significant reduction in the trap density at subsequent GaAs/oxide interfaces was achieved.[25]

3.5 Summary

Semiconductor surfaces are important substrates for the fabrication of quantum structures and advanced devices. Several rules on the formation of surface structures of covalent semiconductors, driven by energy minimization, have been identified by surface scientists. The atomic structures of some famous and intriguing (reconstructed) semiconductor surfaces are now well understood in terms of these rules. How the electronic states of a bulk crystalline solid are modified when the crystal is terminated at a surface is described in this chapter. The conditions for the formation of electronic states that exist only at the surfaces are also discussed. Since the WF at a metal surface may be viewed as a form of energy level alignment at a solid interface, its magnitude is shown to be a direct consequence of the charge distribution of the surface. This chapter illustrates the procedures to predict/explain energy level alignment at a solid interface using the metal WF as the example. As suggested by Chapter 2, a model solid is used as the host of the metal FL, as illustrated in Fig. 3.18 for the Ag(100) surface. The calculated charge distribution in equilibrium (thin solid line) is compared

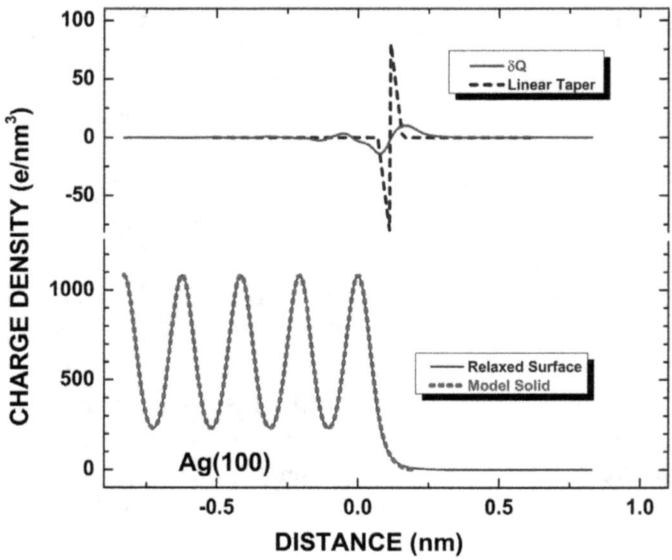

Fig. 3.18. The calculated charge distribution of the Ag(100) surface is shown as thin solid line and compared with that of an NPT model solid of Ag, drawn as thick dotted line, in the lower half of the figure. The difference between these distributions is shown as δQ in the upper part. The effect of δQ is estimated as that due to a linear taper of the average density on the Ag NPT model solid surface over a distance of 0.07 nm, as shown by the dashed line.

with that of an NPT model solid of Ag (thick dotted line) in the lower part of Fig. 3.18. Adequate WF prediction now relies on a reasonable estimation of the effect of the "bath" of this system, which is the difference between the two distributions, shown in magnification (solid line) in the upper part of the figure. Since δQ is only available from supercell calculations, it is considered unknown in charge distribution modeling. However, the total potential energy shift caused by δQ may be modeled as that due to a continuous and linear decrease of the average density of the NPT surface to zero over a distance of 0.07 nm, as shown by the dashed line in the figure. Reasonable estimates of the WFs for elemental metals are obtained by this charge distribution modeling approach.

References

1. G. Chiarotti and P. Chiaradia, eds. *Physics of Solid Surfaces: Subvolume B (Landolt-Börnstein, New Series, 45B)*. 1st Edn. 2018, Springer.

2. D.P. Woodruff, *Modern Techniques of Surface Science*. Third Edn. 2016: Cambridge University Press.
3. E. Bauer, *Rep Prog. Phys.* **57**, 895–938 (1994).
4. T. Sakamoto, N.J. Kawai, T. Nakagawa, K. Ohta, and T. Kojima, **47**, 617–619 (1985).
5. J.H. Neave, B.A. Joyce, P.J. Dobson, and N. Norton, *Appl. Phys. A* **31**, 1–8 (1983).
6. L.C. Feldman, J.W. Mayer, and S.T. Picraux, *Materials Analysis by Ion Channeling*. 1982: Academic Press.
7. G. Binning, H. Rohrer, C. Gerber, and E. Weibel, **49**, 57–61 (1982).
8. C.B. Duke, **96**, 1237–1259 (1996).
9. D. Haneman and A.A. Chernov, **215**, 135–146 (1989).
10. K. Takayanagi, Y. Tanishiro, M. Takahashi, and S. Takahashi, J. *Vac. Sci. Technol. A* **3**, 1502–1506 (1985).
11. C.B. Duke, J. *Vac. Sci. Technol. A* **10**, 2032–2040 (1992).
12. A. Ohtake, **63**, 295–327 (2008).
13. V. Heine, *Phys. Rev.* **138**, A 1689 (1965).
14. P. Martensson, A. Cricenti, and G.V. Hansson, **33**, 8855–8858 (1986).
15. H.B. Michaelson, *J. Appl. Phys.* **48**, 4729–4733 (1977).
16. I. Brodie, *Phys. Rev. B* **51**, 13660–13668 (1995).
17. S. Halas and T. Durakiewicz, *J. Phys.: Condens. Matter* **10**, 10815–10826 (1998).
18. M. Weinert and R.E. Watson, *Phys. Rev. B* **29**, 3001–8 (1984).
19. R.T. Tung and L. Kronik, *Phys. Rev. B* **103**, 035304 (2021).
20. N.D. Lang and W. Kohn, *Phys. Rev. B* **3**, 1215–1223 (1971).
21. N.V. Smith, *Rep. Prog. Phys.* **51**, 1227–1294 (1988).
22. E. Kaxiras, *Phys. Rev. B* **43**, 6824 (1991).
23. P. Skeath, C.Y. Su, I. Lindau, and W.E. Spicer, **57**, 5089–5092 (1985).
24. D. Colleoni, G. Miceli, and A. Pasquarello, *Phys. Rev. B Condens. Matter Mater. Phys.* **92**, 125304 (2015).
25. Y.C. Chang, W.H. Chang, C. Merckling, J. Kwo, and M. Hong, *Appl. Phys. Lett.* **102**, 093506 (2013).
26. Y. Li, W. Long, and R. Tung, *Appl. Surf. Sci.* **284**, 720 (2013).

Chapter 4

BAND ALIGNMENT FUNDAMENTALS: CHARGE DISTRIBUTION AND MODELING

4.1 Foundation of SBH and BO: Interface Charge Distribution

The difference in the CBMs of two semiconductors is known as the conduction band offset (CBO) at a heterojunction interface, while the valence band offset (VBO) is the offset in the VBM. Based on the signs of both band offsets for the same interface, the band alignment at a heterointerface can be distinguished into several types, as illustrated in Fig. 4.1.[1] The kind most commonly found experimentally is the type-I band alignment condition, where the smaller band gap lies entirely inside the band gap of the other semiconductor. When the two band gaps only partially overlap, the interface is said to display the type-II alignment condition. On occasions, one may even find no overlaps between the two band gaps and they lie completely outside of each other, shown as type III in Fig. 4.1. Type-III alignment is only seen at interfaces without intimate chemical bonds, such as those held together by van der Waals forces. At any interface between two insulators, the sum of the VBO and the CBO is a fixed quantity and equals the difference in the band gaps of the two materials. An increase in the VBO is accompanied by a decrease in the CBO, and vice versa. The offset in the energy band across a semiconductor interface plays a critical role in the operations of many advanced devices in microelectronics, photonics, energy harvesting, multiferroics, etc. Therefore, the factors that influence the magnitude of the BO need to be thoroughly understood in order to improve the performance of existing devices or to develop more advanced novel devices. However, investigations of BO formation at semiconductor heterojunctions should not be conducted in isolation because factors controlling BO formation are not fundamentally different from those for band alignment at other types of solid interfaces,

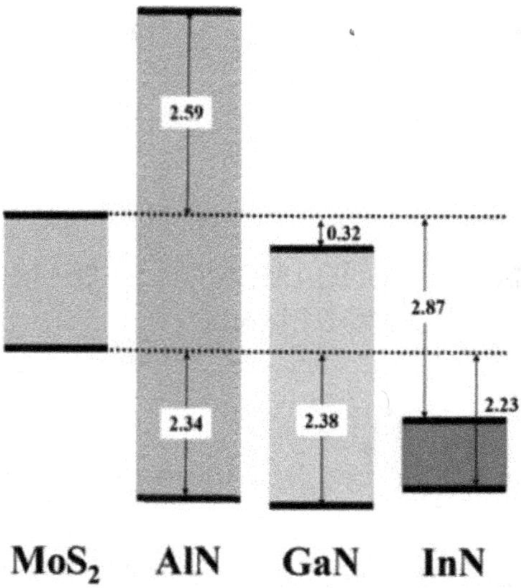

Fig. 4.1. Summary of experimentally measured band diagrams of MoS_2 deposited on three different nitride semiconductors. At the interface with AlN, the entire band gap of MoS_2 is contained within that of AlN, forming the most commonly observed "type-I" band alignment. The band gap of GaN contains only one of the band edges (CBM) of MoS_2, making the band alignment at the MoS_2/GaN interface "type II", while the MoS_2/InN interface has the rare "type-III" band alignment condition because there is no overlap in the band gaps of the two semiconductors.

Source: Reprinted from D. K. Singh[1] with the permission of AIP Publishing.

such as SBH formation at MS interfaces. A broader perspective on the commonalities and distinctions of different types of interfaces is essential to fully understand the band alignment problem.

To understand the factors that ultimately decide the magnitude of the BO, we first focus on a narrow region of material directly influenced by the formation of the interface. In Fig. 4.2, the plane-averaged electric potential energy ($-eV_{\mathrm{Coul.}}$, thin curves) and the energy band edges (thick horizontal lines) at a heterojunction between semiconductors A and B are drawn. Also shown as thick dashed line is the averaged electrostatic potential energy $-e\overline{V_{\mathrm{Coul.}}}$. Boundaries are drawn for an ISR, inside which the electron density is modified from that of the semiconductor solids. The charge distributions left and right of the ISR are nearly identical, respectively, to those of the two bulk semiconductors. In a bulk semiconductor, the energy band positions are fixed with respect to its electric potential energy, from Eq. (2.7). With this information in hand, the energy bands for each

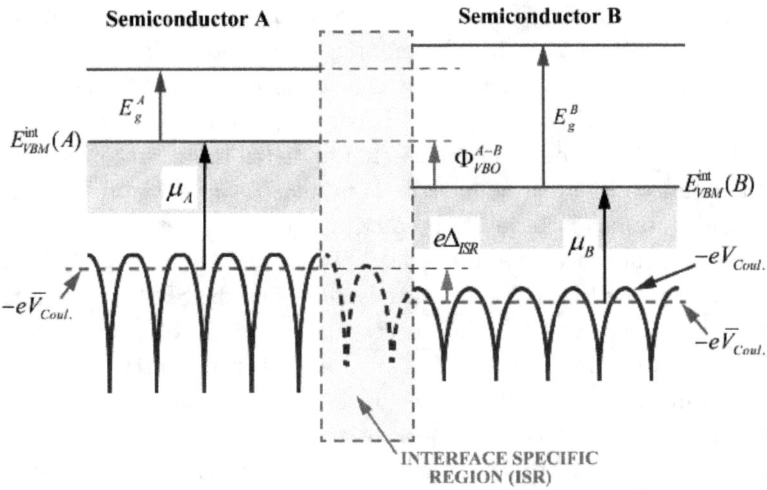

Fig. 4.2. Potential energy and energy band diagram at a semiconductor heterojunction interface. The lateral ($-eV_{\text{Coul.}}$) and macroscopic ($-e\overline{V_{\text{Coul.}}}$) averages of the electrostatic potential energy are shown as solid and dashed thick lines, respectively. The internal ionization potential, μ, is the energy difference between the VBM and the macroscopically averaged electrostatic potential energy. $e\Delta_{ISR}$ is the difference between the bulk values of $-e\overline{V_{\text{Coul.}}}$, $E_g^{A/B}$ is the bandgap of semiconductor A or B, and Φ_{VBO}^{A-B} is the valence band offset between the two semiconductors.

semiconductor can be drawn with respect to the average potential energy on either side of the ISR in Fig. 4.2. The difference between the VBM positions at the interface is the VBO, which can be traced through Fig. 4.2 to consist of three terms:

$$\Phi_{VBO}^{A-B} = E_{VBM}^{\text{int}}(A) - E_{VBM}^{\text{int}}(B) = \mu_A - \mu_B + e\Delta_{ISR}. \tag{4.1}$$

The two bulk terms, μ_A and μ_B, are fixed and can be written down without knowledge of the interface. That leaves the difference between the averaged electrostatic potential energies across the ISR, $e\Delta_{ISR} = -e\overline{V_{\text{Coul.}}^A} + e\overline{V_{\text{Coul.}}^B}$, as the only term on the RHS of Eq. (4.1) that is not known ahead of time. Being purely electrostatic in nature, this term is a direct consequence of, and derivable from, the charge distribution in the ISR. Our ability to explain or predict a BO thus depends on how well the distribution of charge at that heterojunction interface can be modeled. The charge distribution in the interface region naturally depends on the interface atomic structure, which seems to make the prediction of BO from charge distribution a daunting task. However, this turns out not to be a real obstacle thanks to the nearsightedness principle mentioned earlier. In passing, we note that the exact locations of the ISR boundaries may seem unimportant for BO analysis because they do not affect the validity or the conclusion of the discussions above. However, as will become apparent when real attempts to model the BO through interface charge distribution are discussed, the exact locations of the ISR boundaries are crucial. Appropriately chosen ISR boundaries can make the job of predicting the BO easier.

At an MS interface, the difference between the band edge of the semiconductor and the FL of the interface is known as the SBH. MS interfaces/contacts are necessary for the successful operation of nearly all semiconductor devices and the SBH is an important property that controls current transport at the interface. Just as the heterojunction BO, the formation of the SBH is an important topic for both solid-state physics and technology development. To analyze the SBH, we again examine the plane-averaged potential energy (thin curve) and the macroscopically averaged potential energy (thick broken line) drawn in Fig. 4.3 for an MS interface. As before, the region with charge distribution modified by the formation of the interface is marked as the ISR. In the region representing the semiconductor, the band edges (thick straight lines) may be drawn in relation to the average electrostatic potential energy, using Eq. (2.7). Likewise, the Fermi level for the metal can also be drawn. The difference between the CBM and the FL is the n-type SBH, $\Phi^o_{B,n}$, while the p-type

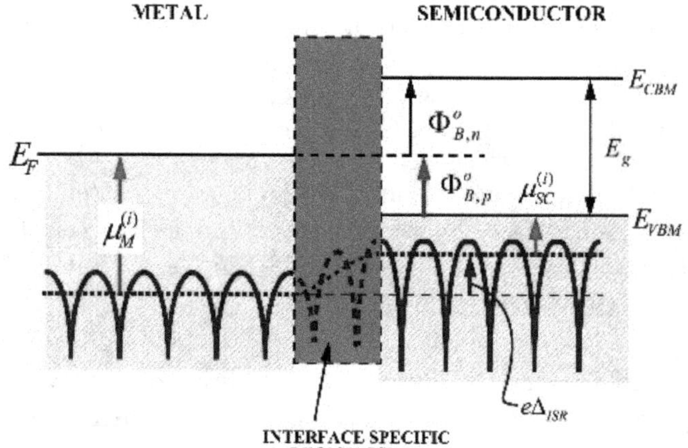

Fig. 4.3. Potential energy and energy band diagram at a metal–semiconductor interface. The electrostatic potential energy is drawn as curvy lines. Dotted lines indicate average electrostatic potential energy.

SBH, $\Phi^o_{B,p}$, is the difference between the VBM and the FL. For interfaces with a uniform SBH, the sum of the n-type and the p-type SBHs is the band gap of the semiconductor. Similar to the heterojunction VBO, the p-type SBH can also be read from Fig. 4.3 to consist of two bulk terms and an interface term:

$$\Phi^o_{B,p} = E_F - E^{\text{int}}_{VBM} = \mu_M - \mu_{SC} + e\Delta_{ISR}. \tag{4.2}$$

Again, $-e\Delta_{ISR}$ is the only term specific to the interface and this dipole term, which is a direct result of charge distribution in the ISR, controls the magnitude of the SBH. A comparison of Figs. 4.2 and 4.3 shows that the SBH and VBO are identical in how they are formed and thus should be understood and modeled the same way.

4.2 Traditional Theories

The Schottky–Mott Theory (SMT) and the Electron Affinity Rule (EAR) were the first theories proposed to explain experimentally observed SBH and BO, respectively.[2–4] They set the precedent of using surface quantities to represent the crystalline solids in the analysis of band alignment at solid interfaces. Following their lead, research investigations have proposed various different models to explain the SBH/BO, although invariably assuming surface quantities to be valid representations of bulk crystals. These theories are collectively known as "traditional theories". The work function, ϕ_M, of an isolated metal surface and the electron affinity, χ_{SC}, of an isolated semiconductor surface are drawn in Fig. 4.4(a) to represent the alignment of bulk bands before an MS interface is formed. At this stage, before contact, the difference between the FL and the CBM gives a virtual n-type SBH of

$$\Phi^o_{B,n} = \phi_M - \chi_{SC}. \quad \text{(Schottky–Mott)} \tag{4.3}$$

The SMT asserts that this relative position of the energy bands when the surfaces are in isolation would still prevail even when an intimate MS interface is formed between these two crystals, as shown in Fig. 4.4(b).[2,3] Equivalently, the p-type SBH is predicted by the SMT to follow

$$\Phi^o_{B,p} = I_{SC} - \phi_M. \quad \text{(Schottky–Mott)} \tag{4.3a}$$

Even though it doesn't mention charge specifically, the SMT is a theory based on the distribution of charge in hiding. The apparent recipe to the SMT seems to be "line up the vacuum levels outside the two surfaces".

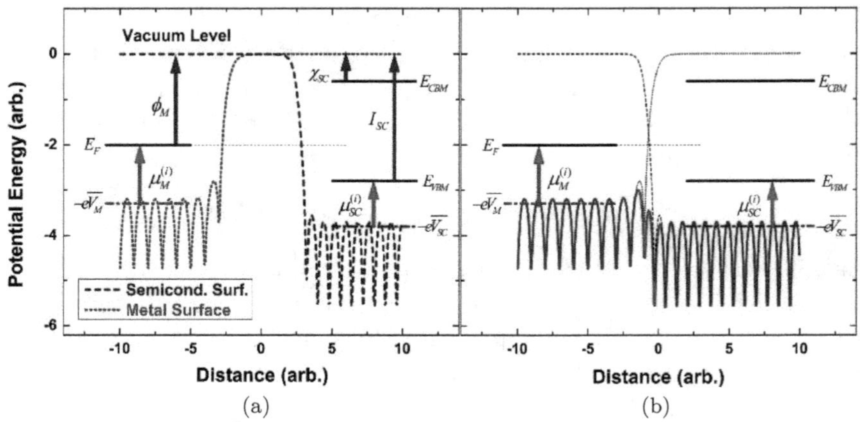

Fig. 4.4. Potential energy for isolated metal surface (dotted) and semiconductor surface (dashed) plotted on the same energy scale. (a) When the two crystals are far apart. (b) When the two crystals are next to each other. The solid line is the sum of the two individual potential energies. Note that the relative position of the semiconductor potential to the metal potential remains unchanged, independent of the distance separating the two crystals.

Yet the underlying reason for Eq. (4.3) is the superposition principle of electrostatic potential. To see this, the potential energy curves from isolated metal and semiconductor surfaces in Figs. 4.4(a) are redrawn in 4.4(b) without a physical gap to find the same difference between the FL and the CBM. The SMT thus describes the SBH of a metaphysical, non-interacting MS interface, put together by combining the charge densities of the two free surfaces, without allowing either the charge density or the atomic positions to relax.

The formation of BO at heterojunction interfaces is nearly identical to the formation of SBH in terms of how such an interface energy level difference may be broken down into bulk and interface contributions. The similarities in the nature of the problem in these two fields prompted the search for mechanisms that could simultaneously explain experimental observations on both the SBH and the BO. In the spirit of the SMT, the EAR proposed that the CBO was the difference between the electron affinities of the two semiconductor surfaces, namely,[4]

$$\Phi_{CBO}^{SC1-SC2} = \chi_{SC1} - \chi_{SC2}. \quad \text{(Electron Affinity Rule)} \tag{4.4}$$

For the VBO, this relationship translates to

$$\Phi_{VBO}^{SC1-SC2} = I_{SC2} - I_{SC1}. \quad \text{(Electron Affinity Rule)} \tag{4.4a}$$

Based on the same assumption as the SMT, the EAR prediction for a heterojunction also corresponds to an unrelaxed interface with its charge distribution a superposition of the distributions at the two isolated semiconductor surfaces. Same as for the SMT, the success of the EAR also rests with the absence of charge relaxation from surface densities.

The traditional SMT and EAR theories relegate the role played by "crystal solids" at an interface to "surfaces of the solids". At the time of their proposal, such replacements were almost inevitable, as there were no known methods for the energy levels of a solid to be referenced to the vacuum level without terminating the solid and exposing a relaxed surface. Whenever the energy level of a solid needed to be compared with those of others, it became automatic to use the surface to represent a solid. Both the SMT and the EAR assume that upon the formation of a solid interface in equilibrium, the charge densities of the two surfaces do not interact with each other. Such a scenario is unrealistic for intimate interfaces for two obvious reasons. First, the electron density spills into the vacuum at a surface to form a surface dipole, with a magnitude that is typically a very significant fraction of either the WP or the IP, as Tables 2.2 and 2.6 show. These large surface dipoles are included in the SMT/EAR and other traditional theories that follow their lead, although they are not part of any intimate interface. On the other hand, the significant dipole as a result of charge rearrangement towards the formation of new interface bonds is not included in the SMT/EAR. The inclusion of unrelated terms and the omission of a critical term prevented the SMT and the EAR from accurately accounting for experimentally observed band alignment results. Instead of the sharp dependence of the SBH on the metal work function predicted by the SMT, a spectacular and poignant FLP phenomenon was observed experimentally decades ago when the formation mechanism of the SBH was very actively pursued. The SBH for covalent semiconductors was found not to depend sensitively on the metal. An example of the FLP phenomenon is shown in Fig. 4.5, where the FL positions in the band gap for various metals evaporated in ultrahigh vacuum on n- and p-type GaAs(100) surfaces are plotted against the work function of the metal.[5] The absence of a clear dependence of the SBH on the metal work function is in sharp contrast to the prediction of the SMT.

The significant departure of the actual charge distribution at an intimate MS interface from that of superimposed charge distributions at the two surfaces is the reason for the SMT to fail. In the absence of a better way to represent the bulk crystals than surface quantities, researchers decades

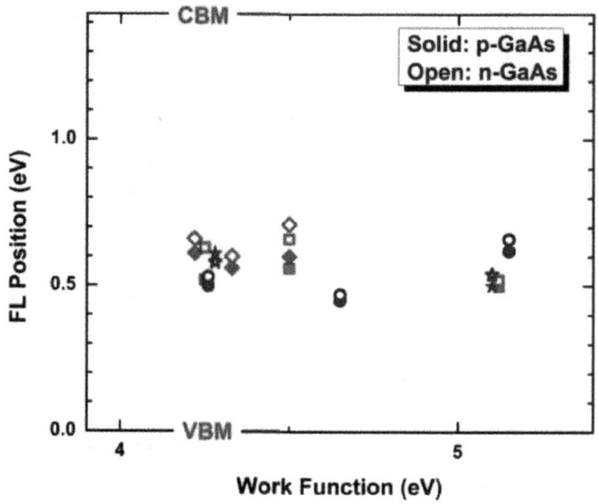

Fig. 4.5. Fermi level position of ideal metal contacts to GaAs. Filled symbols represent SBH measured by I–V method on p-type GaAs, while open symbols are the difference between the band gap and the n-type SBH measured by I–V. Corrections for the effect of image force have been included.

Source: Reprinted from Waldrop[5] with the permission of AIP Publishing.

ago had to keep the SMT framework and look for a correction term to the SMT to make it fit the experimental data. An "interface dipole", D_{int}^{MS}, was added to the SMT:

$$\Phi_{B,n}^{o} = \phi_M - \chi_{SC} - D_{\text{int}}^{MS}, \tag{4.5}$$

where $\Phi_{B,n}^{o}$ is the n-type SBH actually observed from an MS interface. We caution that there are many definitions of "interface dipole" and that any use of this term without an explicit definition is confusing and meaningless. Since the SMT is almost never observed in experiment, it can be concluded that the interface dipole is quite significant for the vast majority of MS interfaces. With the SMT as the starting point, the job to quantitatively explain the experimental SBH becomes tantamount to being able to model D_{int}^{MS} adequately. Since the measured SBH is a result of the actual interface charge distribution, the difference between this and the assumed frozen charge distribution must be electrostatically responsible for D_{int}^{MS}. Discussions above identify two necessary corrections to the frozen charge densities. One is the elimination of the extraneous surface dipoles, D_{surf}^{M} and D_{surf}^{SC}, from the metal and the semiconductor surfaces. The other is to account for the polarization of the bonds that actually form at the interface, eV_{bond}^{MS}. Therefore, the interface dipole contains three identifiable contributions[6]:

$$D_{\text{int}}^{MS} = D_{\text{surf}}^{M} - D_{\text{surf}}^{SC} + eV_{\text{bond}}^{MS}. \tag{4.6}$$

Plugging Eq. (4.6) back into Eq. (4.5) and rearranging, one gets

$$\Phi_{B,n}^{o} = [\phi_M - D_{\text{surf}}^{M}] - [\chi_{SC} - D_{\text{surf}}^{SC}] - eV_{\text{bond}}^{MS}. \tag{4.7}$$

Each set of square brackets above is now the property of only one solid. For the metal, a comparison with Eq. (3.14) shows that the quantity inside the first square brackets is simply ϕ_{ModSol}, or the Fermi level of the model solid of the metal with respect to the vacuum level. Similarly, the second square brackets is the CBM of the model solid of a semiconductor with respect to the vacuum level. Therefore, Eq. (4.7) suggests that a straightforward way to predict the SBH is through the use of model solids, which in fact is the method employed in Chapter 7 to model the SBH. In analogy, the VBO of a heterojunction can also be written in terms of the VBM positions of semiconductor model solids:

$$\Phi_{VBO}^{SC1-SC2} = [I_{SC2} - D_{\text{surf}}^{SC2}] - [I_{SC1} - D_{\text{surf}}^{SC1}] - eV_{\text{bond}}^{1-2}. \tag{4.8}$$

At the time when researchers scrambled to find explanations of the FLP phenomenon, the necessity of surface dipole removal from such analyses and the convenience offered by model solids were not known. The only available path to explain the SBH magnitudes back then was through a study of the systematics of the experimental SBH. Experimentally observed SBHs for different metals on the same semiconductor[7,8] were plotted conventionally against the work function of the metal and fitted linearly to deduce a slope, as is done in Fig. 4.6. Each slope here, ~0.2 and ~0.3, respectively for Si and GaSe, is called the "interface behavior parameter" or the "S parameter" for the semiconductor,

$$S_{\Phi} \equiv \frac{\partial \Phi_{B,n}^{o}}{\partial \phi_M} = 1 + \frac{e\partial D_{\text{int}}^{MS}}{\partial \phi_M}, \tag{4.9}$$

and is a measure of SBH's dependence on metal. SMT would predict an S-parameter of 1 ($D_{\text{int}}^{MS} = 0$). Covalent semiconductors with smaller band gaps, such as Ge, Si, and GaAs, are known as strongly pinned because of their small S-parameters. Compound semiconductors with larger ionicity typically have larger S-parameters and are therefore described as only weakly pinned, although there are exceptions.

The much weaker dependence of the SBH on the work function of the metal than that expected by the SMT is the result of interactions

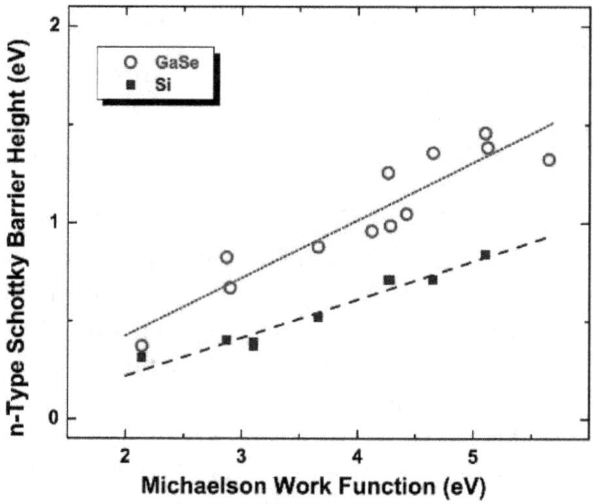

Fig. 4.6. Experimental Schottky barrier height of elemental metal on n-type GaSe and Si.

between the metal and the semiconductor. While it is obvious today that this interaction should be analyzed by first removing the surface dipoles, it wasn't clear how to do this decades ago. An alternative and more direct view of the interface dipole, D_{int}^{MS}, identifies it as the interaction between the two surfaces rather than between the two bulk crystals. From such a vantage point, the only contribution to the D_{int}^{MS} term is the formation of chemical bonds at the interface. Furthermore, these bonds were the result of interactions between the two surfaces and thus are controlled by properties of the surfaces rather than by atomic properties, such as electronegativity, as is the case for bond polarizations in molecules.

A chemical potential equalization (CPE) method[9,10] in molecular chemistry is known to allow the net charges on all the atoms in a molecule to be estimated simultaneously. This method was applied to the MS interface to estimate the amount of charge transfer between the metal and the semiconductor, by allowing the metal and semiconductor "atoms" at the interface to inherit the properties of the respective surfaces. The ionization potential and the electron affinity of each semiconductor "atom" at the interface were thus equated with those for the semiconductor surface. For the metal atoms at the interface, the WF was recognized as both its ionization potential and its electron affinity.[11,12] With these substitutions, the interface dipole was calculated for each MS interface. Indeed, the

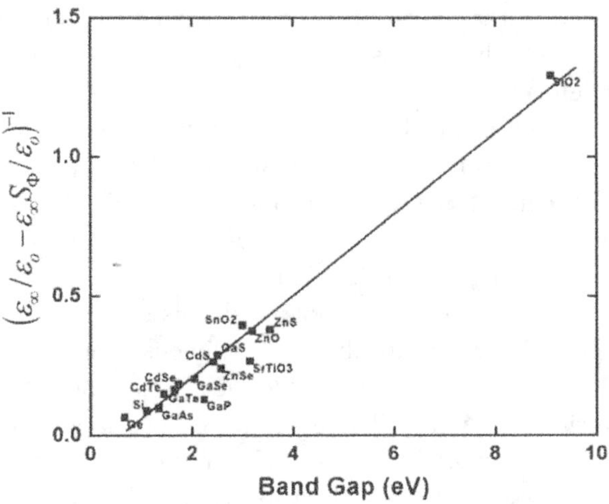

Fig. 4.7. Experimentally observed interface behavior parameter, plotted as $(\varepsilon_0/\varepsilon_\infty)/(1 - S_\Phi)$, against the band gap of the semiconductor.

Source: Reprinted from R.T. Tung.[12] Copyright (2001) American Physical Society.

formation of chemical bonds at the interface reduces the dependence of the SBH on the work function of the metal. The good agreement between the S-parameter predicted by the CPE method and those experimentally observed[13] is shown in Fig. 4.7. The observed linear relationship lends support for chemical bond formation as responsible for the weakened dependence of the SBH on the metal, although this method offers no specific means to quantitatively model individual SBH.

The routine of plotting the experimentally measured SBH against the work function in an effort to study the formation mechanism of the SBH deserves comment. First of all, the surface dipole contribution to the metal WF varies significantly with the metal but in a manner uncorrelated with the WF, as shown in Fig. 3.15. These variations in the surface dipole, which have nothing to do with the MS interfaces under study, become a part of any systematics deduced from SBH analysis. The extraneous effect of the surface dipoles cannot be filtered out by graph fitting, when the WF is used as the x-coordinate. Secondly, the SBHs plotted in Figs. 4.5 and 4.6 are from measurements of interfaces that are polycrystalline in structure. Such interfaces are inhomogeneous, as will be shown explicitly in Chapter 6, and the data plotted in these figures are statistical averages of the local SBHs of the interface. With the x-axis severely affected by surface dipoles unrelated to the interface and the y-axis being statistical averages, systematics from graphs such as Fig. 4.6 do not give information about the SBH at a fundamental level.

Before we leave the subject of, and the problems with, the SMT/EAR, one important exception should be mentioned. It is always the charge distribution at the interface that determines the magnitude of the SBH/BO. While it is nearly impossible for the charge distributions from isolated surfaces not to rearrange when they join to form an intimate interface, the likelihood of that scenario should not be dismissed for all the interfaces. At heterojunction interfaces involving two-dimensional sheet materials or molecules with closed shells, bonds do not form and the interactions between materials are weak. These interfaces are largely held together by van der Waals forces.[1,14] Under such circumstances, the assumption is actually quite reasonable that the total charge distribution of the final interface is a simple sum of the charge distributions of the two original surfaces. The EAR may be quite accurate for the BOs of non-bonded interfaces. But for the vast majority of heterojunctions, new bonds are established during the formation of intimate interfaces, which significantly modify the charge distribution and make EAR predictions inaccurate.

4.3 Empirical BO/SBH Models

Due to the failure of SMT/EAR, empirical models were proposed decades ago in attempts to reproduce the systematic trends of experimentally observed BO/SBH. To understand the motivations behind these empirical models, we take a glimpse of experimental SBH/BO data that were available ~40 years ago. First of all, the FLP phenomenon was the dominant experimental observation from polycrystalline MS interfaces. As shown in Figs. 4.5 and 4.6, this refers to the insensitivity of the SBH measured for zincblende/diamond semiconductors to the work function of the metal and the atomic structure of the interface. As for the heterojunctions, the BO was found experimentally to be largely independent of atomic structure, orientation, and other specifics of the interface for the majority of iso-valent heterojunctions. Furthermore, a "transitivity rule" was often observed to hold among a group of semiconductors. This rule describes the condition that the VBOs measured among three semiconductors A, B, and C approximately satisfy the relationship

$$\Phi_{VBO}^{A-B} = \Phi_{VBO}^{A-C} + \Phi_{VBO}^{C-B}. \quad \text{(Transitivity Rule)} \tag{4.10}$$

The combination of the transitivity rule, the near independence of the BO for iso-valent heterojunction interfaces, and the FLP phenomenon led to the widely held opinion at the time that the formation of the SBH/BO was largely a "bulk effect" of the semiconductor. It was proposed that the SBH/BO magnitudes could be explained in terms of bulk properties of the semiconductor. We clarify, however, that seemingly "bulk-like" behavior for the heterojunction BOs was observed for only some of the semiconductors. For example, the BO for interfaces between hetero-valent semiconductors, e.g. III–V/II–VI and III–V/IV interfaces, varied significantly with the structure and other specifics of the interface. Furthermore, there were many other examples of significant violations of the transitivity rule mixed with BO data in support of the transitivity rule. A valid theory of the BO should explain not only the reason behind the transitivity rule but also why it fails for some. Notwithstanding various evidences of the contrary, the idea of a "bulk" mechanism being responsible for the SBH/BO was the dominant opinion then, leading to the proposal of empirical models.

Traditionally, ideas to explain the heterojunction BO were essentially copied from those proposed for the SBH. Therefore, we focus on the SBH in our discussion of empirical models. The main experimental SBH result that needed explanation was the FLP phenomenon. At the time when FLP dominated SBH research, a phenomenon by the same name was capturing attention in the studies of semiconductor surfaces. As illustrated by Fig. 3.19, the FL of an adsorbate-covered semiconductor surface was measured to be always positioned similarly because of surface states. The role played by the CNL in pinning the surface FL position did not go unnoticed by researchers looking for an explanation of FLP at MS interfaces. The first proposal to involve the surface states in the explanation of both FLP effects came from Bardeen,[15] who suggested that the work function difference between a metal and a semiconductor with surface states could be compensated by an exchange of charge between the metal and the semiconductor surface states. Such a strong pinning mechanism would lead to $S_\Phi = 0$, in disagreement with experiment. From Eq. (4.6), it seems that an interface dipole, eD_{int}^{MS}, that significantly screens out the metal work function would be able to explain the FLP phenomenon. Since a dipole layer is the product of an areal density of charge and a distance, researchers literally looked for such quantities from an MS interface. Bardeen's model was modified by placing the metal not in direct contact with the semiconductor, but at a small gap distance away. The schematic shown in Fig. 4.8 is the "fixed-separation model" of interface

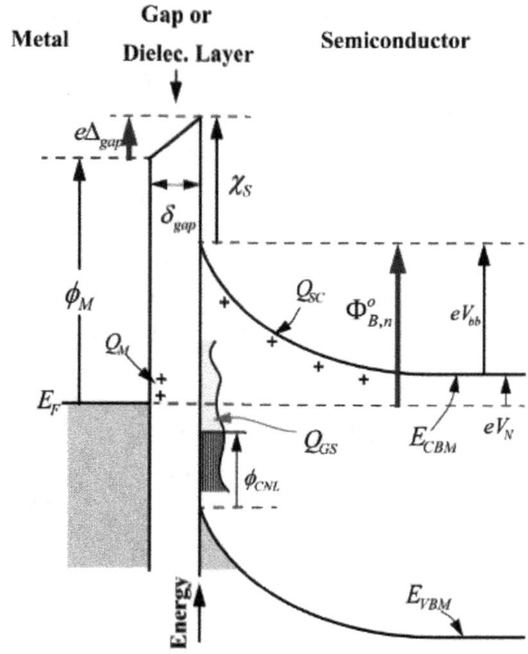

Fig. 4.8. The fixed separation model for SBH formation. Electronic states, with a characteristic CNL and density of states, are assumed to reside on the semiconductor surface, which is held at a fixed distance from the metal.

dipole formation,[16,17] which has been used to explain FLP with not only the surface states of the semiconductor but also other types of interface states, including MIGS, defect states, etc. A thin oxide or dielectric layer that does not affect the surface states of the semiconductor was envisioned to provide the small gap needed. Accepting various assumptions of this model, including that the density of states at the metal FL was high, the n-type SBH would be deduced as[16]

$$\Phi^o_{B,n} = S_{GS}(\phi_M - \chi_{SC}) + (1 - S_{GS})(E_g - \phi_{CNL}), \tag{4.11}$$

where ϕ_{CNL} is the energy of the CNL with respect to the VBM, S_{GS} is a property of the semiconductor, given by

$$S_{GS} = \left(1 + \frac{e^2\delta_{\text{gap}}D_{GS}}{\varepsilon_{\text{int}}}\right)^{-1}, \tag{4.12}$$

which could be identified with the S-parameter experimentally deduced for the semiconductor, and D_{GS}, δ_{gap}, ε_{int} are, respectively, the density of

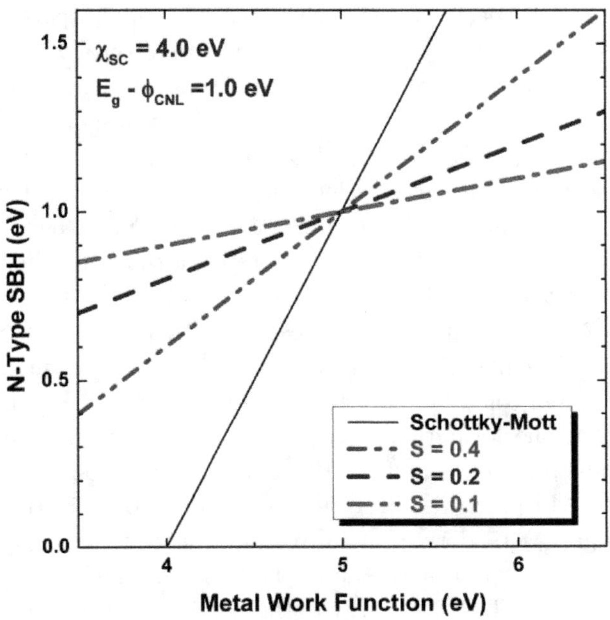

Fig. 4.9. Predictions from SBH model that assumes a fixed separation between interface states and metal. The intercept of lines plotted with different S-parameters is at a metal work function corresponding to the difference between the CNL and the vacuum level, $\chi_{SC} + (E_g - \phi_{CNL})$, which is 5.0 eV for the present example, and at an n-type SBH of $E_g - \phi_{CNL}$, which is 1 eV in this example.

surface states, the thickness of the gap, and the dielectric constant of the gap material. The small separation between the metal and the semiconductor provides some justification for the electronic states on the surfaces of both the metal and the semiconductor not to be modified when an interface is formed and, at the same time, acts as the length over which the charge transfer leads to a dipole. As the form of Eq. (4.8) suggests, the dependence of the SBH on the metal work function is weakened by a factor of S_{GS}^{-1}. When the density of surface states is high, the interface FL is strongly pinned in this model. With two free parameters, D_{GS} and ϕ_{CNL}, to play with, Eq. (4.8) can explain any linear relationship found in SBH vs. work function plots such as Fig. 4.5. The real question is whether the various assumptions of the model are justifiable at real interfaces.

The artificial insertion of a dielectric layer at the MS interface might seem reasonable in the early days of SBH research when the cleanliness of semiconductor surfaces was suspect, but it makes little sense today as microscopic or spectroscopic techniques with high spatial resolution routinely reveal that MS interfaces fabricated with proper care are intimate. Surface states in the band gap of the isolated semiconductor cannot survive the formation of an intimate MS interface[18] because Bloch states are present in the metal. Inside the band gap of the semiconductor, metal wave functions decay a small distance into the semiconductor to become what are known as the metal-induced gap states (MIGS). When MIGS were envisioned to have evolved from surface states of the semiconductor, there was the expectation that the CNL concept would still be applicable to the distribution of MIGSs at MS interfaces.[19,20] To make the CNL concept relevant for FLP explanation, however, it was necessary to justify why the CNL of the MIGS could be independent of the metal. The main argument given was that the formation of gap states was an intrinsic property of the semiconductor and, therefore, the CNL location doesn't change with the metal.[21,22] In particular, Tersoff argued that MIGS in the upper half of the bandgap were derived mainly from the conduction band of the semiconductor and those in the lower half were formed with the valence band states. The CNL is the energy in the band gap where the character of the MIGS changes from valence-band-like to conduction-band-like. These assumptions on the MIGS have all been found to disagree with the results of subsequent first principles calculations, some of which will be discussed in Chapter 7. The necessary dependence of the MIGS on the metal is illustrated in the next section. Due to unrealistic assumptions, there is little scientific basis for models based on the CNL concept. Even though, they were, for a long time, the only models available that could explain some of the SBH/BO data. Charge-density-based theories are now able to explain the entire experimental SBH/BO results, as shown in the remainder of this book.

4.4 Formation of Electronic States at Solid Interfaces

The charge distribution at an interface is from all the occupied electronic states. Energy minimization ultimately determines the charge distribution and all other properties of the interface. In this section, we shall be concerned with only one small aspect of that process, namely, the formation of electronic states at a fully relaxed epitaxial interface of known atomic structure. With either the Hartree–Fock theory or the Kohn–Sham formalism of the DFT, this step involves finding solutions to the single-electron

Schrödinger equation:

$$\left[-\frac{\hbar^2}{2m}\nabla^2 - eV(\vec{r})\right]\varphi_i(\vec{r}) = \varepsilon_i\varphi_i(\vec{r}), \tag{4.13}$$

where the potential $V(\vec{r})$ contains the Coulomb interaction with all charges in the system and also exchange–correlation interactions. Assume that a metal and a semiconductor occupy the space $z < 0$ and $z > 0$, respectively, at a semi-infinite MS interface. An ISR is chosen for $z_{M-ISR} \leq z \leq z_{SC-ISR}$ such that the potential/charge distribution for $z_{SC-ISR} < z$ and $z < z_{M-ISR}$ are bulk-like. As this epitaxial MS interface is assumed to be periodic in x-y directions, the known potential at any z can be expanded in plane waves with the periodicity of the 2D unit cell

$$V(\vec{r}_\parallel, z) = \sum_{\vec{K}_\parallel} U_{\vec{K}_\parallel}(z)\, e^{i\vec{K}_\parallel\cdot\vec{r}_\parallel}, \tag{4.14}$$

where $\vec{K}_\parallel$'s are the 2D reciprocal lattice vectors and note that the Fourier coefficients, $U_{\vec{K}_\parallel}(z)$'s, have explicit z-dependence. All $U_{\vec{K}_\parallel}(z)$'s here are assumed to be known, as our discussion pertains to the very last step of a self-consistent electronic structure calculation. The present discussion is undertaken mainly to demonstrate how the wave functions of the metal and the semiconductor bulk crystals are combined and modified to become solutions for the entire interface and how the bulk states of one crystal may be accommodated in the band gap(s) of the other.

Due to the 2D periodicity, any solution to Eq. (4.13) is characterized by a specific $\vec{k}_\parallel$, from Bloch's Theorem. We now look for parts of a solution $\varphi_{\vec{k}_\parallel}(\vec{r})$ at a specific energy and $\vec{k}_\parallel$, separately for the metal, the ISR, and the semiconductor, before combining them into one complete solution. Since the approach is similar to that already illustrated for the surface, the present discussion for solid interfaces will cover only aspects without analogues with the surface. Like the potential in Eq. (4.14), the wave functions in all regions may be expanded in $\vec{K}_\parallel$. And as before, all sums over $\vec{K}_\parallel$ are assumed to have a finite number (N) of terms. Inside the ISR, wave function with a specific $\vec{k}_\parallel$ can be written as

$$\varphi_{\vec{k}_\parallel}(\vec{r}_\parallel, z) = e^{i\vec{k}_\parallel\cdot\vec{r}_\parallel}\sum_{\vec{K}_\parallel} w^{ISR}_{\vec{k}_\parallel,\vec{K}_\parallel}(z)e^{i\vec{K}_\parallel\cdot\vec{r}_\parallel},$$

$$z_{M-ISR} \leq z \leq z_{SC-ISR}, \tag{4.15}$$

with band and spin indices suppressed. Equations (4.14) and (4.15) can be plugged into Eq. (4.13) to yield coupled equations for the Fourier coefficients, $w^{ISR}_{\vec{k}_\parallel,\vec{K}_\parallel}$:

$$\left[\frac{\hbar^2}{2m}\left(|\vec{k}_\parallel+\vec{K}_\parallel|^2-\frac{\partial^2}{\partial z^2}\right)-\varepsilon_{\vec{k}_\parallel}\right]w^{ISR}_{\vec{k}_\parallel,\vec{K}_\parallel}(z)=e\sum_{\vec{K}'_\parallel}U_{\vec{K}_\parallel-\vec{K}'_\parallel}(z)w^{ISR}_{\vec{k}_\parallel,\vec{K}'_\parallel}(z). \tag{4.16}$$

Since Eq. (4.16) is a second-order differential equation, the solution for the ISR region can always be written down if a total of $2N$ independent boundary values are supplied. When a wave function with this $\vec{k}_\parallel$ and $\varepsilon_{\vec{k}_\parallel}$ is known for the metal, the values and the slopes of its Fourier components at $z = z_{M-ISR}$ provide the $2N$ boundary values needed for the ISR. Likewise, boundary values can also be supplied from the ISR/semiconductor interface to fully define the wave function in the ISR. The ISR thus serves as an intermediary between state(s) with the same energy and $\vec{k}_\parallel$, in the semiconductor and the metal through matching of the boundary values. A complete wave function solution is found only when the matching conditions are satisfied at both the $z = z_{M-ISR}$ and the $z = z_{SC-ISR}$ boundaries. There are four cases concerning the formation of an interface state with specific $\vec{k}_\parallel$ and $\varepsilon_{\vec{k}_\parallel}$ that can be distinguished: (1) there are Bloch states in both the metal and the semiconductor; (2) there are Bloch states in the bulk metal, but no such Bloch states are available in the semiconductor; (3) there are Bloch states in the semiconductor, but not in the metal; (4) there are no Bloch states with the $\varepsilon_{\vec{k}_\parallel}$ and $\vec{k}_\parallel$ in either the metal or the semiconductor. The charge distributions associated with these states are sketched in Fig. 4.10.

We consider what happens when bulk states are available with a $\vec{k}_\parallel$ and energy $\varepsilon_{\vec{k}_\parallel}$, first for the metal. In the metal region, the potential is identical to that of a bulk crystal, and therefore Bloch functions $\psi^{M\_\mathrm{Bloch}}_{\vec{k}_j}$ with the $\vec{k}_\parallel$ ($\vec{k}_j = \vec{k}_\parallel + \vec{k}_{\perp,j}$) and $\varepsilon_{\vec{k}_\parallel}$ automatically satisfy Eq. (4.13) in the metal region. A general form for the wave function in this region is a sum over all such states:

$$\varphi_{\vec{k}_\parallel,\varepsilon_i}(\vec{r})=\sum_{\vec{k}_j}^{\vec{k}_\parallel,\varepsilon_i} b^M_{\vec{k}_j}\,\psi^{M\_\mathrm{Bloch}}_{\vec{k}_j}(\vec{r}),\quad z\le z_{M-ISR}. \tag{4.17}$$

Due to symmetry in the band structure for bulk metal, we assume without loss of generality that only two Bloch states are included in the sum, with

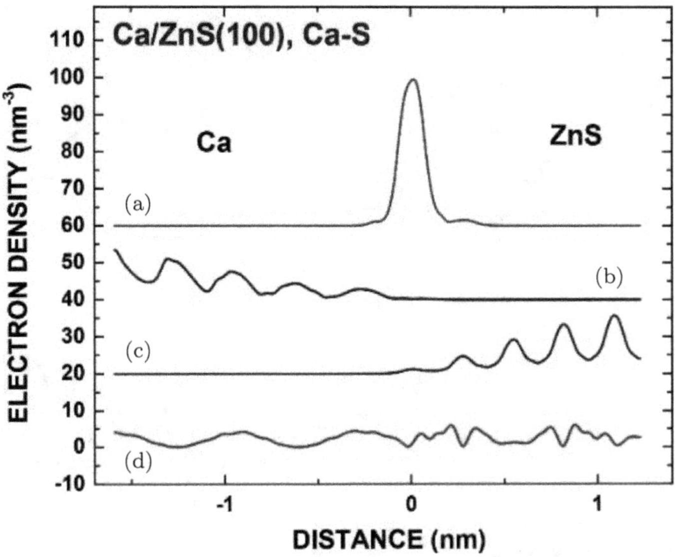

Fig. 4.10. Selected charge density (planar average) of individual state calculated for the relaxed epitaxial Ca/ZnS(100) interface with Ca-S bonds. The mid-point between the Ca and the S planes at the interface marks zero on the horizontal axis. (a) True interface state that decays on both sides. (b) A MIGS that decays toward ZnS. (c) A state from the semiconductor that decays toward a gap in Ca. (d) A state that couples bulk states of the Ca with those of the ZnS.

one "incoming" toward the ISR and one outgoing from the ISR, i.e.,

$$\varphi_{\vec{k}_\parallel,\varepsilon_i}(\vec{r}) = a_{\rm in}^{M}\psi_{\rm in}^{M\_{\rm Bloch}}(\vec{r}) + b_{\rm out}^{M}\psi_{\rm out}^{M\_{\rm Bloch}}(\vec{r}), \quad z \le z_{M-ISR}. \tag{4.18}$$

As these two states form a complete set of the solution to the Schrödinger equation at this energy and wave vector, a combination of these two can always match with any wave function from the ISR at $z = z_{M-ISR}$. The same argument can be made for the semiconductor. When bulk semiconductor states are available at certain $\vec{k}_\parallel$ and $\varepsilon_{\vec{k}_\parallel}$, Bloch states of the semiconductor in the form of

$$\varphi_{\vec{k}_\parallel,\varepsilon_i}(\vec{r}) = a_{\rm in}^{SC}\psi_{\rm in}^{SC\_{\rm Bloch}}(\vec{r}) + b_{\rm out}^{SC}\psi_{\rm out}^{SC\_{\rm Bloch}}(\vec{r}), \quad z_{SC-ISR} \le z, \tag{4.19}$$

are a solution that can match any boundary condition at $z = z_{SC-ISR}$.

The situation of no available bulk states with a $\vec{k}_\parallel$ and energy $\varepsilon_{\vec{k}_\parallel}$ is already encountered during the discussion of surface states formation. Here, we reiterate the case for the energy being in the fundamental band gap of the semiconductor, or any type of gap in the 2DBZ of the semiconductor.

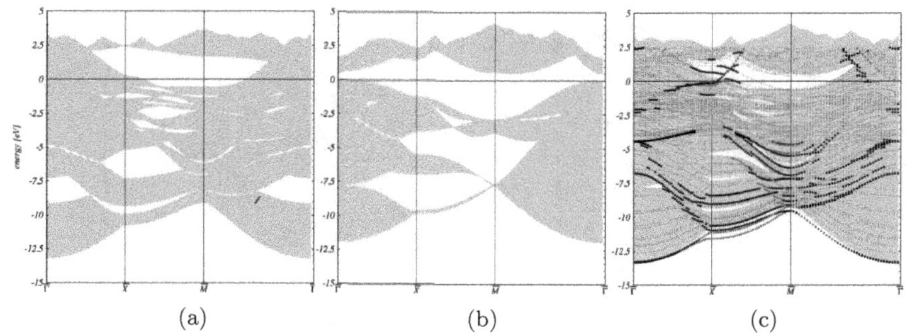

Fig. 4.11. Bulk bands projected along (100) orientation of (a) $CoSi_2$ and (b) Si. (c) Energy bands for the $CoSi_2$/Si(100) interface with the C8 interface type (8-fold coordinated). Shaded area in (c): sum of projected bands of (a) and (b). Interface states localized at positions Si_C(I) (Si interface site of $CoSi_2$ block): black circles; and Si(I) (Si interface site of Si block): grey circles. All other states: small dots. Fermi energy is defined by zero of energy in all cases.

Source: Reprinted with permission from Stadler and Podloucky.[23] Copyright (2000) by the American Physical Society.

Even though there are no states that satisfy the Bloch condition, there are still wave functions at the energy and $\vec{k}_{\parallel}$ that satisfy the Kohn-Sham equation with the potential of the bulk semiconductor. Wave functions with an apparent Bloch form yet with an imaginary $k_{\perp} = iq$ perpendicular wave vector can still satisfy the Kohn-Sham equation. An "evanescent wave" that decays toward the semiconductor ($q > 0$) is the only solution for the semiconductor at a specific $\vec{k}_{\parallel}$ and $\varepsilon_{\vec{k}_{\parallel}}$. Nevertheless, this MIGS is allowed because it can always be matched by a combination of bulk states in the metal. The formation of MIGS is a direct consequence of the presence of bulk metal states at that energy and $\vec{k}_{\parallel}$! In the projected 2D band structures of $CoSi_2$ and Si in the (100) direction in Fig. 4.11, there are bulk states for $CoSi_2$ at the $\bar{M}$ point and at ~5 eV below the Fermi level,[23] but there are no bulk states for the Si; thus serving as an example of Case (2), in addition to the entire fundamental gap of the Si. The present discussion of states in the semiconductor band gap carries over without modification to gaps in the band structure of the metal. A semiconductor bulk state can always decay into gaps of the metal. For example, the gap in bands of $CoSi_2$ at the $\bar{\Gamma}$ point at ~0.85 eV below the Fermi level in Fig. 4.11(a) accommodates evanescent waves from the Si.

At energies and $\vec{k}_{\parallel}$'s with no bulk Bloch states in either the semiconductor or the metal, such as at the $\bar{M}$ point at the FL (or ~10 eV below the FL) in Fig. 4.11, case (4) is encountered. Here, the only acceptable form of the wave function is the kind that decays toward both the metal and the semiconductor. This type of wave function is uncommon because of the overabundance of boundary conditions from both ends. Under rare circumstances, excess boundary conditions turn out to be redundant, and true "interface states" can form that have significant amplitude only in the ISR and decay in both directions.

The formation of the SBH is a result of the entire charge distribution from all four types of electronic states with energies below the Fermi level. Of the charge distribution at an MS interface, such as that in Fig. 4.3, only the part inside the ISR is not known a priori. From the complicated procedures just described for the electronic states to be determined at MS interfaces, it may seem that nothing short of an actual numerical calculation can lead to reliable estimations of the charge distribution and, therefore, SBH predictions. However, the rules on the formation of electronic states in the ISR are not different from those for anywhere else, which leads to similarity in the charge distribution of the ISR with that found in bulk crystals. Even though quantum mechanics clearly governs the formation of charge distribution in the ISR, the latter can actually be modeled adequately without explicit use of quantum mechanics, as explained in the following.

4.5 Interface Charge Modeling Background

The magnitude of the SBH/BO at any solid interface can be experimentally measured. At least in principle, the SBH/BOs of solid interfaces with known atomic structure can also be theoretically calculated, as just described. The availability of these numerical values does not mark the end of the effort to understand the formation mechanism of the SBH/BO but rather the beginning. While one can cavalierly attribute the measured band alignment result to the quantum chemistry of the particular interface, which would be correct technically, there is a desire to conceptually understand the variation in the experimental SBH/BO with specifics of the interface. The conceptual explanation we look for is not an empirical reproduction of the observed systematic SBH/BO trends, but rather a way to arrive at these trends from scientific principles without empirical assumptions. Since the BO/SBH is a direct consequence of the charge distribution at the interface, our approach to understand the BO/SBH is to approximately produce the charge distribution of an interface from physical concepts, without full-fledged numerical calculations.

4.5.1 *Enabler of charge modeling: "nearsightedness" of electron density*

A general property of the density of a many-electron system that essentially enables simulation of the charge distributions at interfaces is the "localness" in the dependence of the electron density on the external potential. In DFT, the total energy and other properties of the ground state of an N-electron system can be expressed as universal functionals of the electron density placed in the presence of "external" potential that arises from the positive charges of the nuclei. Walter Kohn called attention to the "nearsightedness" of the electron density, a reference to the fact that the density of a many-electron system at a particular location depends only on the external potential in its close proximity.[24] Atomic arrangement far away has little influence on the local charge distribution. This property, which obviously reflects how quantum chemistry of many-electron systems works, can now be used directly as a foundation for charge density modeling, without having to bother with the Schrödinger equation and wave functions again. What nearsightedness suggests is that the charge distribution in a volume of the interface with certain local atomic structure is similar to, and therefore may be replaced with, the charge distribution in any other system that we can find with the same local atomic structure. Since only the charge distributions of bulk solid crystals are readily available without specially designed calculations, charge densities of bulk crystals make up the main library to be used to simulate the charge distributions of the interfaces. Properties calculated from the charge density of the bulk crystal are available in Tables 2.5–2.8 for materials commonly used to form solid interfaces for scientific investigations and device applications.

The overall strategy to model the unknown interface charge distribution is to construct it piece by piece with known charge distributions with the same atomic structure found in the bulk crystals. Such an artificially constructed charge distribution, if carried out in the spirit of the nearsightedness principle, should approximately reproduce that of real interfaces. A crucial point to consider is what the "unit" should be for the bulk charge distribution, with which to construct the interface distribution. In other words, what are the size and the shape of each piece of charge distribution in this piece-by-piece replacement, in order to assemble something that resembles the real interface density? The nearsightedness principle, taken at its face value, suggests that the charge distribution at any location of a solid might be dominated by only one of the atoms closest to this location and thus the charge distribution of any solid might be viewed as having been assembled with the charge densities that habitually reside with each of the atoms that make up the structure. If the electron density should be so extremely nearsighted, the charge distribution of an interface would be approximately assembled with the known atomic densities from bulk crystals, and the band alignment conditions would be approximately recreated. To proceed with the concept of individual atoms as units, there is still the important question of how the bulk charge density should be partitioned into atomic densities that are essentially transferable.

4.5.2 *Choosing bulk atomic densities*

Two methods to partition the charge distribution of a bulk crystal into volumes associated with individual atoms/ions, NPT and AIM, are compared in Chapter 2 in terms of the transferability of atomic densities. Atomic transferability among different bulk crystals is the figure of merit for a model solid in interface charge distribution modeling because that likely translates to a replaceability of interface atoms with bulk atoms. NPT is clearly favored over AIM for the analysis of covalent interfaces. Both the NPT and the AIM seem adequate for ionic interfaces, although the latter has a slight edge. Despite the slightly better transferability of Bader atoms among ionic solids, there are restrictions on the use of AIM model solids. NPT is the only choice for covalent interfaces not only because of better transferability in atomic densities but also its connection with the concept of covalent radius. The independence of its average potential energy from the surface termination also makes NPT the only available method to model the BO/SBH for all (polycrystalline) interfaces with completely unknown structure and for interfaces without appropriate AIM model solids.

4.6 Procedures to Model Interface Charge Distribution

The relative band position at a solid interface is a result of the charge distribution of the interface. Any simulation of the band alignment through charge distribution appears to make use of Eqs. (4.1) and (4.2). For example, in order to predict the heterojunction VBO of the InAs/GaSb(100) interface with As-Ga bonds, as shown in Fig. 4.12, it may seem that one would need to produce the $e\Delta_{ISR}$ term to be plugged into Eq. (4.1). In Fig. 4.12, the internal chemical potentials (μ_{SC}) of InAs and GaSb are added respectively to the average electrostatic potentials on the two sides to establish the VBO. However, Eqs. (4.1) and (4.2) are inconvenient

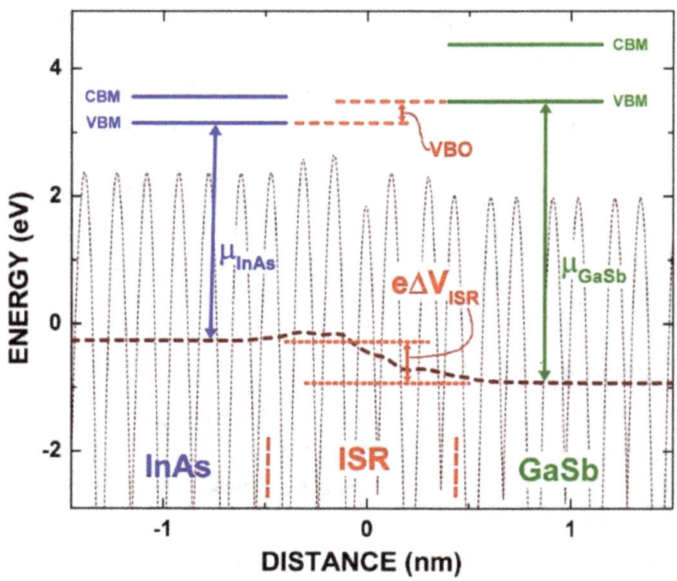

Fig. 4.12. Plane-averaged electrostatic potential energy (dotted line) and macroscopically averaged potential energy (heavy dashed line) at the interface of a semiconductor heterojunction. The bulk band extremes, valence band maximum (VBM) and conduction band minimum (CBM), of each semiconductor are mapped onto the potential energy, using the internal chemical potential, μ_{SC}, calculated for each bulk semiconductor. An interface-specific region (ISR) is identified that separates the two regions, where charge distribution is indistinguishable from that for the respective bulk semiconductor. The difference between the VBM positions for the two semiconductors is the valence band offset (VBO). The actual potential energy distribution from an InAs/GaSb(100) interface with As-Ga interface bonds is used.

for modeling because the $e\Delta_{ISR}$ term, which depends not only on the charge distribution of the ISR but also the bulk charge distributions of both crystals, is nearly impossible to estimate. It is more practical to divide the problem in such a way that the unknown part due to the interface charge distribution reflects a minor correction to what can be reasonably expected from the atomic structure of the interface. The overall plan is to cut the charge distribution of an interface into three parts that represent the two crystalline solids and the ISR. Such a partition strategy would only make sense if the potential distribution for each part could be individually calculated, and the three potential distributions could then be stitched/superimposed directly to approximately recreate the potential distribution of the entire interface. To represent a crystalline solid is exactly what a model solid is created for, in Chapter 2. It is thus clear that the interface charge distribution should be partitioned into two appropriately chosen model solids for the two crystals and an ISR in between, as suggested by Eqs. (4.7) and (4.8).

To illustrate how this general plan may be implemented, we use the same charge distribution of Fig. 4.12 but proceed differently. To represent the two semiconductor bulk crystals in this analysis, NPT model solids are chosen for (100)-oriented InAs and GaSb. From the two-dimensional profile of the charge distribution at this interface, shown as inset in Fig. 4.13, the boundaries of these two model solids are symbolically marked by the dashed zigzag lines. As discussed in Chapter 2, the boundaries of these model solids are not located at the immediate interface but at locations where the charge distributions have reverted back to that of the bulk crystal. The 2D density profile of the interface is literally severed into three parts, with the ISR

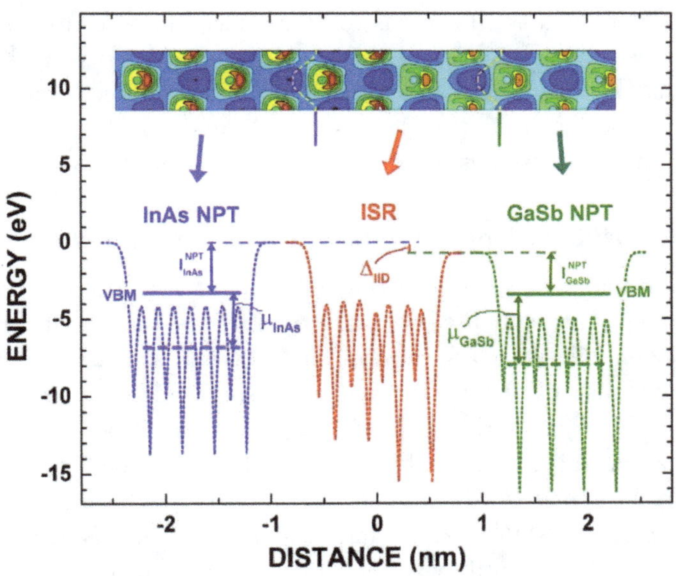

Fig. 4.13. Visualization of the NPT approach to model the VBO of a semiconductor heterojunction. The charge distribution of the actual interface may be partitioned along NPT boundaries, indicated by dashed yellow lines, into three separate parts representing the two semiconductors and the ISR. Since the partitions are along NPT boundaries, the average potential energies of the two semiconductors are precisely known with respect to the vacuum level. The actual charge distribution in the ISR is responsible for Δ_{IID}, as drawn in this diagram. However, for actual NPT modeling, this term will be estimated using methods described in the text.

sandwiched between the two model solids. Due to electrical neutrality, the potential energy curves for the three regions may be individually calculated, as sketched in Fig. 4.13. The VBMs of the two semiconductors can be individually referenced to the vacuum level, as I_{InAs}^{ModSol} and I_{GaSb}^{ModSol}. The numerical values of the ionization potentials of the NPT model solids for zincblende semiconductors are listed in Table 2.7. The two model solids are the hosts and the ISR plays the role of the "bath" at this interface. Note that to the right and the left of the isolated ISR are the empty spaces vacated by the two hosts. Therefore, the potential energy difference across the charge distribution of the ISR, $-eV_{ISR,NPT}^{InAs-GaSb}$, is the difference in the empty potentials. The VBO of this interface can now be read off of Fig. 4.13 as

$$\Phi_{VBO}^{InAs-GaSb} = I_{InAs}^{\text{ModSol}} - I_{GaSb}^{\text{ModSol}} - eV_{ISR,NPT}^{InAs-GaSb}. \tag{4.20}$$

This equation can be recognized as Eq. (4.7), when applied to the present interface. The quantity that needs to be produced in order to predict the VBO is now $-eV_{ISR,NPT}^{InAs-GaSb}$, the net effect of the bath between the model solids. This quantity still seems difficult to estimate directly because the charge distribution inside the ISR is complicated and largely unknown. However, this impression is not quite correct, as we now show.

4.6.1 *Identification of the nature of the corrections needed*

There is an important aspect of the definition of the ISR that should be stressed. Since both cuts used to define the ISR are along boundaries that define model solids, each surface of the ISR has a charge distribution identical to that of the respective model solid surface. Therefore, the $-eV_{ISR,NPT}^{InAs-GaSb}$ contribution depends on the model solids employed for the crystalline solids. Generally, the same type of model solids should be chosen for the two solids of an interface, e.g. both are NPT model solids. To model $-e\Delta V_{ISR}^{1-2}$ at the interface between solid #1 and solid #2 adequately, we first identify what this term is by conducting the following conceptual exercise. The final charge distribution of a relaxed interface with a particular atomic structure may be envisioned to take shape in two steps. *In step one*, two model solids with carefully chosen surface terminations are stacked together to bring all the atoms needed for the formation of a neutral interface with the specified atomic structure. Still in their unrelaxed bulk positions, some atoms in the ISR are not at their final relaxed positions yet. But all atoms should be present, without any missing or surplus ones.

Essentially, step one involves the stacking of bulk atomic densities to create a temporary, unrelaxed charge distribution of the interface. After step one, there is no $-e\Delta V_{ISR}^{1-2}$ and the VBO is simply the difference between the two ionization potentials in Eq. (4.20). *In step two*, the charge density and atomic positions in the ISR are relaxed to reach equilibrium. After step two, the interface may again be partitioned as in Fig. 4.13 and the $-e\Delta V_{ISR}^{1-2}$ term reappears. This exercise shows that the $-e\Delta V_{ISR}^{1-2}$ term is the net effect of charge and lattice relaxations from stacked bulk densities to the equilibrium distribution. Since the first two terms of Eq. (4.20) depend on the method chosen to partition the bulk charge density, the value of $-e\Delta V_{ISR}^{1-2}$ and how this term can be modeled also depend on that choice. For it to be easily estimated from general principles, this term should be a minor correction. In other words, good model solids to use are those that after step one already produce a charge distribution that resembles the final distribution.

4.6.2 *Corrections for charge density difference and interface bonds*

The distribution assembled with frozen bulk atomic densities needs to relax to reach equilibrium. Since each atomic density is already energy minimized for a (bulk) atomic structure that is similar to that at the interface, the rearrangement of charge density within each "atom" is expected to be small and may be ignored. However, when two model solids are stacked together after step one in the above discussion, the average charge densities on the face of the two model solids usually have a finite difference, $\Delta\rho$. In Chapter 2, the average densities of charge on the surfaces of model solids are computed and shown in Tables 2.6–2.8 for commonly studied crystals. Obviously, there should be some transfer of electrons from the model solid with higher face density to the one with smaller density in step two to smooth out the charge distribution and eliminate any finite jumps in charge density. If we assume that the beginning difference $\Delta\rho$ linearly tapers off over a fixed distance Δx, as graphically shown in Fig. 4.14, a contribution to the $-e\Delta V_{ISR}^{1-2}$ term of

$$-e\Delta V_{ISR}^{\Delta\rho} = \frac{2e}{\varepsilon_o}\int_{-\Delta x/2}^{0} dy(y+\Delta x/2)^2\frac{\Delta\rho}{2\Delta x} = 7.54\cdot\Delta\rho(\Delta x)^2\,eV\cdot\text{Å} \tag{4.21}$$

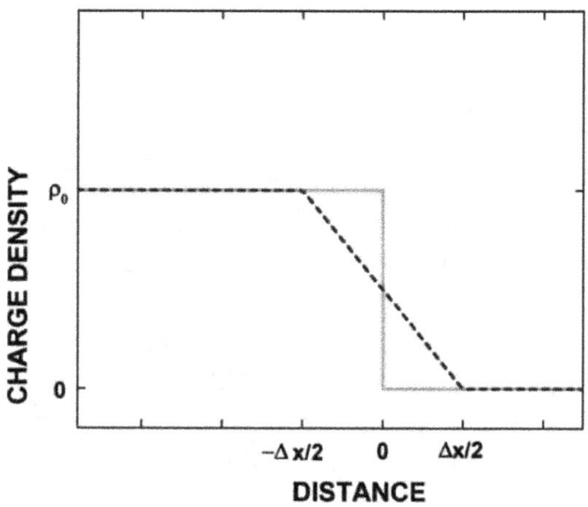

Fig. 4.14. Schematic illustration of an abruptly terminated charge density (solid line) may be assumed to decrease in a linear taper over the distance of Δx, while conserving the total charge.

is required. Since charge smoothing is a universal phenomenon, the distance Δx over which the density difference is dissipated may be determined by comparison with experimental BOs. A systematic study of the measured VBOs and SBHs between binary oxides, described later, suggests that this distance is ~0.07 nm and, therefore, a single formula, $3.76 \cdot \Delta\rho eV \cdot \text{Å}^3$ may be used to account for this effect, regardless of the model solid used. Note that such a taper in the charge density conserves the total charge of the interface.

Another correction for step two becomes necessary when the bonds in the two bulk crystals are dissimilar. When that happens, the new bonds at the interface are not part of either of the bulk crystals, and the local environments for the atoms at the interface are different from the bulk. Considerable adjustments to the stacked atomic densities at the interface are then expected in order to reach equilibrium charge distribution. The main correction required is to account for the polarization of the new interface bonds. In chemistry, the dipole moment of a molecule is often estimated by assigning a net charge to each of the atoms forming the bond and calculating the dipole contributions from point charges placed at atomic positions. This method can be applied to solid interfaces. A baseline to start is a charge distribution with each atom at the interface electrically neutral, which happens to be the situation for an ISR assembled with NPT model solids in step one. From this neutral baseline, the contribution of bond polarization to the potential drop across the ISR, $-e\Delta V_{ISR}^{\text{bond}}$, can be estimated, through the molecular chemistry approach, as

$$-e\Delta V_{ISR}^{\text{bond}} = c\Delta X \cdot L_{\text{bond}} \cdot d_{2D}, \tag{4.22}$$

where ΔX is the difference in electronegativities of the two atoms forming the bond, L_{bond} is the bond length, d_{2D} is the number of bonds per unit area of the interface, and c is a proportionality constant that can be deduced from fitting. This particular correction, which obviously requires a knowledge of the interface atomic structure, is generally unnecessary for interfaces between materials with similar bonds. When the effect of interface bonds cannot be ignored, the total correction for the modeled band alignment is the sum of the two contributions described above, namely,

$$\Phi_{VBO}^{SC1-SC2} = I_{SC1}^{\text{ModSol}} - I_{SC2}^{\text{ModSol}} - eV_{ISR}^{\Delta\rho} - eV_{ISR}^{\text{bond}}. \tag{4.23}$$

4.6.3 *Absolute bulk reference for band position of solid: no bond correction required*

Epitaxial interfaces used in high-performance device applications are often required to have a uniform band alignment condition. This is more easily accomplished by choosing semiconductors of the same type, i.e. both covalent or both ionic, to form the interface. At such solid interfaces, the network of bulk bonds is not interrupted and there is no need to correct for new interface bonds. For these interfaces, only the correction for $\Delta\rho$ is required to model the interface charge distribution. Writing this correction term out explicitly for these interfaces, we have

$$\begin{aligned} E_{VBO}^{1-2} &= E_{V,SC1}^{\text{ModSol}} - E_{V,SC2}^{\text{ModSol}} - k(\rho_{SC1} - \rho_{SC2}) \\ &= E_{VBM,SC1}^{\text{BulkRef}} - E_{VBM,SC2}^{\text{BulkRef}}, \end{aligned} \tag{4.24}$$

where k is a constant ($= 3.76$ eV Å^3) and ρ_{SC} is the density of electrons on the surface of the semiconductor model solid. The bulk reference for each solid, $E_{VBM,SC}^{\text{BulkRef}}$, with respect to the vacuum level is defined entirely from the properties of that crystal,

$$E_{VBM,SC}^{\text{BulkRef}} = E_{V,SC}^{\text{ModSol}} - k\rho_{SC}, \tag{4.25}$$

and remains unchanged for all interfaces of which this crystal is a part. For interfaces with uninterrupted bond networks, charge-density-based predictions of the BO/SBH are thus the difference of the two bulk solids on

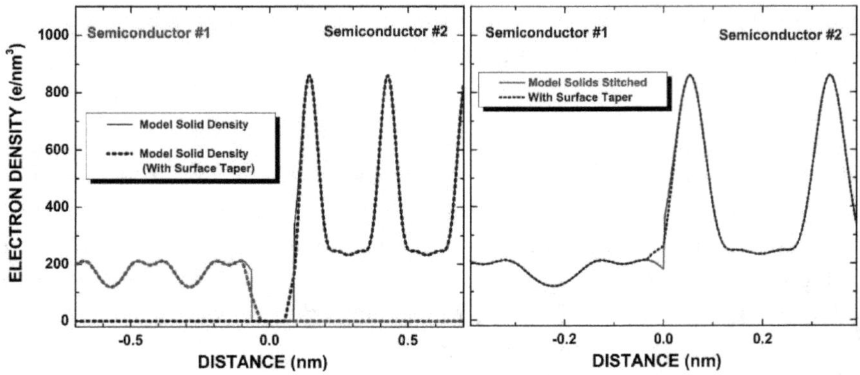

Fig. 4.15. Schematic illustration of the abrupt termination of the charge density of model solid and the need to smooth out the finite difference in charge density across the interface. (a) The abruptly terminated charge density may be tapered down for the model solids. (b) The discontinuity in charge density at the interface from uncorrected model solids may be smoothed out with surface tapers.

a universal scale, as illustrated in Eq. (4.24). The bulk reference $E_{VBM,SC}^{\text{BulkRef}}$ here is the VBM position with respect to the vacuum level, after the model solid has been adjusted for the average density of electrons on its face. As illustrated by the fine lines in Fig. 4.15, the charge density on a model solid ends abruptly on its surface. Stitching of model solid densities of two different crystals usually leads to an abrupt and finite jump in the charge density, such as the fine line in Fig. 4.15(b), which is unphysical. To avoid discontinuities in stitched density, the surface of each model solid may be reshaped preemptively by a linear taper in charge density over a fixed distance of ∼0.07 nm, as schematically shown by dotted lines in Fig. 4.15. When charge densities of two model solids, each with density taper, are stitched together, there are no more jumps in charge density.

A bulk reference of the band edge position of each solid crystal can be established with respect to the vacuum level by analysis of the charge distribution of the crystal. Equation (4.25) puts the band edges of all solids on a common scale and significantly simplifies the procedure to model the interface band alignment. For interfaces without the need for bond corrections, later chapters will demonstrate the amazing accuracy of explaining/predicting the band alignment conditions by simply taking the difference in the bulk reference levels of the two crystals. Since each

SBH/VBO predicted is the difference between two entries on a common scale, the SBH/VBOs for this class of interfaces will display "bulk-like" behavior, namely, they are independent of the orientation and the atomic structure of the interface and, furthermore, they are generally transitive. As mentioned earlier, the band alignment values at many interfaces were indeed found experimentally to be bulk-like. Ignoring the foundation of SBH/VBO in charge distribution, some scientists decades ago chose to explain bulk-like behavior by insisting that a special energy level (CNL) for one bulk semiconductor should align at an interface with that of another bulk semiconductor or the FL of a metal. This empirical CNLconcept conflicted with quantum mechanics and was unnecessary for the explanation of apparently bulk-like behavior. The procedures to model the interface charge distribution show that the SBH/VBO should appear as bulk-like whenever the interface bonds are similar to bulk bonds. They also identify the condition for the interface bonds to be important and for the SBH/VBOs not to appear bulk-like! No matter how differently the SBH/VBO may seem to behave for different types of interfaces, they are all results of, and should be explainable with, the charge distribution of the interface, as is demonstrated in the following few chapters.

4.7 Summary

In this chapter, we begin discussions on band alignment at solid interfaces. How the SBH/VBO of a solid interface is a direct consequence of the charge distribution is the first thing illustrated and the most important point of this chapter. From that perspective, the most logical and sure-footed way to understand/predict the band alignment at an interface seems to be through its charge distribution. However, at the time when band alignment concepts were developed from scratch decades ago, approaches based on the charge distribution were largely ignored. Traditional theories represented the band positions of solid crystals with their surface properties, which was a mistake that prevented the explanation of experimentally observed SBH/VBOs. The failure of the traditional theories prompted the proposal of various empirical models specifically designed to explain the FLP phenomenon and the appearance that some experimental data were bulk-like. Unconnected with charge distribution, empirical models relied on assumptions of the energetic distribution of interface states that disagree with later ab initio calculations. In reality, the entire experimental SBH/VBO data encompass a wide range of dependences in addition to the small set that appear as bulk-like. So, the band alignment problem has always been beyond the scope of empirical models and in need of a comprehensive treatment with a broad perspective.

This chapter emphasizes the importance of relying on the charge distribution of the interface, which is the foundation of the band alignment problem, to understand and explain the entire gamut of different SBH/VBO results found in the literature. To be able to do this, we need to understand the rules of charge distribution and be able to predict how the charge distribution changes with interface structure and bond. Since electron density depends very locally on the atomic structure, it is feasible to simulate the charge distribution of an interface by assembling with charge distributions found in bulk crystals with identical atomic structures. An overall plan, along with detailed steps, to approximately produce the charge distribution at a solid interface is discussed in this chapter. An example of how the SBH/VBO at a solid interface may be modeled with bulk charge densities is shown graphically in Fig. 4.16. The calculated electron density at a Rh/ZnS(100) interface with Rh-S bonds is partitioned into three parts: the Rh, the ZnS, and the ISR. To represent the regions of the interface where the charge distribution is essentially bulk-like, model solids of Rh and ZnS are used, as shown in the middle panels of the figure. The model solids drawn here are both defined by the NPT method, which requires that the two boundaries of the ISR are also of this type. For each model solid, the electric potential energy and the energy band position(s) with respect to the vacuum level can be calculated in isolation with formulas derived in Chapter 2, as shown at the bottom of Fig. 4.16. In general, the electric potential of the ISR is difficult to compute explicitly. However, this chapter identifies that the net effect of the ISR is from charge relaxation of the bulk charge density at the interface. There are two main contributions to the potential drop across the ISR: smoothing out the difference in average densities of the two model solids and the formation of polarized bonds at the interface. Since the magnitudes of these two contributions can be independently estimated, there is no need for detailed computations for the ISR. The effect of charge relaxation is estimated and inserted between the potential energies of the two model solids, as shown in the bottom panel, to produce the final band alignment prediction for the interface. Figure 4.16 illustrates the steps that will be used in subsequent chapters to model the SBH/VBO for various solid interfaces.

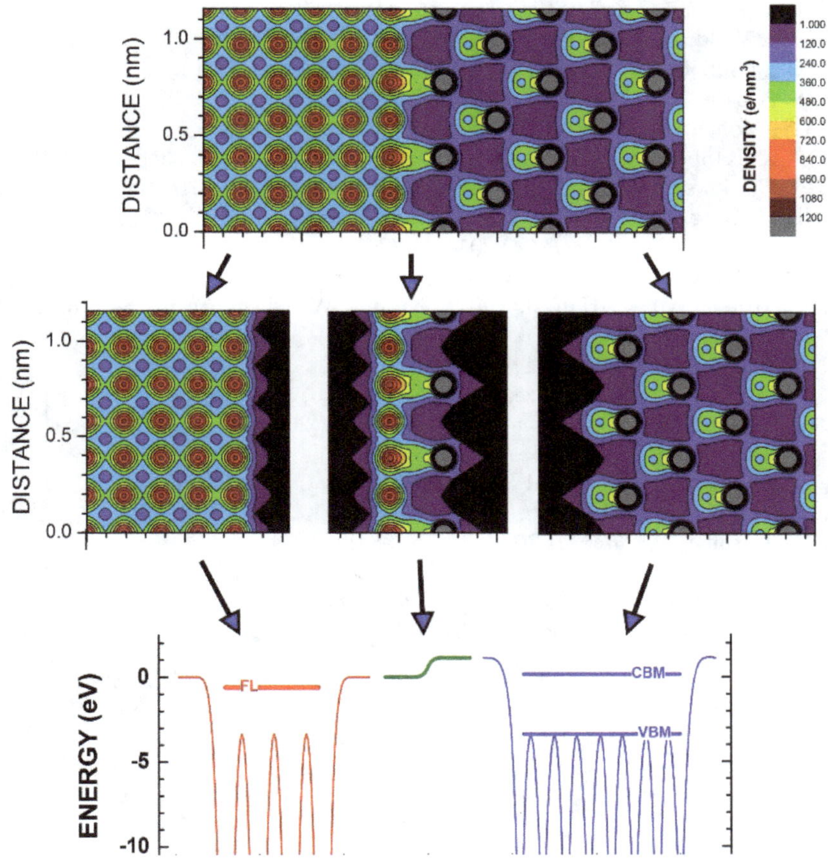

Fig. 4.16. Top: [110] view of the distribution of electron density at the interface of a Rh/ZnS(100) interface with Rh-S bonds. Middle: To model the SBH of this interface, the charge distribution of the interface is partitioned into a Rh model solid on left, ISR in the middle, and a ZnS model solid on the right. Bottom: The electric potential and the energy band positions of the Rh and the ZnS model solids can be individually calculated with respect to the vacuum level. The contribution of the ISR, which plays the role of the bath in this analysis, is estimated from the known density difference of the two model solids and the interface bond characteristics and used to rigidly shift the relative band positions established by the model solids.

References

1. D.K. Singh, B. Roul, R. Pant, A.M. Chowdhury, K.K. Nanda, and S.B. Krupanidhi, *Appl. Phys. Lett.* **116**, 252102 (2020).
2. W. Schottky, *Z. Physik* **113**, 367 (1939).
3. N.F. Mott, *Proc. Roy. Soc. (London)* **A171**, 27–38 (1939).

4. R.L. Anderson, *Solid-St. Electron.* **5**, 341 (1962).
5. J.R. Waldrop, *Appl. Phys. Lett.* **44**, 1002–1004 (1984).
6. C. Mailhiot and C.B. Duke, *Phys. Rev. B* **33**, 1118 (1986).
7. S. Kurtin, T.C. McGill, and C.A. Mead, *Phys. Rev. Lett.* **22**, 1433 (1969).
8. W. Monch, *Phys. Rev. Lett.* **58**, 1260 (1987).
9. A.K. Rappe and W.A. Goddard, III, *J. Phys. Chem.* **95**, 3358–63 (1991).
10. J. Cioslowski and B.B. Stefanov, *J. Chem. Phys.* **99**, 5151–62 (1993).
11. R.T. Tung, *Phys. Rev. Lett.* **84**, 6078–81 (2000).
12. R.T. Tung, *Phy. Rev. B* **64**, 205310 (2001).
13. M. Schluter, *Phys. Rev. B* **17**, 5044–7 (1978).
14. Z. Hu, Z. Wu, C. Han, J. He, Z. Ni, and W. Chen, *Chem. Soc. Rev.* **47**, 3100–3128 (2018).
15. J. Bardeen, *Phys. Rev.* **71**, 717 (1947).
16. A.M. Cowley and S.M. Sze, *J. Appl. Phys.* **36**, 3212 (1965).
17. R.T. Tung, *Mater. Sci. Eng. Rep.* **35**, 1–138 (2001).
18. V. Heine, *Phys. Rev.* **138**, A 1689 (1965).
19. C. Tejedor, F. Flores, and E. Louis, *J. Phys. C* **10**, 2163–77 (1977).
20. J. Tersoff, *Phys. Rev. Lett.* **52**, 465–8 (1984).
21. J. Tersoff, *Phys. Rev. B* **30**, 4874–7 (1984).
22. J. Tersoff, *Phys. Rev. B* **32**, 6968–71 (1985).
23. R. Stadler and R. Podloucky, *Phys. Rev. B* **62**, 2209–19 (2000).
24. W. Kohn, *Phys. Rev. Lett.* **76**, 3168–3171 (1996).

Chapter 5

HETEROJUNCTION BAND OFFSETS EXPLAINED WITH CHARGE DISTRIBUTION

Heterojunction between semiconductors with different band gaps is a structure required for a large number of solid-state devices that have revolutionized the semiconductor industry in recent decades. Our lives have improved tremendously because of modern technologies employing heterojunctions in different fields of application. Even though the main interest of this book is not in device applications, it would be remiss not to mention the importance of heterojunctions in the development of advanced devices such as lasers, light-emitting diodes, solar cells, heterojunction bipolar transistors, and high electron mobility transistors that have directly impacted our daily lives. At the same time, heterojunctions are a key structure that underpins the ongoing investigations and explorations of various quantum effects for science and technology. Due to the importance of heterojunctions, the formation of band offset at heterojunction interfaces needs to be thoroughly understood. After brief descriptions of how heterojunction BOs may be experimentally measured and theoretically calculated, this chapter illustrates how these BOs may be quantitatively explained through modeling of interface charge distribution.

5.1 Band Offset Measurement

(a) Surface Photoemission Spectroscopy

Photoemission spectroscopy is a versatile technique used extensively to measure the BO of semiconductor heterojunctions.[1] Photoemitted electrons can only escape from a shallow depth beneath the surface without inelastic scattering. Since at least some parts of each semiconductor forming the interface must be visible in order for the relative band position to be determined by this technique, the surface semiconductor (over-)layer

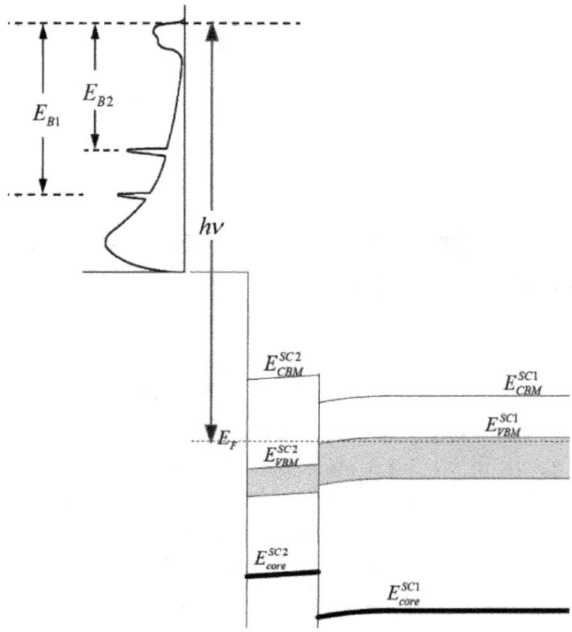

Fig. 5.1. Energy bands and core levels of a heterojunction comprising a thin semiconductor #2 layer on a substrate of semiconductor #1. The energy difference between the two core levels can be measured from the photoelectron spectrum and used to determine the offset in the VBM or the CBM of the two semiconductors.

is limited to less than ~10 nm in thickness. To avoid oxidation and contamination, heterojunction interfaces to be studied by photoemission are usually fabricated in situ in the photoemission analysis chamber. For accurate measurements, the semiconductors should not be heavily doped to minimize the band-bending that contributes to measurement errors. The availability of a wide range of photon energies from different light sources allows any occupied level to be energy analyzed with respect to the FL of the system. It is possible to directly detect the energy difference between the VBMs of the two semiconductors when the difference is large enough for both band edges to be individually analyzed. More frequently, however, the BO of a heterojunction interface is inferred from measurement of the core levels of the two semiconductors. A schematic of how core levels are used to measure the band offset of the insulator interface is shown in Fig. 5.1. The energy difference between the VBM and the core level of a constituent atom, $E_{\mathrm{VBM}} - E_{\mathrm{core}}$, is a fixed quantity for a semiconductor, known from calculations or photoemission studies of the bulk semiconductor. Photoemission measured binding energies of chosen core levels of the two semiconductors, against the same FL, can now be converted into the VBO of the heterojunction. For example, the binding energies, E_{B1} and E_{B2}, of appropriately chosen core levels of the two semiconductors forming a heterojunction can be measured in a single photoemission experiment, as shown in Fig. 5.1. The difference in the measured core level binding energies can be converted to the VBO of the heterojunction

$$\Delta E_{\rm VBO}^{\rm SC1-SC2} = E_{\rm B2} - E_{\rm B1} + (E_{\rm VBM}^{\rm SC2} - E_{\rm core}^{\rm SC2}) - (E_{\rm VBM}^{\rm SC1} - E_{\rm core}^{\rm SC1}), \quad (5.1)$$

using the known fixed quantities (in parentheses) for the two semiconductors. An explicit demonstration of the photoemission technique is shown in Fig. 5.2 for the VBO at a MgSe/$Zn_{0.42}Cd_{0.58}Se$(001) interface.[2] The Mg 2p core level at the interface is measured to shift downward by $0.57 \pm 0.05\,eV$ with respect to the bulk standards, while the Cd 3d level is kept unchanged.

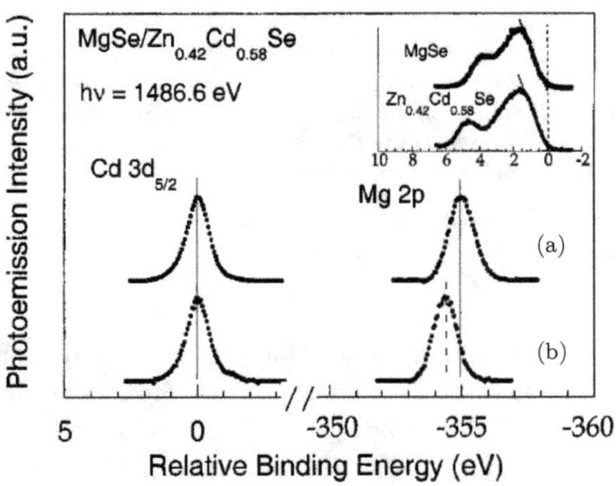

Fig. 5.2. Photoemission determination of the valence band offset for a MgSe/$Zn_{0.42}Cd_{0.58}Se$(001) heterojunction. Inset: valence band emission from $Zn_{0.42}Cd_{0.58}Se$ (bottom) and MgSe (top) epitaxial standards. The zero of the binding energy scale is at the linearly extrapolated position of the valence band maximum E_v. (a) Mg 2p and Cd 3d emissions from the bulk standards. The apparent core energy separation is that expected from a hypothetical heterojunction with zero valence band offset. (b) Mg 2p and Cd 3d emissions from a MgSe/$Zn_{0.42}Cd_{0.58}Se$(001) heterojunction which includes a 3-nm-thick MgSe overlayer. The variation in the core separation ΔE_{cl} relative to (a) was independent of overlayer thickness in the thickness range of interest (1–5 nm) and gives directly the valence band discontinuity $\Delta E_v = 0.57 \pm 0.05\,eV$.

Source: Reprinted from Pelucchi *et al.*[2] with the permission of AIP Publishing.

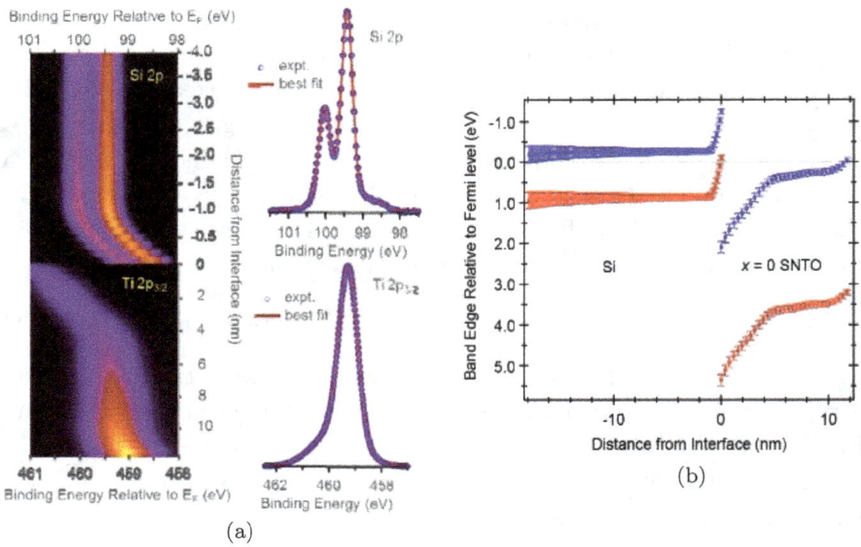

Fig. 5.3. (a) Contour intensity plot of layer-resolved Si 2p and Ti $2p_{3/2}$ spectra versus distance from the interface resulting from fitting the spectra from the STO/Si heterojunction that takes into account built-in fields. Overlays of the sums of all experimental spectra are shown on the right. (b) Band edge profiles for 12 nm n-$SrTiO_3$/i-Si(001) heterojunction taken from fits shown in (a). The conduction band edge profiles are simply the valence band edge profiles less the band gap for the appropriate material.

Source: Reprinted with permission from Z. H. Lim *et al.*[3] Copyright (2019) by the American Physical Society.

As the VBMs of the bulk standards for both semiconductors have been artificially set as equal, the measurement now translates to a VBM of the MgSe lower than the VBM of $Zn_{0.42}Cd_{0.58}Se$ by 0.57 eV.

Accurate measurement of the VBO through the determination of core level peak positions usually relies on the lack of significant band bending in the semiconductor overlayer. The relative core level peak positions reflect their alignment at the interface under such a favorable condition. When the bands of either semiconductor are not flat within the escape depth of emitted photoelectrons, core levels at different depths appear at different energies in the PES signal, and the apparent peaks in the recorded spectrum do not necessarily indicate their positions at the interface. Under such circumstances, the experimental determination of the band offset only from the peak core level positions may be in error. However, band bending does not prevent band offsets from being accurately measured with the photoelectron spectroscopy technique. Even with significant band bending, photoemission spectrum of a single core level peak, recorded with high-resolution electron energy analyzer and highly monochromatized light source, can be fitted with multiple peaks originating from different surface depths to high precision. A single broad peak is deconvoluted into a spatially resolved band-bending diagram which allows the core level position at the interface to be accurately determined, as shown in Fig. 5.3. Simply from the photoelectron peak positions of the Si 2p and the Ti $2p_{3/2}$ core levels, the VBM of the $SrTiO_3$ overlayer would have been deduced to be ~2.5 eV below the VBM of the Si for a 12 nm $SrTiO_3$/Si(001) heterojunction, which corresponds to a type II band offset situation. However, a careful analysis of the details of these peaks reveals significant band bending on both sides within a short distance of the interface.[3] At the interface, the VBM of STO is >5.5 eV below that of the Si, constituting a type III heterojunction according to this analysis. This somewhat extreme example of an interface between covalent and ionic semiconductors serves as a reminder of possible pitfalls of the photoelectron technique and, at the same time, a demonstration of the kind of information that can be extracted with careful and advanced analysis.

(b) Transport Measurement

The band diagram of a heterojunction interface between two *n*-type doped semiconductors A and B is shown in Fig. 5.4 without applied bias. The band bending drawn in this diagram is that needed to support an offset in the conduction band and, at the same time, maintain overall neutrality of the interface. As semiconductor B has a larger band gap and a higher CBM position than semiconductor A in this figure, the space charge region at the interface resides essentially on the side of semiconductor B. As indicated by the small downward band bending, semiconductor A is negatively charged at the interface. The rectifying potential barrier for the transportation of electrons is approximately the CBO, just as the VBO would be for the transport of majority carriers (holes) at an isotype *p*-type heterojunction. In Fig. 5.4, the accumulation of electrons on the semiconductor A side of the interface has the consequence that the electrochemical potential there should remain largely unchanged, when a small bias potential of either sign is applied. Accordingly, the flow of electrons from semiconductor B to A is "easy" and represents the "forward" direction. Electron transport in the opposite direction is more difficult. It is thus possible to measure the heterojunction band offset magnitude

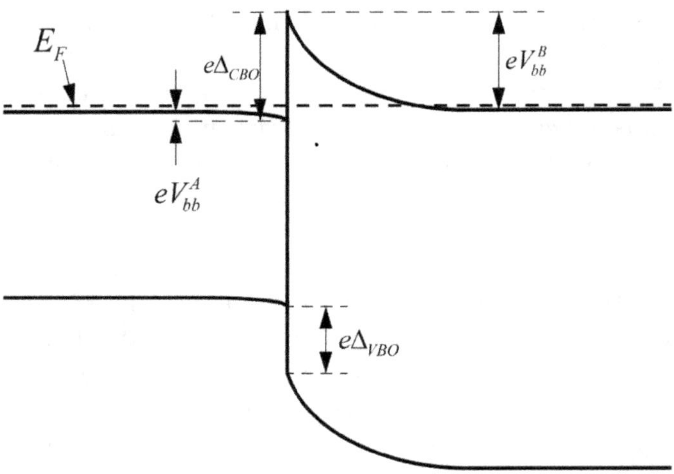

Fig. 5.4. Band diagram of an n-type heterojunction interface.

from the characteristics of transport perpendicular to the interface. Both current-voltage and capacitance-voltage methods can be used for CBO measurement, although reliable measurements can only be performed with adequate, low-resistance contacts to both semiconductors. Except during the initial period of BO research, however, transport measurements have not been relied upon for accurate BO determination.

(c) Internal Photoemission (IPE) Measurement

Photon-induced conduction of carriers, or "photo-response", can be employed to measure the offset in the energy band at a heterojunction interface. The current flow in the short-circuited junction is measured as a function of the incident photon energy. The physical process utilized for the measurement is the photoexcitation of carriers over the potential barrier. The minimum energy required for any current flow is obviously the magnitude of the BO, while the photon energy should not exceed the band gap of either semiconductor to stay transparent to the semiconductors. These constraints set the range of photon energies to be used. Since the magnitudes of the induced currents are small, good ohmic contacts to the two semiconductors are needed for internal photoemission, as drawn in the structure fabricated for CBO measurement at a $CdCr_2Se_4$/AlGaAs heterojunction shown in Fig. 5.5.[4] To improve the signal-to-noise ratio, measurements are typically conducted in the ac mode, with the use of chopped light beams and lock-in amplifiers. There is some uncertainty in

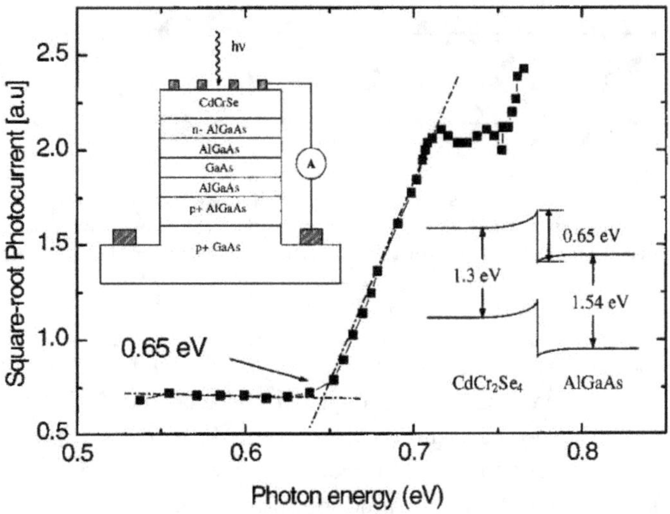

Fig. 5.5. IPE spectrum at room temperature from unbiased heterojunction $CdCr_2Se_4$/AlGaAs. The inset shows the cross-section of the device structure and the band alignment.

Source: Reprinted from Zhao *et al.*[4] with the permission of AIP Publishing.

how the internal photoemission signal should be analyzed to render the height of the rectifying barrier. A square root line shape is traditionally used to determine the threshold energy of internal photoemission when the density of states is high at the Fermi level, similar to the determination of SBH at MS interfaces. For the heterojunctions, the thin two-dimensional electron (or hole) gas at the interface forms discrete energy levels due to quantum confinement, although the square root dependence may still be used.

5.2 Numerical Calculation of VBO

Computer hardware and computational software have advanced to the point where carefully conducted first-principles calculations routinely reproduce experimentally measured properties of metal-semiconductor and heterojunction interfaces.[5] These capabilities make numerical computation not only a trustworthy companion in the interpretation of experimental results but also a valuable tool in exploring novel materials for improved performance or new functionality in devices. To calculate the properties of interfaces, there have been mainly two geometries commonly used: a

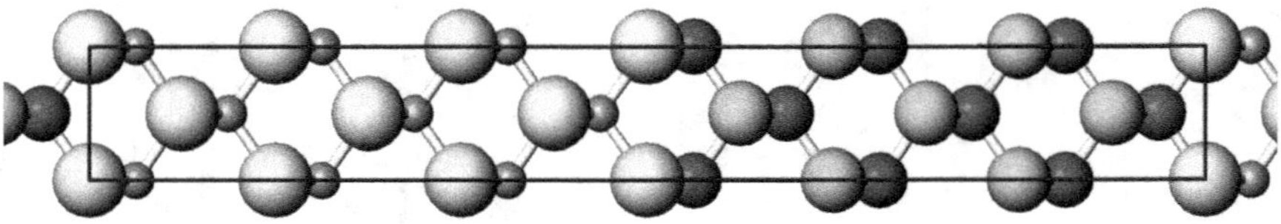

Fig. 5.6. Ball-and-stick models of structures used in the calculation of a (100) supercell between zinc blende semiconductors. Larger and smaller spheres represent anions and cations, respectively. The outlines of the repeating unit used for the calculation, which contains 12 anions and 12 cations, are marked.

supercell that contains back-to-back interfaces and a slab containing a single interface. For the latter, the two exposed surfaces should be terminated properly to be chemically inert. More commonly, interface calculations employ the supercell geometry, an example of which is shown in Fig. 5.6, with the full unit cell of a heterojunction interface marked. The presence of two interfaces in one supercell should be noticed. The size of the supercell has to be large enough for the characteristics of the bulk solids, such as the distribution of charge density, to be recovered in regions away from the interface. There is always the question of how many layers of each material should be included in a supercell to make the calculation meaningful. Obviously, longer supercells are safer as one interface in a supercell would not know the presence of the other, although more resources and time would be required for the calculations. This issue depends on the types of bonds in the two crystals. Between materials of similar crystal structures and chemical bonds, not many layers of atoms are required. As the network of bulk bonds continues across the interface, the formation of an interface is a minor disturbance to the bulk crystal. In contrast, the creation of an interface between materials of dissimilar crystal structures and electronic properties, e.g. between covalent and ionic compounds, considerably disrupts the network of bonds and thus requires longer supercells. The underlying issue is the nearsightedness of the electron density, which may be tested through specially designed calculations, as shown in the following.

The magnitude of the VBO is a direct result of the distribution of all occupied electronic states in the interface region. However, core-level electrons are usually ignored through the use of pseudopotentials without affecting the accuracy of the calculations because they do not contribute to the VBO. DFT has been the preferred method to calculate interfaces with large supercells. With relatively modest demand on computational resources, local and semi-local functionals are routinely used as the workhorse to perform DFT calculations.[6] For the majority of interfaces, these highly efficient calculations adequately produce the charge and potential distributions of the interface. To alleviate possible problems with band gaps, the (final) calculations for band offset determination usually employ functionals with some non-local interactions, which are known to produce band gaps in reasonable agreement with experiment.[7] Although it comes at the price of a much heavier demand on the computational capabilities, the use of these hybrid functionals is not prohibitively difficult for interface calculations of moderate size. It is almost routine nowadays for accurate and reliable calculations involving insulators to be conducted with functionals that include non-local interactions.

With the size of the supercell selected for an interface calculation, the dimensions of its Brillouin zone are also set. As the supercell is elongated in the direction perpendicular to the interface, the Brillouin zone of the supercell is squashed in that direction into the shape of a thin pancake. To sample the entire Brillouin zone, a single sheet of k-points, such as $6 \times 6 \times 1$ or $8 \times 8 \times 1$, usually provides enough Kohn–Sham orbitals to produce the interface properties accurately. As the orbitals are conveniently constructed with plane waves, a cutoff in energy is specified to ignore fine spatial details of little scientific consequence. A calculation invariably works in iteration cycles until a preset self-consistency criterion is reached between the calculated charge density and the density input. A supercell calculation is actually only one of the three calculations that are needed to allow the VBO of a heterojunction interface to be determined, as illustrated by the three separate terms of Eq. 4.1. The potential energy distribution from a supercell calculation provides the difference in the average potential energies, $e\Delta_{\text{ISR}}$, in the bulk regions of the two semiconductors across the interface. The internal chemical potentials, μ_A and μ_B, of the two semiconductors obtained from separate bulk calculations are then added to render the valence band offset of the heterojunction. For accuracy and consistency, all three calculations should employ the same density functionals and have fine enough sampling of the individual Brillouin zones.

We use the plane-averaged "local potential energy", $-eV_{\text{loc}}$, obtained from a GaAs/AlAs(100) heterojunction and plotted as dash-dotted line in Fig. 5.7, to show explicitly how the calculated VBO is arrived at for this interface. A difference of 0.22 eV is found between the "macroscopically averaged"[8] local potential energies for GaAs and AlAs, shown as bold dash lines. The differences between the VBM and $-e\overline{V_{\text{loc}}}$ on either side of the heterojunction, 4.11 eV and 3.43 eV, respectively, obtained from separate bulk calculations of GaAs and AlAs are used to mark the VBM positions in Fig. 5.7 as thin solid lines. Through Eq. 4.1, these values

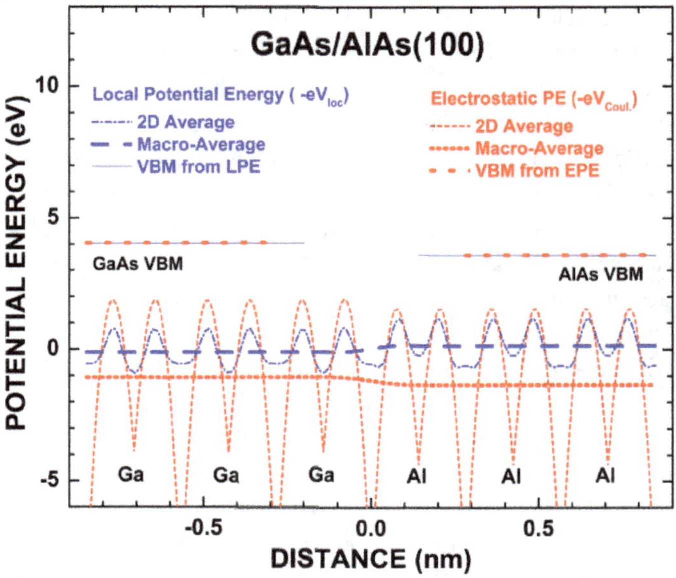

Fig. 5.7. The lateral 2D-average and the macroscopic 3D-average of two types of potential energy are plotted along the length of a 6 × 6 GaAs/AlAs(100) supercell. The local potential energy ($-eV_{\text{loc}}$) contains pseudopotential contributions. EPE ($-eV_{\text{Coul.}}$) is the pure electrostatic (Coulomb) potential energy calculated from the charge density output. In the GaAs region of the supercell (left), the VBM position is computed and shown as thin solid line, using the average $-e\overline{V}_{\text{loc}}$ in this region and the known difference (VBM + $e\overline{V_{\text{loc}}}$) for bulk GaAs. Independently, the VBM in the GaAs region is also determined and shown as coarse dotted line, using the average EPE ($-e\overline{V_{\text{Coul.}}}$) in this region and relative position established for bulk GaAs. VBM is also determined for the AlAs region (right) of the supercell with both methods, using relative positions established for bulk AlAs.

Source: Reprinted with permission from Tung and Kronik.[9] Copyright (2016) by the American Physical Society.

lead to a VBO of 0.49 eV for the heterojunction. While well suited for accurate extraction of the VBO, the three calculations above (a supercell and two bulk calculations) employ pseudopotentials and, therefore, should not be directly compared with potential energies that are based entirely on charge density, such as those for model solids. The ion core contributions missing from the pseudopotential have to be added back to facilitate such a comparison. The calculated charge distribution of the GaAs/AlAs(100) interface is used to generate the electrostatic (Coulomb) potential energy, $-eV_{\text{Coul.}}$, and its average, $-e\overline{V}_{\text{Coul.}}$, also shown in Fig. 5.7. The offset between $-e\overline{V}_{\text{Coul.}}$ and $-e\overline{V}_{\text{loc}}$ can be established from the bulk calculation for each semiconductor, which is then used to map the VBM to the average Coulomb potential energy in Fig. 5.7. There is no difference between the VBMs extracted from $-e\overline{V}_{\text{Coul.}}$ and $-e\overline{V}_{\text{loc}}$ is on either side of the heterojunction, confirming the equivalence of VBO determination from either charge distribution or pseudopotential energy distribution of the interface. With this extra step, results from supercell calculations can be directly compared with band positions based on the charge distributions of model solids.

5.3 Epitaxial Heterojunctions between Covalent Semiconductors

The excellent lattice-matching conditions of many combinations of semiconductors with the zincblende and diamond structure (ZB/D) provide researchers and engineers with a wide selection of epitaxial heterojunction interfaces. Device physicists have been able to use the different band gaps of these semiconductors and the continuously tunable band gaps of their alloys to design band diagrams and then use advanced technologies to fabricate such structures. The plot of the band gaps of some common

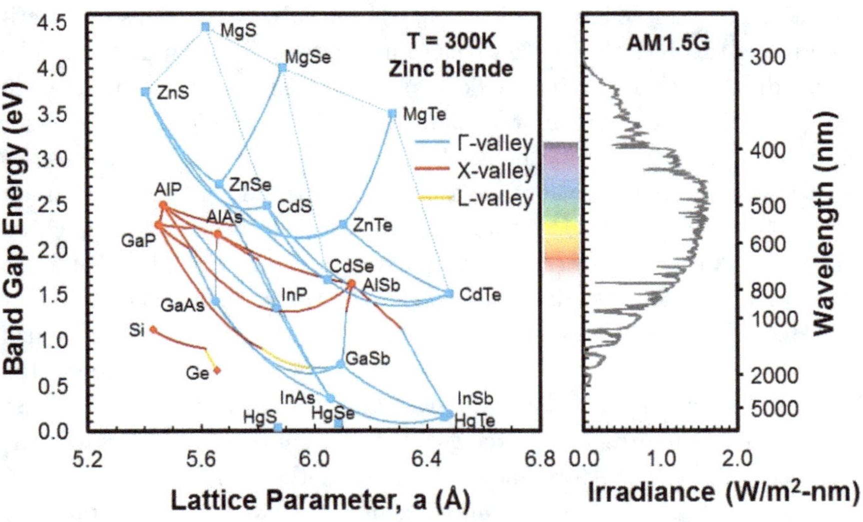

Fig. 5.8. Band gap energies and lattice parameters of semiconductors with the Zincblende and diamond lattice structure. Solid lines connecting semiconductors indicate alloys between the semiconductors. Downloaded from the University of California Davis' website, with permission from Prof. J. Woodall.

zincblende semiconductors against the lattice parameters in Fig. 5.8 can serve as a blueprint for such "band-gap engineering" of semiconductor heterostructures for device applications. Between lattice-matched ZB/I semiconductors, uniform epitaxial heterojunction interfaces without extended defects may be routinely fabricated. Besides applications in advanced devices, these interfaces form well-controlled systems for heterojunction band offset to be investigated experimentally and theoretically. Lattice-matched semiconductors may be of the same valence, e.g. GaAs and AlAs are both III-V semiconductors, in which case an iso-valent heterojunction may be formed between them. Or they can come from different groups, e.g. ZnSe and GaAs, and thus form hetero-valent interfaces. Heterojunctions of either kind can be fabricated with high structural quality. However, the electronic nature of the interface and the factors that determine the band offset are very different for iso-valent and hetero-valent heterojunctions. Therefore, these two types of interfaces will be discussed separately.

5.3.1 *Iso-valent Zincblende Heterojunctions*

(a) Band Offsets of Lattice-Matched Heterojunctions

Quite a few iso-valent interfaces between zincblende semiconductors with excellent lattice matching conditions have been subject to extensive experimental and theoretical investigations to understand their band offsets. The tetrahedral bonds in a zincblende crystal are polar and are directed in the {111} directions. Among the common crystallographic orientations for the interface, the (110) is a non-polar orientation because equal densities of cations and anions are contained in each plane. Along two other major directions, (001) and (111), atomic planes alternate between cationic and anionic, making these orientations "polar" for the zincblende crystal. At each polar direction, two different atomic structures are possible for planar interfaces. For example, at a (100) interface between two lattice-matched semiconductors, InAs and GaSb, the two crystals could be connected through As-Ga bonds or In-Sb bonds. One can convert between these two atomic structures by replacing a layer of cations (or anions) at the interface with cations (or anions) from the other semiconductor. Such a replacement does not significantly change the overall electrostatics of the interface because iso-valent cations (or anions) have the same formal charges. Similarity in electrostatics is the main reason that the BO for iso-valent heterojunctions has been found not to depend significantly on the orientation and the atomic structure of the interface in both experimental

Table 5.1. Valence band offset at iso-valent heterojunctions of zincblende semiconductors.[9] A positive value indicates that the VBM of semiconductor #1 is positioned higher than that of semiconductor #2. Results from DFT calculations and experimental measurements are listed. The VBOs predicted by modeling with bulk charge distribution, which are the differences in NPT bulk references for the two semiconductors, are also listed.

Semicond #1	Semicond #2	Calc. VBO (eV)	Expt. VBO (eV)	NPT (eV)
GaP	AlP	0.51	0.55	0.44
GaAs	AlAs	0.49	0.55	0.40
BeTe	ZnSe	0.86	0.90	0.62
InGaAs	InP	0.42	0.37	0.25
InGaAs	InAlAs	0.20	0.20	0.20
InP	InAlAs	0.27	0.23	−0.05
AlSb	InAs	0.19	0.09	0.19
GaSb	AlSb	0.42	0.40	0.38
GaSb	InAs	0.53	0.53	0.56
ZnTe	CdSe	0.68	0.64	0.56
HgTe	CdTe	0.44	0.36	0.26

and theoretical investigations. Experimentally measured VBOs for iso-valent interfaces are listed in Table 5.1 and can be seen to be well reproduced by theoretical calculations. This agreement illustrates the reliability and accuracy of modern computational tools.

(b) Nearsightedness of Electron Density

For experimentally measured VBOs between zincblende semiconductors, e.g. Table 5.1, the question naturally arises as to whether these can be understood through charge distribution modeling. Our ability to accurately model the charge distribution of the interface depends on the nearsightedness of the electron density.[10] With the detailed charge densities revealed by calculations, how nearsighted the electron density really is at these iso-valent interfaces can be examined through side-by-side comparisons. The top panel of Fig. 5.9(a) shows the calculated electron density of a (100)-oriented heterojunction interface between GaAs and AlAs, which is among the simplest interfaces one can find.[9] The zero on the horizontal axis marks the location of the arsenic plane common to both semiconductors. The atomic structure on one side of this plane is identical to the bulk GaAs, while it has the AlAs structure on the other side. An obvious thing to try is to place the bulk densities of the GaAs and the AlAs on the opposite

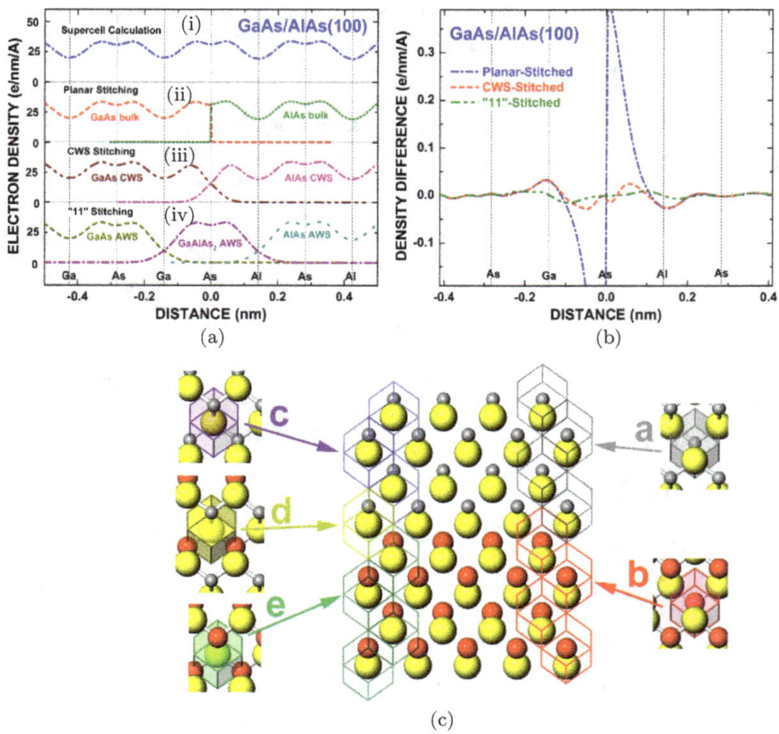

Fig. 5.9. (a) Electron density, per nm and per area of the 2D unit cell ($0.16\,\text{nm}^2$), of a GaAs/AlAs(100) supercell. (i) obtained by ab initio calculation, (ii) from stitching bulk GaAs and bulk AlAs densities at the interface As plane, (iii) from stitching bulk GaAs and AlAs densities terminated at CWS boundaries, (iv) from stitching the bulk GaAs AWS density, the bulk AlAs AWS density, and the density of an AWS cell from a "1 1" $GaAlAs_2$(100) cell (see text for details). (b) Difference between the equilibrium electron density calculated for a GaAs/AlAs(100) supercell and that constructed for the heterojunction by the various stitching methods. (c) Schematic illustration of two stitching methods for the simulation of charge density at GaAs/AlAs(100) heterojunction. The atomic arrangement of the heterojunction is drawn in the center diagram, with As shown as large spheres, Al as small spheres, and Ga as mid-sized spheres. In the CWS-stitching method shown on the right, the volume of the heterojunction is divided into CWS's and the charge density is stacked CWS-by-CWS with charge densities calculated for "a" bulk AlAs and "b" bulk GaAs. In the "1 1"-stitching method shown on the left, the volume of the heterojunction is divided into AWSs and the charge density is stacked AWS-by-AWS with charge densities calculated for "c" bulk AlAs and "e" bulk GaAs, except for sites of the interface As common to both semiconductors. On these interface sites, AWSs are filled with charge density calculated for "d" $AlGaAs_2$ oligo-cell.

Source: Reprinted with permission from Tung and Kronik.[9] Copyright (2016) by the American Physical Society.

sides of this planar boundary, which corresponds to employing the plane terminated model solids discussed in Chap. 2. Unfortunately, such a "planar stitched" charge distribution deviates from that calculated for the supercell, as shown in Fig. 5.9(b). Upon closer examination, significant discrepancies are confined only to the region between the closest Ga and Al planes, which happens to have a local atomic structure different from that of either the bulk AlAs or GaAs. This shows that the local electron density for these semiconductors is nearsighted to structural changes beyond a bond length away.

The planar stitching strategy just described is chemically unwise because the charge distributions associated with the individual bonds at the interface are not preserved as a unit. From the perspective of nearsightedness, it seems that agreement with the actual charge distribution at ZB/I interfaces should improve if the integrity of the individual bond is protected. From Chapter 2, the charge distribution associated with each bond is essentially confined to the quadrant with the cation and the anion of the bond at the apices. A bond-conserving method to create a trial charge distribution is to insert a quadrant of GaAs bulk density between every Ga and As nearest neighbors of Fig. 5.9, and bulk AlAs density in every Al-As quadrant. This is equivalent to placing a GaAs model solid assembled with CWS cells next to an AlAs model solid similarly assembled. Shown in Fig. 5.9(c) is a schematic of such a stitching strategy based on CWS model solids, leading to much improved agreement with the actual density. The electron distribution within each quadrant of a semiconductor is thus largely unaffected by the structure of its neighboring quadrants, at least in this overall iso-valent chemical environment. This nearsightedness of the electron distribution to the ions of its own quadrant is explicitly demonstrated in Fig. 5.10(b), which compares, as a function of the distance from the cation to the anion of a quadrant, the electron densities of quadrants in different chemical environments. For example, the integrated electron density in a Ga-As quadrant is little changed from its bulk distribution when this quadrant is placed at any of the GaAs/AlAs(100), (110), and (111) heterojunctions. Similarly, Al-As quadrants also remain essentially "bulk-like" when placed at these iso-valent heterojunction interfaces. The good agreement of CWS-stitched density in Fig. 5.9 may be further improved by using quadrants with exactly the same chemical environment as that of the interface. In a ternary $GaAlAs_2$ compound, marked "d" in Fig. 5.9(c), with alternating Ga and Al cation planes, each As anion is in the same chemical environment as the As at the

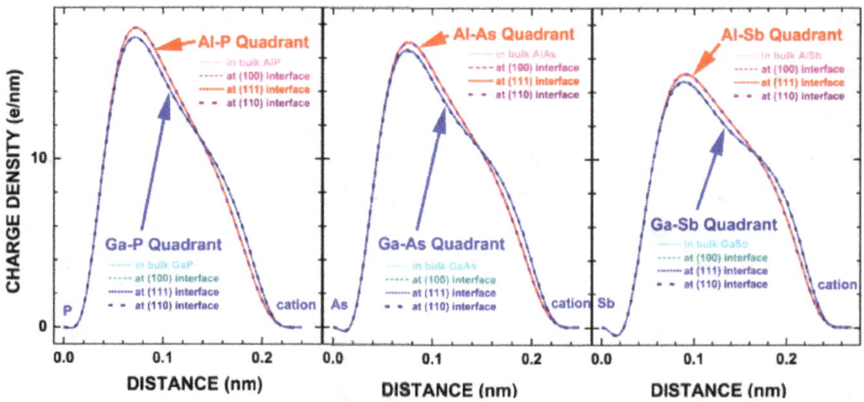

(a) (b) (c)

Fig. 5.10. Superimposed plots of charge density (per nm) of quadrants, integrated in the direction perpendicular to the center axis, {111}, of Ga-anion and Al-anion quadrants, as extracted from the (100), (110), and (111) common anion III-V heterojunction interfaces, as well as from the bulk semiconductor. At the interfaces, the cation in the quadrant is bonded with both Ga and Al. (a) Ga-P and Al-P quadrants, (b) Ga-As and Al-As quadrants, (c) Ga-Sb and Al-Sb quadrants.

Source: Reprinted with permission from Tung and Kronik.[9] Copyright (2016) by the American Physical Society.

interface. An AWS cell of such a "one-one" ("1 1") oligo-cell provides charge density matched to the interface chemical environment, leading to nearly perfect agreement with the supercell equilibrium density, as Fig. 5.9(b) reveals. The charge distributions of oligo-cells used to stack the interface charge distribution are dubbed "local electron densities generated in oligo-cells" (LEGO).

In the example of charge distribution at the InAs/GaSb(100) interface with As-Ga interface bonds of Fig. 5.11(a), there is a sliver of material at the interface with a local structure (As-Ga) not found in either InAs or GaSb. A reasonable charge distribution of this interface may be produced by using an InAs model solid constructed with CWS cells, a GaSb model solid from AWS cells, and a single layer of quadrants from GaAs calculations. The density stitched together in this fashion, shown as single-quadrant layer ("1Q") stitching, approximately reproduces the equilibrium density obtained in supercell calculation, as shown in Fig. 5.11(b), but the fit is mediocre. Using the same argument just made for the GaAs/AlAs interface, we seek the charge density of the As atom from an $InGaAs_2$ "11" cell and that of the Ga atom from a Ga_2AsSb "11" LEGO to adjust the chemical

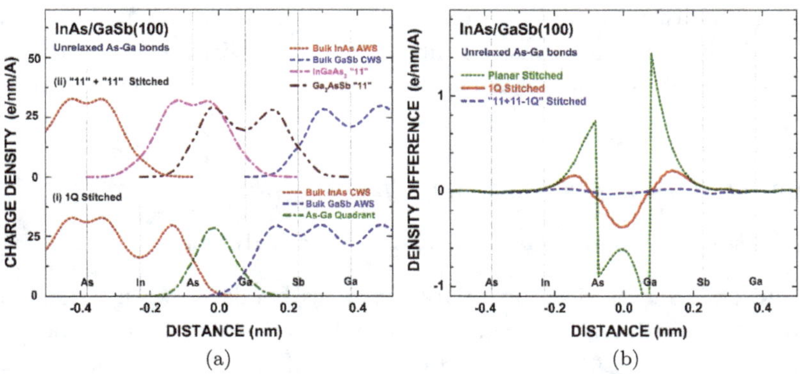

Fig. 5.11. (a) Electron densities, per nm and per area of the 2D unit cell ($0.185\,\text{nm}^2$), of bulk semiconductors and oligo-cells that are stitched together to simulate the charge density at the InAs/GaSb(100) heterojunction with As-Ga interface bonds. (b) The difference between the equilibrium charge density calculated for the unrelaxed InAs/GaSb(100) supercell with As-Ga interface bonds and the simulated LEGO-stitched charge density for this heterojunction, based on two stitching methods.

Source: Reprinted with permission from Tung and Kronik.[9]

environment. Taking both corrections is equivalent to stitching the densities of InAs(AWS), "11" InAs-GaAs, "1 1" GaAs-GaSb, and GaSb(CWS) together, as illustrated in Fig. 5.11. The charge density of a layer of As-Ga quadrants, which is double-counted in the process, is then subtracted to complete the "11 + 11-1Q"-stitching. As shown in Fig. 5.11, such a LEGO-stitching strategy puts together a density in excellent agreement with the equilibrium charge density of the supercell, demonstrating a near-perfect replaceability of electron density at semiconductor interfaces when the chemical environment is matched.

(c) Iso-valent VBOs Modeled with Bulk Densities

With bonds at the interface similar to those found in bulk crystals, the VBOs of iso-valent zincblende heterojunctions can be modeled without considering interface bonds. Earlier analysis also showed that model solids defined by NPT were well suited for these interfaces. The difference between the VBM positions of the two NPT bulk references is the predicted VBO between the two semiconductors. For example, the difference between $-3.45\,\text{eV}$ and $-3.89\,\text{eV}$, respectively, for GaP and AlP from Table 2.7, is the NPT prediction of $0.44\,\text{eV}$ for the VBO of GaP/AlP heterojunction. This and other predictions from NPT bulk references agree well with VBOs experimentally measured or theoretically calculated. This success is the result of the excellent transferability of charge densities between the bulk and the interface, a feat that can be attributed to the nearsightedness of electron density.

5.3.2 *Heterovalent Zincblende Heterojunctions*

(a) Hetero-valent bonds and interface metallicity

A hetero-valent interface necessarily involves bonds between cations and anions that ordinarily do not form bulk binary semiconductors. For example, at the interface between GaAs and ZnSe, either Zn-As or Ga-Se bonds must be present. In bulk III-V semiconductors, isolated group II and VI atoms on substitutional III and V lattice sites form shallow acceptor and donor defects, respectively. Conversely, group III and V atoms are donors and acceptors, respectively, for II–VI semiconductors. Therefore, Ga-Se and Zn-As bonds are donor-like and acceptor-like, respectively, at a hetero-valent GaAs/ZnSe interface. By the same token, Ga-Ge and As-Ge bonds behave as acceptors and donors, respectively, at III–V/IV interfaces. If a hetero-valent interface could be fabricated in a polar orientation to be planar and sharp, it would have a sheet of either donor or acceptor hetero-valent bonds at the interface. Such an interface would be electrically charged and locally metallic. Metallic interfaces have little use for practical devices, and the long-range electric fields make the band offset at such interfaces not meaningful scientifically. Fortunately, such charged interfaces are uncommon in nature. An analysis of the formation energies shows that these charged interfaces are unstable against decomposition into patches that are individually neutral, such as the neutral interfaces with mixed-cations or mixed-anions to be discussed in the following. Due to clear thermodynamic disadvantages, charged planar interfaces are not expected to be important under equilibrium conditions and they should not contribute much to the VBOs measured at hetero-valent heterojunctions. Therefore, our discussion of the hetero-valent interfaces will be limited to those that are electrically neutral.

(b) Atomic and Electronic Structures of "Neutral" Hetero-valent Interfaces

What makes a hetero-valent interface structure electrically neutral? The electrical activities of hetero-valent bonds suggest a simple rule for stable and neutral hetero-valent interfaces is to have equal densities of donor and

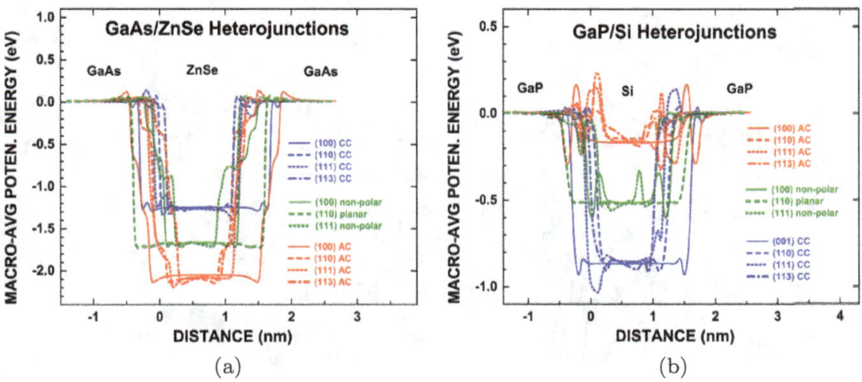

Fig. 5.12. Macroscopically-averaged potential energy profile across an entire supercell of neutral III-V/II-V and III-V/IV heterojunctions with different interface structures. (a) Unrelaxed GaAs/ZnSe, (b) unrelaxed GaP/Si. Potential energy profiles for (100), (110), (111), and (113) interface orientations are shown as solid, dashed, dotted, and dash-dotted lines, respectively. Blue, red, and green curves are used to designate interface structures that are cation-centered (CC), anion-centered (AC), or non-polar (see text for a detailed definition of these structures).

Source: Reprinted with permission from Tung and Kronik.[11] Copyright by John Wiley & Sons, Inc.

acceptor bonds. We begin by examining planar heterojunctions with the nonpolar (110) orientation. With the densities of donor and acceptor hetero-valent bonds being equal, a planar (110) hetero-valent interface is expected to be electrically neutral. This is indeed borne out from the calculated GaAs/ZnSe(110) and GaP/Si(110) heterojunctions of Fig. 5.12. The wide and flat sections of averaged potential energy profiles between interfaces are evidence for neutrality of perfectly compensated interfaces.

How can neutral structures be constructed for hetero-valent interfaces with the polar (100) and (111) orientations?[11] This can be illustrated through attempts to assemble these interfaces with WS cells, analogous to Fig. 5.9. Two highly symmetric WS cells for the ZB crystal are the CWS and the AWS. When a GaAs/ZnSe(100) interface is assembled with CWS cells of GaAs and ZnSe, each atomic site at the plane of this interface contains ½ As and ½ Se. This translates to a cation-centered (CC) GaAs/ZnSe(100) heterojunction with mixed-anions, namely, a 50-50 (1:1) mixture of As and Se atoms, on the interface plane, forming a stacking sequence of – As – Ga –$As_{0.5}Se_{0.5}$ –Zn – Se – in atomic planes. Similarly, an anion-centered (AC) interface with a stacking sequence of – Ga – As –

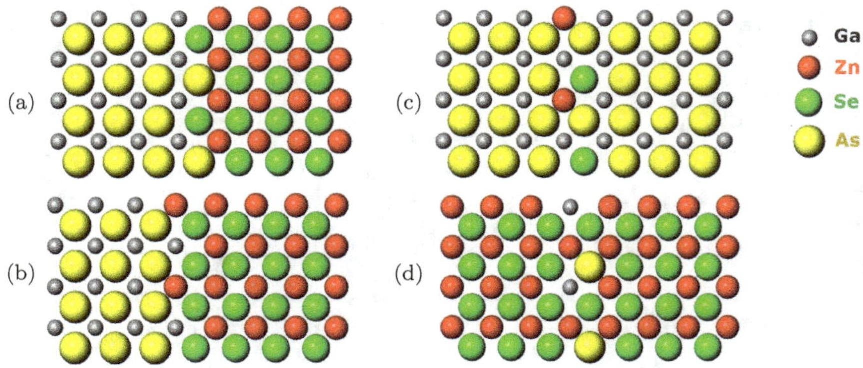

Fig. 5.13. Structural models related to the neutral GaAs/ZnSe(100) interfaces. The spheres represent, in ascending order of the size of the sphere, Ga, Zn, Se, and As. (a) Cation-centered interface with a single 1:1 mixed-anion plane at the interface, (b) anion-centered interface with a single 1:1 mixed-cation plane at the interface. If the pattern of (a) is symbolically subtracted from (b), the net effect is that at the interface one-half monolayer of As needs to be replaced by Se and one-half monolayer of Ga needs to be replaced by Zn. When these replacements are applied to two adjacent (110) atomic layers of bulk GaAs, the structural diagram drawn in (c) is obtained. Alternatively, one can subtract (b) from (a) to perform the opposite atomic replacements. When those replacements are applied to two adjacent (110) atomic layers of bulk ZnSe, the structural diagram drawn in (d) is obtained. Diagrammatically speaking, (c) = (b) − (a) + "bulk GaAs" and (d) = (a) − (b) + "bulk ZnSe".

$Ga_{0.5}Zn_{0.5}$ – Se – Zn – in atomic planes would be assembled with AWS cells. As can be checked from the drawings in Fig. 5.13, the densities of donor and acceptor bonds are equal for either CC or AC interfaces. Both structures are indeed electrically neutral, as evidenced by the wide, flat sections seen in the averaged potential energy profiles of Fig. 5.12(a). However, the VBOs of these two (100) interfaces are significantly different. With respect to the VBM of GaAs, the VBM of ZnSe is higher by ~0.9 eV for the CC interface than that for the AC interface. The VBO for the planar (110) interface is approximately midway between those two interfaces. For all hetero-valent heterojunctions studied, the neutral AC and the CC interfaces have nearly identical formation energies and are favored over interfaces that are charged. Therefore, even without deliberate effort to control their structures, hetero-valent interfaces are likely neutral in equilibrium, but with laterally inhomogeneous VBO.

The reason that the VBOs are so different for the three neutral interfaces between the same two materials, GaAs and ZnSe, can be traced to the relative position of the positively charged donor bonds with respect to the negatively charged acceptor bonds. For the AC GaAs/ZnSe(100) interface, the donor (As-Zn) bonds are on the GaAs side of the interface, while the acceptor (Ga-Se) bonds are on the ZnSe side. For the CC structured interface, these bond positions are reversed. For the planar (110) interface, both types of bonds reside at the plane of the metallurgical interface. The average potential energy plots shown in Fig. 5.12(a) agree with the signs of the dipoles due to exposed dopants. Regardless of the interface orientation, moving the positive donor bonds toward the ZnSe (or bulk GaAs) tends to lower the relative VBM position of that semiconductor. Following these clues, it becomes obvious how a "non-polar" neutral structure may be created at a normally polar GaAs/ZnSe(100) interface. The distinguishing character of a non-polar interface is "a symmetry in the distribution of the donor bonds and the acceptor bonds about a common location". This can be accomplished by having a layer of donor bonds sandwiched between two half-layers of acceptor bonds, or vice versa. With a stacking sequence of – Ga –$As_{0.5}Se_{0.5}$ –Zn – Se –, the CC (100) structure has $1/2$ monolayer of donor bonds on the left and $1/2$ monolayer of acceptor bonds on the right. A monolayer of bonds here is defined as two bonds per atom on a (100) plane. One now moves $1/4$ monolayer of the original donor bonds to the right of the acceptor bonds by using the sequence – Ga –$As_{0.75}Se_{0.25}$ – $Ga_{0.25}Zn_{0.75}$ – Se –. About the midway point between the mixed-atoms planes, the distribution of either donor bonds or acceptor bonds is symmetric, while their densities remain equal. Such a neutral interface is verified by Fig. 5.12 to be non-polar, and electrically equivalent to the planar (110). While the actual fabrication of structures requiring such levels of atomic control challenges even today's advanced growth capabilities, this idea of "non-polarity" is nevertheless proven valid through computations. It is possible to create different structures for the non-polar GaAs/ZnSe(100) interface, not shown redundantly in Fig. 5.12, that have nearly identical VBOs. The creation of the CC, AC, and non-polar structures is not limited to the (100) orientation; CWS cells and AWS cells may be used to produce CC and AC interfaces with any orientation. All hetero-valent interfaces with the same type of structure have nearly identical VBOs, independent of the interface orientation.

(c) Lattice Relaxation and Nearsightedness Principle

The VBO of a hetero-valent interface is largely decided by the relative placement of the donor and the acceptor bonds. As the interface structure relaxes to minimize energy, the movement of the atomic positions should shift the locations of these bonds and change the VBO. However, the VBO for a relaxed interface has been found by calculation to be only slightly modified ($\leq 0.2\,\text{eV}$) from that for the unrelaxed interface, for the vast majority of hetero-valent interfaces theoretically studied. The systematic trends discussed above for the VBOs of hetero-valent interfaces thus remain essentially intact. Directly impacting the VBO, the movement of atoms at a hetero-valent interface during lattice relaxation should be understood. To determine how the entire charge distribution of an interface is affected by the relaxation of its atomic structure, theoretical calculations can be used to analyze the contributions from specific shifts in the atomic positions. Investigations have been conducted on the detailed relaxation process of several hetero-valent interfaces, and the most important finding has been a lack of "collaborative" effort on the charge distribution from the relaxation of multiple atomic positions.[11] To illustrate this interesting aspect, the relaxation of the GaAs/Ge(110) interface is shown through its charge distribution in steps. First, properties of this interface with the optimized, slightly expanded, interface distance but without relaxation of any of the atoms at the interface planes are calculated. With respect to the density of this unrelaxed structure, the density change resulting from the movement of each different atomic type (e.g. Ga) to its final relaxed position is separately shown in Fig. 5.14 (e.g. marked as "Only Ga"). Each such density change can be seen to be local, being essentially confined to the width of the WS cell about the shifted atom. Furthermore, the change in electron density associated with the movement of each atom is found to be independent of the positions of other atoms. This is shown in the lower part of Fig. 5.14, where the entire charge difference between the fully relaxed structure and the expanded yet unrelaxed structure is shown. There is little difference between the density change induced by the simultaneous movement of four different atoms to the sum of the four difference curves associated with the independent individual atomic moves, shown in the same part of the figure. One can conclude from Fig. 5.14, which is representative of results obtained at several hetero-valent interfaces, first of all that the dependence of the electron density on atomic structure is extremely localized. Secondly, the density change associated with an atomic movement is uncorrelated with that due to the movement of any other atom. Both of these characteristics are expected from, and serve as vivid demonstrations of, the nearsightedness principle of electron density.

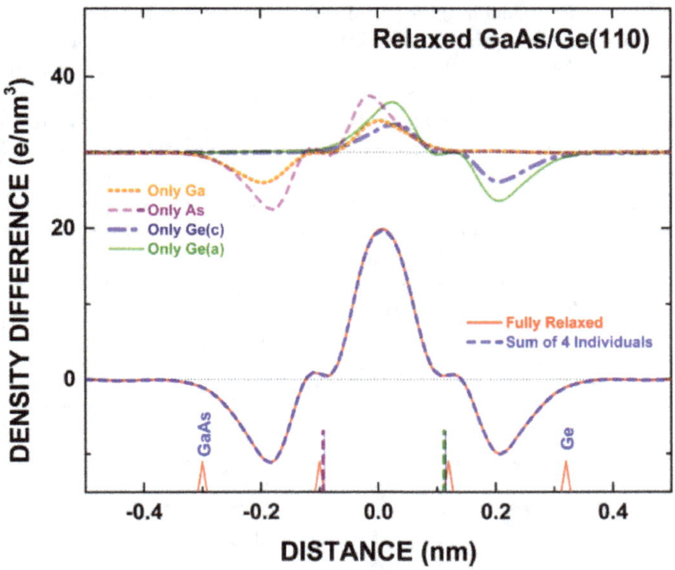

Fig. 5.14. Charge density difference associated with movements of atoms, collectively or individually, for the partially and fully relaxed GaAs/Ge(110) interfaces. The baseline density is that of the expanded (or contracted) yet unrelaxed interface, for which the positions of atomic planes are marked by chevrons at the bottom. Upon full relaxation, the charge density differs from the original unrelaxed density by the line marked "Fully Relaxed". The line marked "Only X" is the difference between the baseline density and that calculated for an interface where only one type of atom, "X", is placed at its final relaxed position, with the rest remaining at the unrelaxed positions. "X" here stands for Ga, As, Ge(c), and Ge(a), where (c) and (a) denote a cationic and anionic site, respectively. Positions of relaxed planes are marked by vertical lines (thin solid lines for cations and dashed lines for anions) at the bottom. "Sum of 4 Individuals" denotes the numerical sum of the four separate lines involving single-atom movement. For clarity, different curves for individual movement are vertically shifted.

Source: Reprinted with permission from Tung and Kronik.[11] Copyright by John Wiley & Sons, Inc.

(d) Modeling the Hetero-valent VBOs with Bulk Densities and Screened Dopant Dipoles

Calculated VBOs of several lattice-matched hetero-valent systems with hybrid functionals are listed in Table 5.2 in different columns for three interface structures: the non-polar planar (110), the AC (100), and the CC (100). For each heterojunction system, there is a significant difference

Table 5.2. VBOs calculated for lattice-matched, fully relaxed, hetero-valent, and heterojunctions with three different interface structures: the non-polar planar (110), the AC (100) with mixed-cation interface layer, and the CC (100) with mixed-anion interface layer. A positive VBO value indicates that the VBM of semiconductor #1 is higher compared to that of semiconductor #2 at that interface. Hybrid HSE functionals are used in these calculations. Available experimental VBOs are also shown. VBOs modeled with (1) charge densities of bulk semiconductors through the NPT approach and, when necessary, (2) screened dipoles due to hetero-valent bonds at the interface for these interface structures are listed as "NPT" for comparison.

Semicond. #1	Semicond. #2	(110) VBO (eV)		AC (100) VBO (eV)		CC (100) VBO (eV)		Expt. VBO (eV)
		Calc.	NPT	Calc.	NPT	Calc.	NPT	
Ge	AlAs	1.12	1.08	1.40	1.39	0.80	0.78	0.95
Ge	GaAs	0.56	0.68	0.80	0.96	0.28	0.40	0.52
AlAs	ZnSe	0.61	0.69	1.01	1.11	0.22	0.28	0.40
GaAs	ZnSe	0.91	1.10	1.27	1.47	0.57	0.73	0.98
Ge	ZnSe	1.48	1.78	2.10	2.41	0.87	1.14	1.40
Si	AlP	1.06	0.92	1.40	1.31	0.68	0.54	1.35
Si	GaP	0.42	0.49	0.58	0.85	−0.13	0.12	0.80
AlP	ZnS	0.77	0.88	1.25	1.32	0.26	0.43	
GaP	ZnS	1.05	1.31	1.50	1.73	0.58	0.90	0.83
Si	ZnS	1.67	1.80	2.39	2.60	0.86	1.00	

between the VBOs for the AC and the CC interfaces. This difference, which may potentially be limited by the band gaps of the semiconductors, can be as much as 1.5 eV for heterojunctions between group IV and II–VI semiconductors, and is ∼0.5–0.9 eV for heterojunctions between III–V and II–VI or group IV and III–V semiconductors. As discussed, these significant differences in the VBO are correlated with the relative positions of the donor and acceptor bonds for different interface structures. Ultimately, each VBO is a direct result of, and therefore should be understood from, the charge distribution of that interface. For hetero-valent interfaces, the interface bonds play a dominant role and cannot be neglected. Fortunately, the electrostatic effects of these elements, which are routinely employed as shallow dopants in semiconductors, are well described by semi-classical theories, with which the dipolar effect of hetero-valent bonds at the interface can be modeled.

To consider how the VBO may be modeled with bulk densities, one begins with the simplest among hetero-valent interfaces, the non-polar planar (110). At this interface, hetero-valent donor and acceptor bonds are located at the same atomic plane. Even though their magnitudes may be significant, lateral transfers of charge do not modify the VBO appreciably because the net dipole moment vanishes. Corrections for interface bonds are unnecessary and the VBOs for these non-polar hetero-valent interfaces can be modeled the same way as for iso-valent interfaces. As before, an NPT predicted VBO for the (110) interface in Table 5.2 is simply the difference between the NPT bulk references. The good agreement with calculated VBOs confirms the negligible effect of interface bonds for non-polar interfaces.

For the polar AC or CC interfaces, interface bonds cannot be ignored. From the perspective of the effective charge for an isolated shallow dopant in a semiconductor, each donor and acceptor contribute a point charge of $+e/\varepsilon_S$ and $-e/\varepsilon_S$, respectively, with ε_S being the dielectric constant. This well-known result provides the basis to model the charge distribution at hetero-valent interfaces.[8] However, there's ambiguity in the assignment of charged dopants at an interface due to the presence of two semiconductors. At a hetero-valent interface, neither the donor nor the acceptor layer is entirely buried in one semiconductor. Such a dilemma may be resolved by tracking the changes in atomic structure from one type of interface to the other.[11] We again use the GaAs/ZnSe(100) interface as the example. Starting from a CC (100) interface with a stacking sequence of – As – Ga – $As_{0.5}Se_{0.5}$ – Zn – diagrammed in Fig. 5.13(a), one notices that "replacing $1/2$ ML of As with $1/2$ ML of Se" in the first interface plane and "replacing $1/2$ ML of Ga with $1/2$ ML of Zn" in the second interface plane lead to a sequence of – As – $Ga_{0.5}Zn_{0.5}$ – Se – Zn –, which is precisely the AC (100) structure of Fig. 5.13(b). These two replacements may be enforced in any compound with Ga and As, e.g. a bulk GaAs crystal. The result is a plane with $1/4$ ML of substitutional Se adjacent to a plane containing $1/4$ ML of substitutional Zn in GaAs, shown in Fig. 5.13(c). Through similar analysis, the non-polar (110) is found to be halfway between the AC and CC structures, i.e. with 1/8 ML of substitutional replacements. Alternatively, the AC to CC conversion may be implemented by embedding $1/2$ ML GaAs in bulk ZnSe, as shown in Fig. 5.13(d). Calculated potential energy curves of GaAs and ZnSe embedded with these dipole layers are shown in Fig. 5.15. The shift in average potential energy between the AC and the CC structures is essentially independent of the interface orientation, in agreement with the result of Fig. 5.12. These shifts in potential energy calculated from embedded dopants are in good agreement with those estimated from the dielectric constant of the semiconductor. The classical idea of dielectric

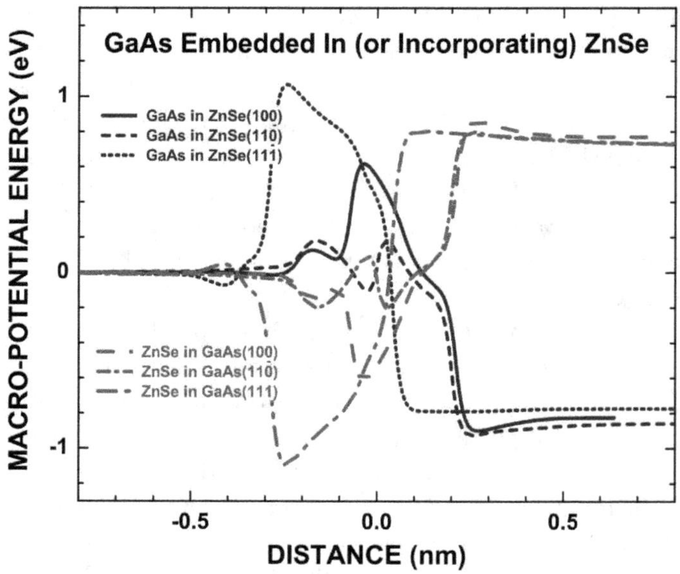

Fig. 5.15. Macroscopically averaged potential energy of GaAs crystal embedded with a fraction of a monolayer of ZnSe (red) and ZnSe crystal embedded with GaAs (blue), calculated by DFT with relaxed atomic positions. For the (100) and (110) orientations, the density of embedded pairs is 1/2 monolayer. For the equivalent supercell with (111) orientation, 3/4 monolayer of embedded pairs is used within a bi-layer while 1/4 monolayer of embedded pairs is used across bi-layers. The embedded anions and cations are slightly negative and positive, respectively, on the horizontal axis.

Source: Reprinted with permission from Tung and Kronik.[11] Copyright by John Wiley & Sons, Inc.

screening thus appears valid for interfaces and can be used to model the VBO dependence on interface structure.

Since the atomic structure of a non-polar interface is midway between the AC and the CC structures, its VBO also lies midway between the AC and the CC interfaces, as already noted from Fig. 5.12. With this knowledge, the VBOs of the CC, the AC, and the non-polar interfaces can all be modeled with the NPT model solids of the semiconductors. The difference between the VBOs of the non-polar interface and the AC (or CC) interface corresponds to the potential energy shift across a layer of dopant dipoles (e.g. 1/4 ML in (100) direction) embedded in the semiconductor. Dielectric screening can be used to estimate these potential energy shifts, with the dielectric constant assumed to be the average of the two semiconductors.

For heterojunctions between II-VI and group IV semiconductors, a hetero-valent atom on a lattice site is treated as a depleted double-donor or double-acceptor, allowing VBOs for all three interfaces to be modeled. Adding the corrections for interface bonds estimated from dielectric screening to the NPT predicted VBOs for the non-polar (110) interfaces, VBOs for the AC (100) and the CC (100) interfaces are modeled as the "NPT" values shown in Table 5.2. A very good agreement is seen between VBOs calculated quantum mechanically and those modeled by NPT for all the hetero-valent interfaces investigated. With the help of simple electrostatics, the entire dependence of the VBO on the interface structure at hetero-valent heterojunctions can be accounted for by the modeled charge distribution of the interface.

5.4 Epitaxial Heterojunctions between Perovskite Oxides

Oxides are materials that are abundant in nature, can occupy numerous crystal structures and compositions, and have been used in various capacities for applications. Therefore, the band alignment at interfaces between oxides and those between oxides and other materials is of great technological importance and scientific interest. The vast majority of oxides for practical applications are either polycrystalline or amorphous. How the band alignment at polycrystalline interfaces may still be handled through charge density is the topic of the following section. Here, a small number of oxide interfaces that could be formed epitaxially are examined and how the VBOs of these interfaces could be modeled through bulk densities is discussed. The cubic perovskite structure is a common structure for ionic compounds, and a large family of oxides are found in the cubic or distorted perovskite structures, many with similar lattice constants. In principle, single-crystal interfaces between selected perovskite oxides that are closely matched in their lattice parameters may be fabricated in major interface orientations. The magnitude of the VBO of a perovskite interface may seem amenable to the same kind of analysis employed to handle zincblende VBOs successfully. However, ionic screening makes these ionic interfaces fundamentally different from the covalent interfaces just discussed.

5.4.1 *Iso-valent and Hetero-valent Epitaxial Interfaces*

A large metal cation (A) occupies the corner site and a transition metal cation (B) occupies the body-center site of the cubic unit cell of a perovskite oxide, ABO_3, with oxygen atoms at all the face-center positions. Some properties of the insulating perovskite oxides can be found in the upper half of Table 2.8. Due to close lattice matching conditions, a few oxides that ordinarily occupy the orthorhombic crystal structure are also studied in the cubic structure to broaden the scope of the investigation. Perovskite oxides included in Table 2.8 belong to three groups: "II-IV-Ox", "I-V-Ox", and "III-III-Ox", based on the formal charge states of the cation occupying the A and the B sites. Lattice matching conditions are found for both iso-valent perovskite heterojunctions, e.g. between two II-IV-Ox oxides, and hetero-valent heterojunctions, e.g. between one II-IV-Ox oxide and an oxide from either the I-V-Ox or the III-III-Ox group. Due to the cubic structure, matched lattice parameters allow two perovskite oxides to form epitaxial interfaces in all major orientations, (100), (110), and (111).

At a planar (110) interface, atomic planes alternate between layers of oxygen and layers that contain equal densities of both cations of the same perovskite compound. This stacking sequence represents a non-polar arrangement of the atoms. Therefore, planar (110) interfaces are always neutral even between hetero-valent perovskite semiconductors. With polar (100) and (111) orientations, there are two possible atomic arrangements for planar interfaces. For example, between $CaHfO_3$ and $SrSnO_3$, there could be an interface with corner cations (CC) of the first oxide, Ca, and body-centered (BC) metal atoms of the second oxide, Sn, on adjacent planes, along with stoichiometric oxygen atoms. Such an interface is distinct from one with BC Hf atoms and CC Sr atoms occupying adjacent atomic planes. These two types of interfaces are referred to below as CC1-BC2 and BC1-CC2 interfaces, respectively. Between iso-valent oxides, interfaces are neutral in all orientations. However, planar hetero-valent perovskite oxide interfaces in polar directions are charged. As before, our examination of band alignment at perovskite interfaces focuses on neutral interfaces with well-defined BOs. Neutral structures for hetero-valent perovskite interfaces with polar orientations can be assembled with WS cells, in steps analogous to those described for zincblende interfaces. Interfaces assembled with WSC centered on the body-centered cation have a 1:1 mixture of the two corner cations on a common interface plane separating the two perovskite oxide crystals for either the (100) or the (111) orientation. For example, a stable (100) interface between $BaTiO_3$ and $KTaO_3$ has a stacking sequence of - BaO-TiO_2-$(Ba_{0.5}K_{0.5})O$-TaO_2-KO- and is known as the "mixed BaK" interface. On the other hand, construction with corner-cation-centered WSC leads to a stacking sequence of -Ti-BaO_3-$Ti_{0.5}Ta_{0.5}$-KO_3-Ta-, in a mixed-TiTa (111) interface. Both can be verified to be neutral from valence counting of the atoms at the interfaces. It is also possible to create non-polar interface structures along polar directions, between hetero-valent perovskite oxides.

5.4.2 *Ionic Screening: Effect of Lattice Relaxation*

Since the lattice relaxation process has a dominant effect on the properties of ionic structures, the band alignment conditions of the perovskite oxide interfaces with neutral structures are calculated first in the unrelaxed structure, with all atoms clamped to their bulk positions, before allowing the atomic positions to relax. The VBO of unrelaxed iso-valent perovskite interfaces is found to be nearly independent of the interface atomic arrangement and orientation, and it does not change significantly when the interface structure is allowed to relax. As shown in Fig. 5.16, the

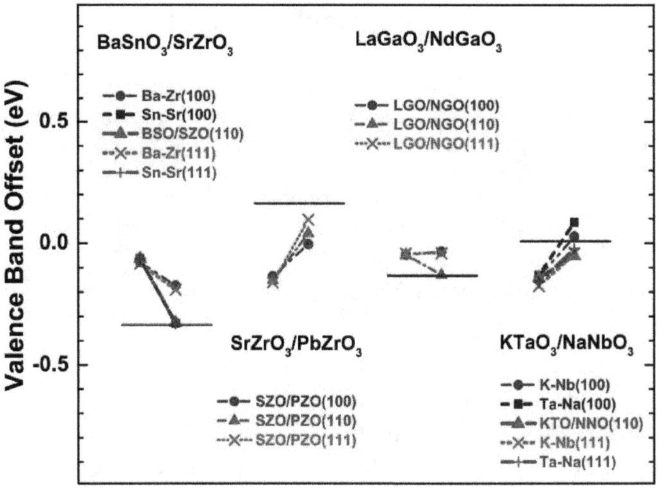

Fig. 5.16. Valence band offsets calculated for iso-valent heterojunctions of perovskite oxides. The orientation and the atomic arrangement of the interface are marked by the identities of atoms on interface planes, where distinction is necessary. For example, under the BaSnO3/SrZrO3 heading, the Ba-Zr(100) is a (100) interface with Ba and Zr on adjacent planes, and the BSO/SZO(110) is a planar (110) interface between these two oxides. Within each group, the data shown on the left are VBOs calculated for unrelaxed, clamped interfaces and those shown on the right are from the final, relaxed interfaces. The VBO predicted by the NPT for each heterojunction is drawn as a horizontal solid bar, while the Bader model solid prediction is shown as broken line.

Source: Reprinted with permission from Tung and Kronik.[12] Copyright (2019) by the American Physical Society.

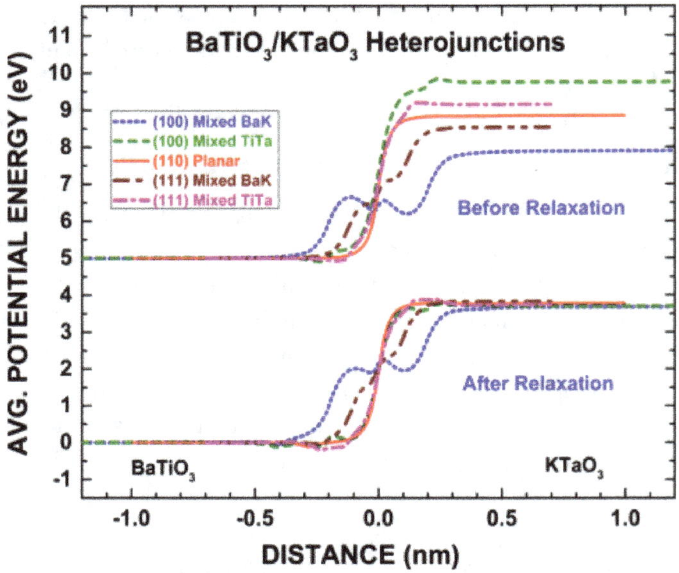

Fig. 5.17. Macroscopically averaged potential energy distributions calculated for neutral $BaTiO_3/KTaO_3$ interfaces. The orientation and the atomic arrangement of the interface are as marked. Curves are vertically shifted for clarity. For each interface, the potential distribution before and after lattice relaxation is shown in the upper and lower parts of the figure. The significant dependence of the potential energy offset on interface orientation and atomic arrangement before relaxation becomes negligible after the lattice is relaxed.

Source: Reprinted with permission from Tung and Kronik.[12] Copyright (2019) by the American Physical Society.

absolute values of the VBOs found for the relaxed iso-valent heterojunctions are small, <0.4 eV, resulting in the conduction band offset accounting for much of the difference between the two band gaps. Unrelaxed hetero-valent interfaces behave differently. Calculated average potential energies of the unrelaxed $BaTiO_3/KTaO_3$ junctions with various interface orientations and compositions are shown in the upper part of Fig. 5.17. A significant dependence of the VBO on the atomic structure and orientation is observed, similar to hetero-valent zincblende interfaces. And as before, this VBO dependence may be attributed to the relative positions of the donor- and acceptor-like hetero-valent elements at the interface.[12] Before lattice relaxation, the VBOs of hetero-valent perovskite oxide interfaces behave similarly to zincblende heterojunctions.

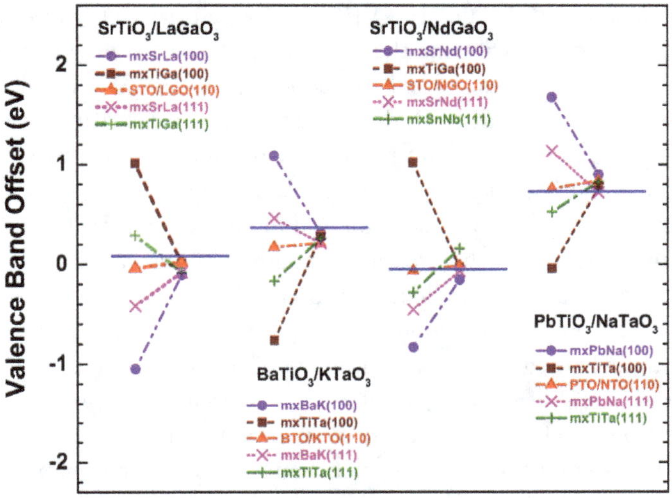

Fig. 5.18. Valence band offsets calculated for hetero-valent (a) II-IV-Ox/I-V-Ox, (b) II-IV-Ox/III-III-Ox heterojunctions of perovskite oxides. The orientation and the atomic arrangement of the interface are marked by atoms on interface plane, when necessary. For example, from the $BaTiO_3/KTaO_3$ group, "mxBaK(100)" is a (100) interface with a 50–50 mixture of Ba and K, and an appropriate number of oxygen atoms, on the interface plane. Within each group, the data shown on the left are VBOs calculated for unrelaxed, clamped interfaces and those shown on the right are from the final, relaxed interfaces. The VBO predicted for each heterojunction by the NPT with corrections for density on polyhedra face is drawn as a horizontal solid bar. In (a), the NPT predictions corrected according to the intermediate density are also shown, as broken horizontal bars.

Source: Reprinted with permission from Tung and Kronik.[12] Copyright (2019) by the American Physical Society.

Lattice relaxation, which has a negligible effect for zincblende heterojunctions, overwhelms the band alignment at hetero-valent perovskite interfaces! The sharp dependence of the VBO on the orientation and structure of hetero-valent perovskite interfaces is wiped out by lattice relaxation, as shown in the lower part of Fig. 5.17. The shifts in the average potential energy across the interface, widely scattered for the unrelaxed interfaces, collapse into a narrow range upon relaxation. Figure 5.18 summarizes these changes in the VBOs, with unrelaxed interfaces plotted on the left of each group and the relaxed VBOs on the right. Lattice relaxation renders the VBO for neutral hetero-valent heterojunction between two perovskite oxides essentially independent of interface orientation and initial atomic arrangement.[12] Since the VBO for the neutral (110) interface does not change significantly during the relaxation, ionic screening converges the VBOs of polar interfaces toward the non-polar value. The implications of Figs. 5.16–5.18 are striking. Independent of the orientation, the atomic structure, or whether the interface is iso-valent or hetero-valent, the equilibrium VBO is that for the non-polar interface!

The effectiveness of ionic screening is further demonstrated by embedding a sub-monolayer of hetero-valent polar pairs into a perovskite oxide, i.e. the creation of an "artificial" dipolar interface in an otherwise bulk structure. The embedded hetero-valent pairs are chosen to be nominally lattice-matched to the hosts, to minimize effects due to strain. Profiles of the macroscopically averaged electrostatic potential energy distributions obtained from such "embedded homojunctions" are given in Fig. 5.19. Clearly, the as-embedded hetero-valent pairs induce large potential energy offsets (>1.5 eV) in the bulk oxide, in directions consistent with the polarity of the embedded dipole. However, here too these large potential energy shifts all but disappear after lattice relaxation. This behavior

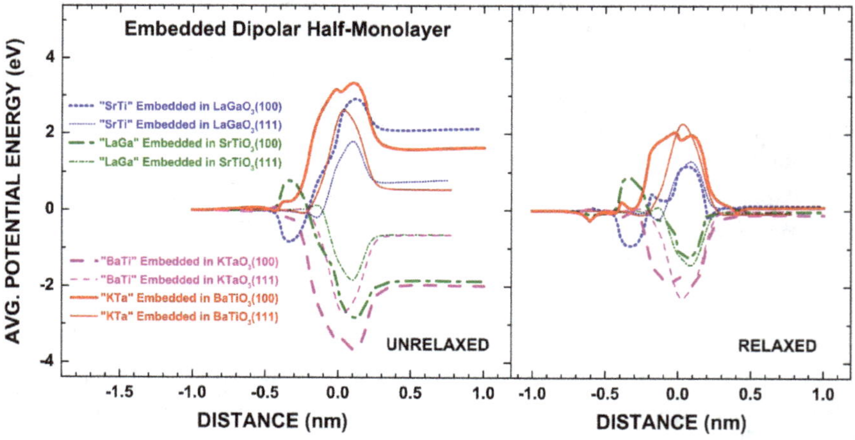

Fig. 5.19. Macroscopically averaged electrostatic potential energy distributions calculated for perovskite oxide embedded with one-half monolayer of hetero-valent pairs in the ⟨100⟩ direction. Before relaxation, the results for SrTi embedded in $LaGaO_3$, LaGa embedded in $SrTiO_3$, BaTi embedded in $KTaO_3$, and KTa embedded in $BaTiO_3$ are shown as dash-dot-dot, dashed, dash-dotted, and dotted lines, respectively. Potential distributions after relaxation are shown as thin solid lines.

Source: Reprinted with permission from Tung and Kronik.[12] Copyright (2019) by the American Physical Society.

for ionic compounds is markedly different from that found for covalent semiconductors. As shown in Fig. 5.15, the potential shifts due to embedded hetero-valent pairs in zincblende semiconductors remained largely intact after relaxation and were the explanation of the observed VBO dependence. Unlike the tunability in BO demonstrated for covalent semiconductors, ionic screening makes it impossible to adjust the VBO of a perovskite heterojunction by any means.

5.4.3 *VBO Modeled with Bulk Densities: NPT*

The disappearance of the sharp VBO dependence on interface structure highlights the dominant role played by lattice relaxation for ionic interfaces. Between unrelaxed and relaxed structures, the main difference is that each atom in the latter occupies an optimized position with vanishing Hellmann–Feynman forces,[13] whereas forces are not managed for the former. The nulling of electric fields at atomic sites has negligible effect for covalent heterojunctions. Yet, it has such a dominant effect on the charge distribution at perovskite heterojunctions that the equilibrium BO may even be deduced simply from the requirement for the electric field to vanish! To examine the relationship between charge distribution and electric fields, we first look backward at how a charge distribution with vanishing electric fields at atomic sites might have been constructed in the first place. It is known that within each neutral polyhedron, the charge distribution is symmetric, and the electric field from its own charge vanishes at the atomic position. Therefore, when symmetric neutral cells, not all of which are necessarily identical, are used to build a volume of charge distribution, the electric field on any internal atomic site vanishes because neither the charge distribution within its own cell nor the charge distribution from any neighboring cell provides a non-zero contribution. Such an assembled charge density induces no external electric field either and has no influence on the potential distribution outside of its confines. Therefore, if a certain volume is known to be stackable from symmetric neutral cells, the charge distribution inside this volume can be ignored altogether for band alignment considerations, regardless of the position or chemical identity of the symmetric neutral cells. For the perovskite heterojunction interfaces, ionic screening effectively renders the structure and the charge distribution of the ISR irrelevant. With no bond correction needed, the BOs of fully relaxed perovskite heterojunctions are determined by bulk properties. The difference in the NPT bulk reference VBM positions of the two perovskite

Table 5.3. The average of VBOs calculated for lattice-matched, fully relaxed, perovskite heterojunctions with up to five different interface structures: (100) and (111) orientations with mixed-cation and mixed-anion planes and the planar (110). A positive VBO value indicates that the VBM of oxide #1 is higher than oxide #2 at that interface. Hybrid HSE functionals are used in these calculations. VBOs modeled with charge densities of bulk semiconductors are marked as "NPT".

Ox #1	Ox #2	Calc. VBO (eV)	NPT (eV)
LaGaO3	NdGaO3	−0.068	−0.133
KTaO3	NaNbO3	0.000	0.007
CaHfO3	SrSnO3	0.227	0.251
CaZrO3	SrHfO3	0.099	0.007
BaSnO3	SrZrO3	−0.270	−0.335
BaSnO3	PbZrO3	−0.145	−0.172
SrZrO3	PbZrO3	0.044	0.163
SrTiO3	LaGaO3	−0.054	0.074
SrTiO3	NdGaO3	−0.086	−0.059
PbTiO3	NaTaO3	0.802	0.718
BaTiO3	KTaO3	0.245	0.355
BaTiO3	NaNbO3	0.149	0.362
CaSnO3	KNbO3	0.175	0.384

oxides is the modeled VBO of their interface, irrespective of the orientation and the structure. The average of VBOs calculated for lattice-matched, fully relaxed, perovskite heterojunctions with up to five different interface structures, (100) and (111) orientations with mixed-cation and mixed-anion planes and the planar (110), are listed in Table 5.3. These are seen to be in very good agreement with those predicted by the difference on the NPT bulk scale.

5.5 Modeling Experimental VBO of Polycrystalline Interfaces

Single-crystal interfaces have been the preferred venue to investigate the fundamentals of the band alignment phenomenon. The roles played by various parameters in deciding the band offset can be clearly identified when the interfaces are investigated one at a time. However, for many practical applications such as energy conversion, the size of the active interface area needs to be macroscopic, and it is rare to find materials with the desired properties that are also lattice-matched. Therefore, polycrystalline interfaces are widely used in various advanced technologies, with the band offsets at these interfaces providing the functionality needed for the operations of some devices. Due to their importance, the factors that control the efficiency and quality of polycrystalline interfaces and determine the band alignment conditions also need to be understood. Even though the structures of these interfaces are non-uniform, the alignment of band positions still must be a result of the local distribution of charge. To explain and predict the band alignment at polycrystalline interfaces, modeling the interface charge distribution remains the only scientifically proven approach. Polycrystallinity seems to make charge modeling difficult, but it does not prevent a density-based approach from accurately predicting the BO/SBH for certain polycrystalline interfaces because of common rules that all interfaces should follow. In any crystal with polar bonds, it is energetically favorable for the cations to stay away from each other and be surrounded by anions, and *vice versa* for the anions. Following any path on the skeletal bond diagram of a bulk crystal, one expects to find cations and anions to alternate. This general rule, observed for all bulk compounds, makes a strong case for the network of alternating polarized bonds to continue across any interface, irrespective of the interface orientation. Some consistency in the arrangement of atoms may therefore be expected simply based on energetics, even for polycrystalline interfaces with supposedly unknown atomic structures.

Between some materials, e.g. binary oxides, this rule of alternate arrangement of cations and anions alone allows adequate modeling of the BO/SBH. For other interfaces, variations in the BO/SBH may arise depending on which of several available cations and anions actually form bonds at the interface. This concern is particularly relevant for compounds with cations or anions of different valences, such as perovskite oxides. For these interfaces, additional assumptions should be made about the interface structure to make more realistic predictions. Then there are still other interfaces, e.g. those between covalent and ionic compounds, where the polarization of the interface bonds and thus the BO/SBH should depend on the interface structure. Due to competition among possible structures, the actual structure(s) of these interfaces are not predictable from simple energetic arguments. A successful modeling of band alignment conditions at these interfaces becomes feasible only when the effect of variations in bond polarization is averaged out from the randomness in the polycrystallinity interface structure. These different types of interfaces are discussed separately.

NPT and AIM model solids are the main choices for charge modeling of polycrystalline interfaces. A factor that affects this choice is whether the interface structure is completely unknown, as may be assumed to be the case for some polycrystalline interfaces, or whether it is either known or may be reasonably inferred. Due to the charged nature of Bader atoms, the average potential energy of an AIM model solid depends sharply on its surface termination. Therefore, only when the interface structure is known and can be handled by the two specific model solids discussed earlier can AIM be used for the analysis. NPT is the main method to model the BO/SBH for most of the polycrystalline interfaces, first because the same NPT model solid is applicable to all (including unknown) interface structures, and second because the interatomic distance at the interface is already approximately optimized which minimizes the need for further lattice relaxation. Since the effect of the interface bonds cannot be considered for polycrystalline interfaces, the difference in the bulk references, $E_{\mathrm{VBM,SC}}^{\mathrm{BulkRef}}$, of the two semiconductors is the modeled VBO.

5.5.1 *Interfaces of Binary Oxides*

Each binary oxide has one cation species and one anion (oxygen). Since oxygen is the common anion throughout, the bonding sequence for an interface between any two binary oxides is essentially predictable. For example, between HfO_2 and SiO_2, one expects to find -Hf-O-Si- at the interface, regardless of the orientation. Here, the band alignment conditions for these interfaces are modeled by NPT and compared with those experimentally measured.[14] The good agreement between experiment and theory is shown in Fig. 5.20, with the data for the individual interfaces listed in Table 5.4. This agreement suggests that, under the condition of a continuous network of bonds, the effect of lattice and charge relaxation from stacked bulk densities is small and arises mainly from the smoothing out of charge density differences. Without the need for interface bond corrections, NPT predicts the band alignments at binary oxide interfaces to be "bulk-like", i.e. independent of interface orientation and structure. Furthermore, among a group of materials, the band alignment should be transitive, as can be verified for some interfaces of Tables 4.4–4.6. Ultimately, both the bulk-like behavior and the success of the NPT analysis can be attributed to the absence of new bonds at these polycrystalline binary oxide interfaces.

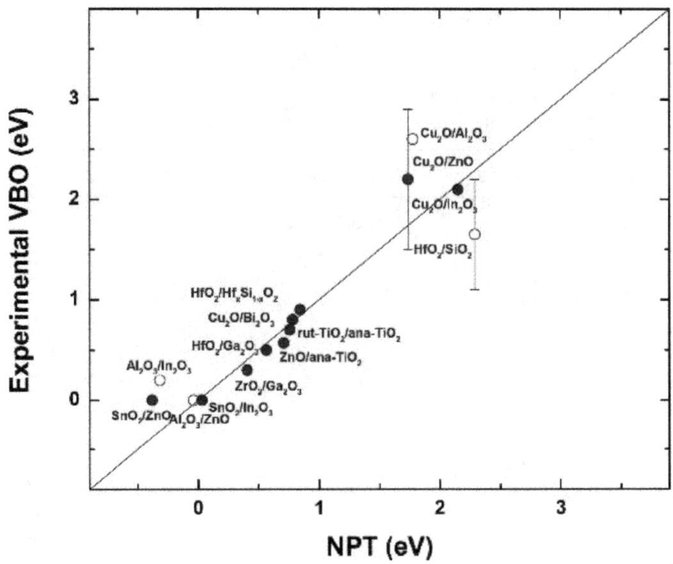

Fig. 5.20. Experimentally measured VBO are shown against the values expected from the NPT, with corrections for charge density difference. The data shown in Table 5.4 are used for this plot. Empty symbols represent experiments conducted with amorphous oxides but theoretically modeled with a crystalline oxide of the same composition.

Source: Reprinted from open-access article Tung.[14]

5.5.2 *Polycrystalline Interfaces with Perovskite Oxides*

Each perovskite oxide has two cations, usually of different formal charges. A simple application of the rule of alternating cation and anion does not lead to a unique choice of the interface structure. Which of the two cations of the perovskite is actually at the interface with another material may affect the band alignment condition appreciably. Therefore, to accurately model the BO/SBH for perovskite interfaces, the effect of polarized bonds should not be ignored, if possible. As the average potential energy of an NPT model solid is independent of its surface structure, the NPT band offset does not change with interface structure. An NPT prediction for a perovskite interface thus amounts to an interface average, in the absence of a dominant atomic structure. NPT-modeled band offsets of all the present perovskite interfaces, summarized in Table VI, show varied degrees of agreement with those experimentally measured.

Table 5.4. Experimentally measured valence band offset, VBO between binary oxides. A positive value indicates that the band edge position of oxide #1 (first column) is higher than that of oxide #2. The NPT values include corrections for charge density differences. The experimental data are obtained from various investigations listed in Ref. 14.

Oxide #1	Oxide #2	VBO/p-SBH (eV)	NPT (eV)	References
rut-TiO2	ana-TiO2	0.7	0.76	15, 16
HfO2	SiO2	1.1–2.2	2.29	17, 18
Cu2O	Al2O3	2.6	1.78	19
Cu2O	Bi2O3	0.9	0.84	19
Cu2O	Sn:In2O3	2.1	2.15	20
Cu2O	ZnO	1.5–2.9	1.74	21–23
HfO2	HfxSi1-xO2	0.8	0.78	24
HfO2	β Ga2O3	0.5	0.56	25
ZrO2	β Ga2O3	0.3	0.40	25
Al2O3	Sn:In2O3	0.2	−0.3	26
SnO2	In2O3	0	0.03	27
SnO2	ZnO	0	−0.38	28
Al2O3	ZnO:Al	0	−0.04	28
ZnO	ana-TiO2	0.57	0.71	29

Since Bader atoms have better transferability among perovskite oxides than neutral polyhedra, their use when appropriate is expected to give more accurate predictions of the band offsets. For some of the interfaces shown in Table 5.5, the interface structure may actually be reasonably deduced. For example, at the $BaTiO_3/SrTiO_3$ interface, one expects the two crystals to be joined at a common TiO_x layer. We could use group II-centered (A-centered) Bader model solids, listed in Table II, to represent both crystals of this interface. This strategy recreates the "shared" TiO_2 layer that still remains overall neutral. Much improved VBO prediction over the NPT model is seen for the Bader model for this interface. Other interfaces that have been modeled with corner-cation-centered Bader model solids are $PbTiO_3/SrTiO_3$ and $SrTiO_3/BaSnO_3$. The AIM projections for these interfaces are much closer to the experiment than those modeled with NPT. Some interfaces between perovskite oxides and binary oxides have structures that are either deducible or may be narrowed down to a few choices with similar band offset consequences. These include all the yet-undiscussed interfaces in Table 5.5 that are marked with predictions from the Bader method. Experimental band alignment results from Table 5.5 are plotted in Fig. 5.21 against either the predictions of AIM analysis when available or the NPT projections.[14] A very good agreement is found for most

Table 5.5. Valence band offset, VBO, between perovskite oxides and materials. A positive value indicates that the band edge position of material #1 is higher than that of #2. Estimates from NPT contain density difference corrections. The Bader estimates, which contain the same density difference corrections, may use a specific termination of the Bader model solid. See text for explanations and justifications. When available, the Bader approach is preferred as it provides a closer representation of charge distribution at the actual interface. The experimental data are obtained from various investigations listed in Ref. 14.

Material #1	Material #2	VBO (eV)	NPT (eV)	Bader (eV)
BaTiO3	SrTiO3	0	−0.22	0.24
PbTiO3	SrTiO3	1.1	0.28	0.97
LaAlO3	SrTiO3	0.15	0.42	
SrTiO3	BaSnO3	0.28	0.89	−0.11
LaAlO3	BaSnO3	0.5	1.31	
SrTiO3	SrZrO3	0.5	0.66	
(Ba,Sr)TiO3	a-TiO2	0	1.05	−0.06
(Ba,Sr)TiO3	Ta2O5	0	1.23	−0.15
LaAlO3	SiO2	2.5	3.14	
SrTiO3	SnO2	0.85	1.06	
ana-TiO2	SrTiO3	0.2	−1.39	0.06
SrTiO3	MgO	0	0.30	
BiFeO3	SrTiO3	1	0.26	1.24
GaAs	SrTiO3	2.5	1.57	
Si	SrTiO3	2.1	2.32	
Si	BaTiO3	2.4–2.7	2.54	
Ge	SrHfO3	3.27	2.82	
Sn:In2O3	(Ba,Sr)TiO3	0.1	−0.75	0.46
Pb(Zr,Ti)O3	Sn:In2O3	1.1	0.74	1.02
Al2O3	(Ba,Sr)TiO3	0.15	−0.38	0.20
RuO2	BiFeO3	0.72	0.72	
RuO2	(Bi,Na)TiO3	1.05	1.34	0.94
RuO2	(K,Na)NbO3	2.15	1.81	
RuO2	PbZrO3	0.7–0.8	1.36	0.64
RuO2	(Ba,Sr)TiO3	2.35	1.32	2.35
RuO2	Pb(Zr,Ti)O3	1	1.34	

of the interfaces, thereby confirming the validity of density-based analysis of polycrystalline interfaces.

5.5.3 *Interfaces between Oxides and Zincblende/Wurtzite Semiconductors*

The experimental VBOs between oxides and zincblende/wurtzite/diamond semiconductors are compared with the NPT model in Fig. 5.22. As more

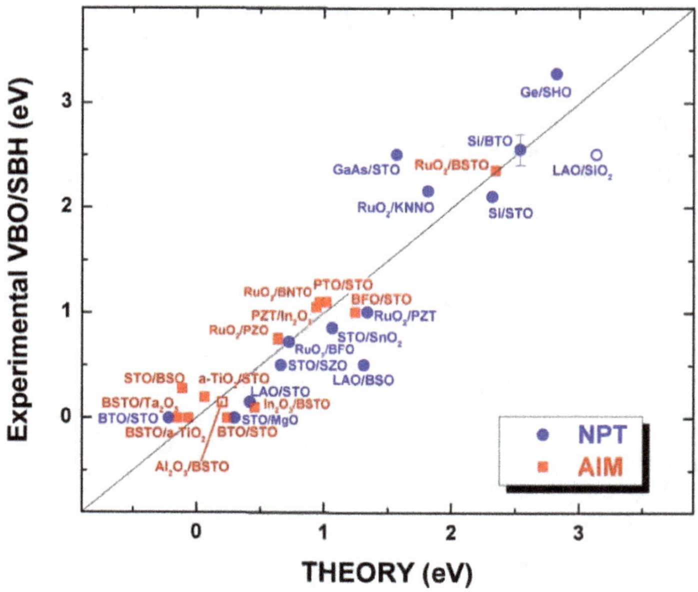

Fig. 5.21. Experimentally measured VBO and *p*-type SBH of perovskite oxides, shown in Table 5.5, are plotted against the theoretical values expected from either the NPT (circles) or, when available, the AIM theory of Bader (squares). Corrections for charge density difference are included in the theory. Empty symbols represent experiments conducted with amorphous oxides but theoretically modeled with a crystalline oxide of the same composition.

Source: Reprinted from open-access article Tung.[14]

than one cation and more than one anion are available for every interface, the VBO is expected to depend on the atomic structure for most of these interfaces. With a few exceptions, there is no strong energetic argument to favor a particular interface structure over another, without actual calculation. Any agreement of the experimental VBO with the NPT model here is thus an indication that the effect of polar bonds at the interface has been rendered insignificant by averaging over polycrystallinity or ionic screening. An examination of Fig. 5.22 shows that the agreement between the NPT and experiment is quite reasonable for most of these interfaces.[14] There are a few interfaces with very significant disagreements that should be examined more closely. Of all interfaces the largest errors, ~2 eV, are found for all forms of the SiC semiconductors. SiC is typically fabricated at high temperatures and investigations have found signatures of C–C bonds and evidence for graphitic structure at SiC interfaces.[48] Based on these reports,

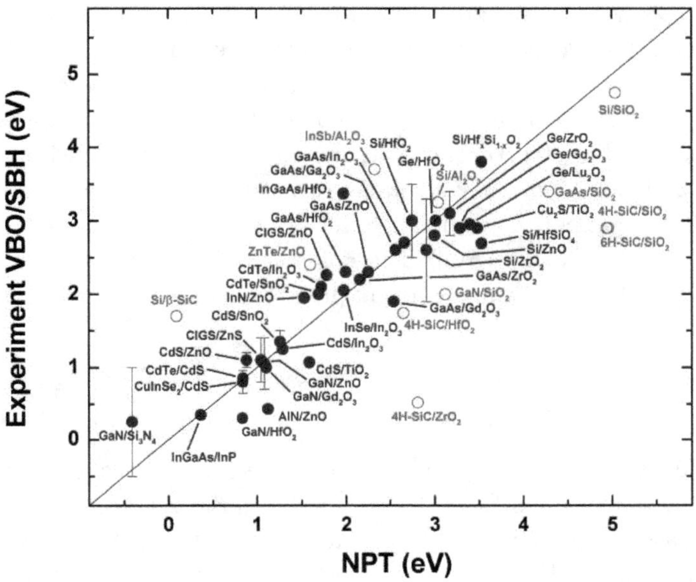

Fig. 5.22. Experimentally measured VBOs from interfaces between compound semiconductors and oxides are plotted against the NPT predictions with corrections for charge density difference. Empty symbols represent special cases discussed in the text or experiments conducted with amorphous oxides but theoretically modeled with a crystalline oxide of the same composition.

Source: Reprinted from open-access article Tung.[14]

it seems reasonable to assume the presence of a Si–C dipole, associated with the defects, at the interface with the positive pole pointing away from the interface and toward the silicon carbide, independent of other details of the interface. The large magnitude of this dipole would significantly lower the VBM for the SiC bulk crystal and move the NPT values significantly in the direction of the experimental VBO for all SiC interfaces studied.

An examination of Figs. 5.20–5.22, which span a wide range of materials with different types of bonds, shows that, for the most part, the band alignment condition can be well understood in terms of the distribution of charge even for polycrystalline interfaces. However, there is no single approach to model the charge distribution that is right for all the interfaces. Attention should be paid to what chemistry tells us about the bonds at each interface. At interfaces where the bulk network of polar bonds is expected to continue without interruption, there is no need to consider the polarization of interface bonds. Both the NPT and the AIM, if proper model solids may be constructed, should be accurate. At interfaces between dissimilar materials, a knowledge of the interface structure is required to accurately predict the effect of the interface bonds. With the atomic structure known, AIM is the preferred technique to model the effect of interface bonds. Without a known structure, one resorts to the NPT approach and relies on the randomness in a polycrystalline structure to average out the effect of interface bond polarization. These recipes for band alignment prediction are built on the sound scientific basis of charge distribution. As such, they serve as guidelines for band alignment prediction for yet unexplored or novel interfaces.

5.6 Summary

Band offsets at heterojunction interfaces are of much academic interest and technological importance. Due to their causal relationship, the band offset of a heterojunction interface can be explained/predicted whenever the charge distribution at its interface can be reasonably modeled. The procedures to model the charge distribution at a solid interface, discussed in Chapter 4, are applied to model the VBO of various kinds of heterojunction interfaces in this chapter. These procedures for VBO modeling are different for heterojunctions with interface bonds that are similar or dissimilar to those in the bulk crystals. With this simple rule, this chapter demonstrates that the VBOs measured and/or calculated for epitaxial heterojunctions with known atomic structures can all be accurately explained/predicted. In addition, VBOs of the majority of polycrystalline heterojunctions of unknown interface structures may be reasonably modeled. How the VBOs of different types of heterojunctions are modeled in this chapter is summarized in Table 5.6. Whether a correction is needed for the effect of the bath depends on the nature of the bonds at the interface. Interface bonds that are a continuation of the network of bulk bonds do not need bond corrections. The VBOs for such heterojunctions can be accurately modeled as the difference of two bulk reference values. Due to this property, these VBOs will display bulk-like behavior, including orientation independence and transitivity. When the interface bonds are dissimilar to the bulk bonds in either solid, their effect cannot be ignored. For epitaxial hetero-valent zincblende heterojunctions, bond polarizations are accurately modeled by screened point charges from exposed shallow dopants at the interfaces. The effect of bonds at polycrystalline interfaces between perovskite oxides and other oxides is adequately handled by using AIM model solids, when

Table 5.6. Methods used to model VBOs of different types of epitaxial and polycrystalline heterojunctions.

Heterojunction Type	VBO Modeling Method	Hosts	Bath
Epitaxial Iso-valent Zincblende Heterojunctions	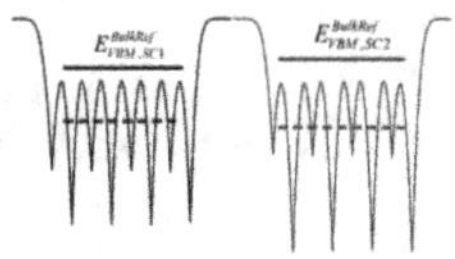	NPT Bulk References	No
Epitaxial Hetero-valent Zincblende Heterojunctions	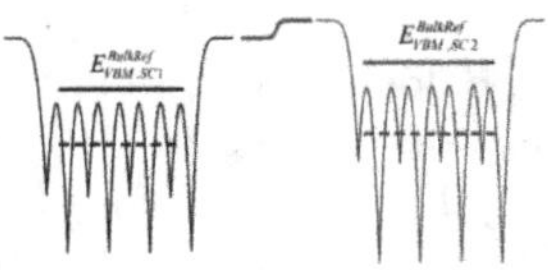	NPT Bulk References	Screened Dopant Point Charges
Relaxed Epitaxial Perovskite Heterojunctions	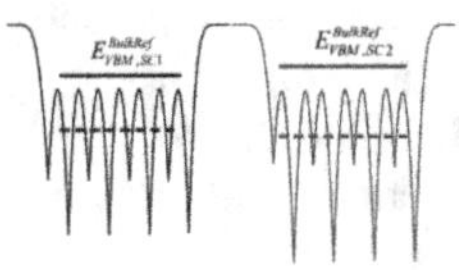	NPT Bulk References	No
Polycrystalline Binary Oxide Heterojunctions and Oxide/Zincblende Heterojunctions	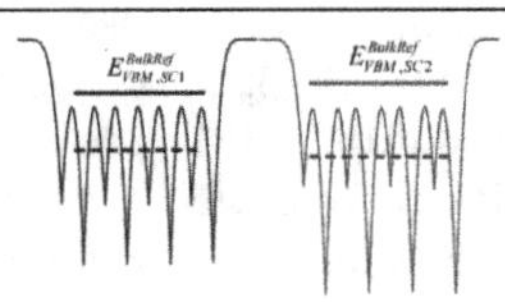	NPT Bulk References	No
Polycrystalline Perovskite/Oxide Heterojunctions	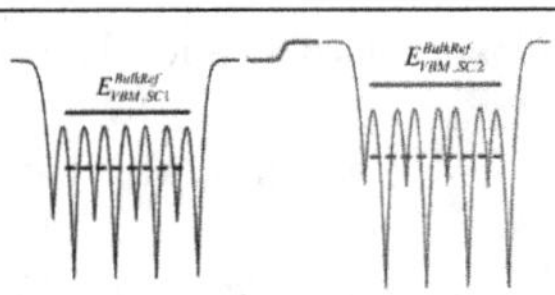	NPT Bulk References	AIM Model Solids, When Appropriate

appropriate. No longer expressible as the difference between two bulk references, the VBOs of this group of interfaces generally do not behave as bulk-like, i.e. they are not transitive and they may depend on interface orientation. Whether the VBO of a heterojunction is bulk-like or not is attributable to the nature of its interface bonds! This chapter demonstrates that charge distribution is indeed the foundation of band offset formation.

References

1. S.A. Chambers, *Surf. Sci. Rep.* **79**, 100638 (2024).
2. E. Pelucchi, S. Rubini, B. Bonanni, A. Franciosi, and M. Peressi, *Appl. Phys. Lett.* **78**, 1574–1576 (2001).

3. Z.H. Lim, N.F. Quackenbush, A.N. Penn, M. Chrysler, M. Bowden, Z. Zhu, J.M. Ablett, T.L. Lee, J.M. Lebeau, J.C. Woicik, P.V. Sushko, S.A. Chambers, and J.H. Ngai, *Phys. Rev. Lett.* **123**, 026805 (2019).
4. H.B. Zhao, Y.H. Ren, B. Sun, G. Lüpke, A.T. Hanbicki, and B.T. Jonker, *Appl. Phys. Lett.* **82**, 1422–1424 (2003).
5. G. Kresse and J. Hafner, *Phys. Rev. B* **49**, 14251 (1994).
6. J.P. Perdew, K. Burke, and M. Ernzerhof, *Phys. Rev. Lett.* **77**, 3865–3868 (1996).
7. J. Heyd, G.E. Scuseria, and M. Ernzerhof, *J. Chem. Phys.* **118**, 8207–8215 (2003).
8. M. Peressi, N. Binggeli, and A. Baldereschi, *J. Phys. D* **31**, 1273–1299 (1998).
9. R.T. Tung and L. Kronik, *Phys. Rev. B* **94**, 075310 (2016).
10. W. Kohn, *Phys. Rev. Lett.* **76**, 3168–3171 (1996).
11. R.T. Tung and L. Kronik, *Adv. Theo. Simul.* **1**, 1700001 (2017).
12. R.T. Tung and L. Kronik, *Phys. Rev. B* **99**, 115302 (2019).
13. R.P. Feynman, *Phys. Rev.* **56**, 340–343 (1939).
14. R.T. Tung, *J. Appl. Phys.* **136**, 165301 (2024).
15. P. Deák, B. Aradi, and T. Frauenheim, *J. Phys. Chem. C* **115**, 3443–3446 (2011).
16. V. Pfeifer, P. Erhart, S. Li, K. Rachut, J. Morasch, J. Brötz, P. Reckers, T. Mayer, S. Rühle, A. Zaban, I. Mora Seró, J. Bisquert, W. Jaegermann, and A. Klein, *J. Phys. Chem. Lett.* **4**, 4182–4187 (2013).
17. L. Xie, Y. Zhao, and M.H. White, *Solid-State Electron.* **48**, 2071–2077 (2004).
18. O. Renault, N.T. Barrett, D. Samour, and S. Quiais-Marthon, *Surf. Sci.* **566–568**, 526–531 (2004).
19. J. Deuermeier, E. Fortunato, R. Martins, and A. Klein, *Appl. Phys. Lett.* **110**, 051603 (2017).
20. J. Deuermeier, J. Gassmann, J. Brötz, and A. Klein, *J. Appl. Phys.* **109**, 113704 (2011).
21. M. Ichimura and Y. Song, *Jpn. J. Appl. Phys.* **50**, 051002 (2011).
22. M. Yang, L. Zhu, Y. Li, L. Cao, and Y. Guo, *J. Alloys Compd.* **578**, 143–147 (2013).
23. S. Siol, J.C. Hellmann, S.D. Tilley, M. Graetzel, J. Morasch, J. Deuermeier, W. Jaegermann, and A. Klein, *ACS Appl. Mater. Interfaces* **8**, 21824–21831 (2016).
24. S. Toyoda, J. Okabayashi, H. Kumigashira, M. Oshima, K. Ono, M. Niwa, K. Usuda, and N. Hirashita, *J. Electron Spectrosc. Relat. Phenom.* **137–140**, 141–144 (2004).
25. V.D. Wheeler, D.I. Shahin, M.J. Tadjer, and C.R. Eddy, Jr., *ECS J. Solid State Sci. Technol.* **6**, Q3052–Q3055 (2017).
26. Y. Gassenbauer, A. Wachau, and A. Klein, *Phys. Chem. Chem. Phys.* **11**, 3049–3054 (2009).
27. A. Klein, *Thin Solid Films* **520**, 3721–3728 (2012).
28. S. Li, F. Chen, R. Schafranek, T.J.M. Bayer, K. Rachut, A. Fuchs, S. Siol, M. Weidner, M. Hohmann, V. Pfeifer, J. Morasch, C. Ghinea, E. Arveux,

R. Günzler, J. Gassmann, C. Körber, Y. Gassenbauer, F. Säuberlich, G.V. Rao, S. Payan, M. Maglione, C. Chirila, L. Pintilie, L. Jia, K. Ellmer, M. Naderer, K. Reichmann, U. Böttger, S. Schmelzer, R.C. Frunza, H. Uršič, B. Malič, W.B. Wu, P. Erhart, and A. Klein, *Phys. Status Solidi Rapid Res. Lett.* **8**, 571–576 (2014).
29. D.Q. Fang and S.L. Zhang, *J. Chem. Phys.* **144**, 014704 (2016).
30. R. Schafranek, S. Li, F. Chen, W. Wu, and A. Klein, *Phys. Rev. B Condens. Matter Mater. Phys.* **84**, 045317 (2011).
31. S.A. Chambers, M.H. Engelhard, V. Shutthanandan, Z. Zhu, T.C. Droubay, L. Qiao, P.V. Sushko, T. Feng, H.D. Lee, T. Gustafsson, E. Garfunkel, A.B. Shah, J.M. Zuo, and Q.M. Ramasse, *Surf. Sci. Rep.* **65**, 317–352 (2010).
32. S.A. Chambers, T.C. Kaspar, A. Prakash, G. Haugstad, and B. Jalan, *Appl. Phys. Lett.* **108**, 152104 (2016).
33. R. Schafranek, J.D. Baniecki, M. Ishii, Y. Kotaka, K. Yamanka, and K. Kurihara, *J. Phys. D* **45**, 055303 (2012).
34. V.V. Afanasev, A. Stesmans, C. Zhao, M. Caymax, T. Heeg, J. Schubert, Y. Jia, D.G. Schlom, and G. Lucovsky, *Appl. Phys. Lett.* **85**, 5917–5919 (2004).
35. C. Ke, W. Zhu, Z. Zhang, E.S. Tok, and J. Pan, *Surf. Interface Anal.* **47**, 824–827 (2015).
36. A.C. Tuan, T.C. Kaspar, T. Droubay, J. J. W. Rogers, and S.A. Chambers, *Appl. Phys. Lett.* **83**, 3734–3736 (2003).
37. P. Cásek, S. Bouette-Russo, F. Finocchi, and C. Noguera, *Phys. Rev. B* **69**, 854111–8541111 (2004).
38. R. Schafranek, J.D. Baniecki, M. Ishii, Y. Kotaka, and K. Kurihara, *New J. Phys.* **15**, 053014 (2013).
39. Y. Liang, J. Curless, and D. McCready, *Appl. Phys. Lett.* **86**, 082905 (2005).
40. S.A. Chambers, Y. Liang, Z. Yu, R. Droopad, J. Ramdani, and K. Eisenbeiser, *Appl. Phys. Lett.* **77**, 1662–1664 (2000).
41. F. Amy, A.S. Wan, A. Kahn, F.J. Walker, and R.A. McKee, *J. Appl. Phys.* **96**, 1635–1639 (2004).
42. M.D. McDaniel, C. Hu, S. Lu, T.Q. Ngo, A. Posadas, A. Jiang, D.J. Smith, E.T. Yu, A.A. Demkov, and J.G. Ekerdt, **117**, (2015).
43. S. Li, C. Ghinea, T.J.M. Bayer, M. Motzko, R. Schafranek, and A. Klein, *J. Phys. Condens. Matter* **23**, 334202 (2011).
44. F. Chen, R. Schafranek, S. Li, W.B. Wu, and A. Klein, *J. Phys. D* **43**, 295301 (2010).
45. S. Li, A. Wachau, R. Schafranek, A. Klein, Y. Zheng, and R. Jakoby, *J. Appl. Phys.* **108**, 014113 (2010).
46. S. Li, J. Morasch, A. Klein, C. Chirila, L. Pintilie, L. Jia, K. Ellmer, M. Naderer, K. Reichmann, M. Gröting, and K. Albe, *Phys. Rev. B* **88**, 045428 (2013).
47. R. Schafranek, J. Schaffner, and A. Klein, *J. Eur. Ceram. Soc.* **30**, 187–192 (2009).
48. V.V. Afanasev, M. Bassler, G. Pensl, and M. Schulz, *Phys. Status Solidi A* **162**, 321–337 (1997).

Chapter 6

SCHOTTKY BARRIER HEIGHT MEASUREMENT

The measurement of the SBH of an MS interface sounds like a boring routine that has been conducted millions of times before, without fanfare or much worth talking about. But the truth is that, for decades, transport data from SBH measurements of routinely formed polycrystalline MS interfaces always contained exciting clues about the formation mechanism of the SBH. These clues, which were hidden in plain sight in old data, were unrecognized by researchers for a long time, until a theory describing the potential distribution and current transport at inhomogeneous MS interfaces was developed.[1] That analytic theory showed what experimental data from polycrystalline MS interfaces long tried to reveal: the SBH was generally inhomogeneous. This was an important step in SBH research, as scientists were forced to confront clear evidence that the SBH depended on the local atomic structure, in direct conflict with what the FLP phenomenon seemed to suggest at the time. Due to the importance of SBH data to be interpreted correctly, this entire chapter is devoted to SBH measurement. We begin with the distribution of potential in the space charge region of an MS interface, followed by carrier transport across the Schottky barrier, and conclude with the many different forms SBH inhomogeneity can manifest itself in experimental data.

6.1 Space Charge Region

The n-type and p-type SBHs are the energy barriers for the transport of electrons and holes, respectively, across the MS interface. Since the CBM at an n-type MS interface is at a significantly higher energy ($\Phi^o_{B,n}$) above the electrochemical potential (FL), dopants near the interface are depleted of electrons. These positively charged dopants bend the CBM, over the width of a space charge region, back to its equilibrium position in the bulk semiconductor. Similarly, the exposed negative charge of acceptors in the depletion region near the interface bends the VBM at a p-type MS interface. The formation of depletion regions at MS interfaces is similar to that at the surface of a semiconductor already discussed. At a uniform MS interface, the actual form of the band bending in the space charge region is unimportant and does not play a significant role in carrier transport. However, when the SBH is inhomogeneous, the current transport may be overwhelmingly controlled by the potential distributions in the space charge region. A thorough understanding of the potential distribution in the space charge region is thus a prerequisite to adequate interpretation of transport phenomenon observed at inhomogeneous MS interfaces. For that reason, the potential distribution in the space charge region of an MS interface is discussed in detail, but only for n-type semiconductors. Similar behavior is expected at the space charge region of p-type MS interfaces.

6.1.1 *Ideal case: Homogeneous SBH*

Since each energy band edge is at a fixed energy with respect to the average potential energy of a semiconductor, band bending is directly linked to variations in the long-range potential energy in the semiconductor. In general, the electrostatic potential is derivable from the charge distribution through Poisson's equation. To provide a backdrop for later discussions, the SBH $\Phi^o_{B,n}$ is here assumed to be uniform, which reduces the electrostatics of the interface to a one-dimensional problem. Assuming that the MS interface is the $z = 0$ plane and the n-type semiconductor, with a doping level of N_D, occupies the space $z > 0$, one writes

$$\frac{\partial^2}{\partial z^2}\bar{V}_{\text{semicond}}(z) = -\frac{\rho(z)}{\varepsilon_S}, \quad z \geq 0, \tag{6.1}$$

where ε_S is the semiconductor permittivity and $\rho(z)$ is the density of local free charge. Being different only by a rigid shift, $\varepsilon_{CBM}(z)/e$ can be inserted in place of $\bar{V}_{\text{semicond}}$ in the above Poisson's equation, with only modifications needed in the boundary conditions supplied. Here, $\varepsilon_{CBM}(z)$ is defined as the CBM position with respect to the FL of the metal.

This is a familiar one-dimensional problem with its solution graphically shown in Fig. 6.1. Since the band diagram here is eventually used to describe current transport under bias, it is drawn with a small external potential, V_a,

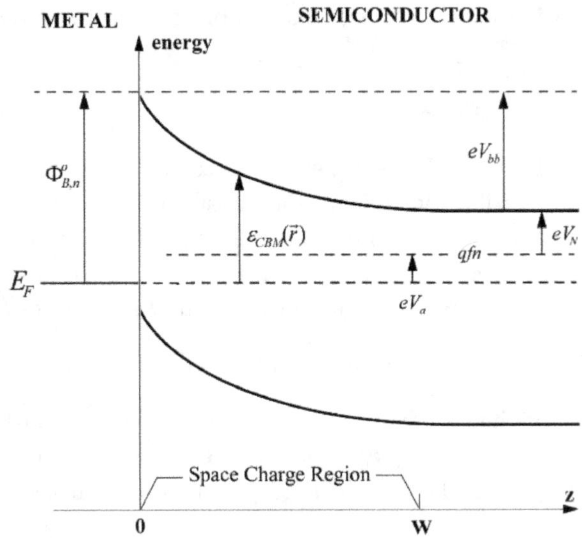

Fig. 6.1. Band diagram for a homogeneous n-type SB with an applied potential V_a.

applied to the semiconductor with respect to the metal to accommodate discussion. Otherwise, the potential distribution here is identical to that for semiconductor surfaces. Again, the rigorous Fermi–Dirac statistics is abandoned in favor of the more transparent depletion approximation, which assumes that all donor impurities are completely unoccupied when the electrochemical potential lies above the donor level and are fully occupied when it is below. The density of free charge becomes a step function with a non-zero (positive) density only within a distance of the MS interface:

$$\rho(z) = \begin{matrix} eN_D, & 0 < z < W \\ 0, & z > W \end{matrix}, \tag{6.2}$$

where W, the depletion region width, is given by

$$W = \lambda_D \sqrt{\frac{2eV_{bb}}{k_B T}}, \tag{6.3}$$

and λ_D is the Debye length of the doped semiconductor:

$$\lambda_D = \sqrt{\frac{\varepsilon_S k_B T}{e^2 N_D}}. \tag{6.4}$$

The CBM position with respect to the FL of the metal is given by

$$\varepsilon_{CBM}(z) = \begin{matrix} eV_{bb}[1-(z/W)^2] + eV_N + eV_a, & 0 < z < W \\ eV_N + eV_a, & z > W \end{matrix} . \tag{6.5}$$

This analytic solution to Poisson's equation captures all important aspects of the potential distribution and current transport close to the MS interface. In Eqs. (6.3) and (6.5), eV_N is the difference between the CBM and the electrochemical potential in a neutral n-type semiconductor given by Eq. (2.6) and eV_{bb} is the total band bending at the MS interface given by

$$eV_{bb} = \Phi^o_{B,n} - eV_N - eV_a. \tag{6.6}$$

Equation (6.5) obviously satisfies the boundary condition on both sides of the space charge region, $\varepsilon_{CBM}(0) = \Phi^o_{B,n}$ and $\varepsilon_{CBM}(z > W) = eV_N + eV_a$, as shown in Fig. 6.1. The total charge per unit area in the space-charge region is

$$Q_{SC} = eN_DW = \sqrt{2\varepsilon_S N_D eV_{bb}}, \tag{6.7}$$

and that the maximum electric field near the MS interface is

$$E_{\max} = \sqrt{\frac{2eN_DV_{bb}}{\varepsilon_S}}. \tag{6.8}$$

For moderately doped semiconductors, this field is only of the order of 10^4–10^5 V cm^{-1}, which leads to a potential change of 1 volt in 50 nm or longer. The small electric fields justify the present semi-classical treatment of band bending. At low temperatures and doping levels, the depletion approximation is quite accurate. The errors may become appreciable when the temperature is high and the carriers at the edge of the depletion region make a contribution. Analytic expressions can still be written down, with the replacement of the V_{bb} in Eqs. (6.5)–(6.8) with $(V_{bb} - k_BT/e)$.

6.1.2 *General case: Inhomogeneous SBH*

When the SBH varies locally at an MS interface, Eq. (6.1) needs to be solved in three dimensions and the above results have to be modified to accommodate lateral inhomogeneity. Whereas the boundary condition for ε_{CBM} at $z > W$ remains unchanged $(= eV_N + eV_a)$, the boundary values at $z = 0$ now reflect the lateral variations of the SBH:

$$\varepsilon_{CBM}(x, y, 0) = \Phi^o_{B,n}(x, y) = \Phi^{oo}_{B,n} + e\delta(x, y), \tag{6.9}$$

where $e\delta(x, y)$ is the difference between the local SBH and a "mean" SBH, $\Phi_{B,n}^{oo}$. For an arbitrary distribution of the SBH, these boundary values can be used to solve for the band bending profile numerically in the space charge region. However, a "dipole layer approach" allows the potential to be accurately described with simple analytic formulae.[1] In this approach, the variations of the potential due to the presence of the SBH inhomogeneity are treated as a perturbation. It is obvious that the potential due to a dipole layer with a varying dipole moment per area, $2\varepsilon_S\delta(x, y)$, should be used as the perturbative term, since it satisfies the Laplace equation and reproduces the desired boundary condition at the MS interface. In other words, the potential in the depletion region of the semiconductor ($0 < z < W$) is approximately described by

$$\varepsilon_{CBM}(x, y, z) = eV_{bb}\left(1 - \frac{z}{W}\right)^2 + eV_N + eV_a + \iint dx_1 dy_1 \frac{ze\delta(x_1, y_1)}{2\pi\left[z^2 + (x_1 - x)^2 + (y_1 - y)^2\right]^{3/2}}, \qquad (6.10)$$

where eV_{bb} is the band bending corresponding to $\Phi_{B,n}^{oo}$. For a rapidly fluctuating SBH or for any isolated variation of the SBH on a small lateral length scale, Eq. (6.10) gives a nearly perfect account of the potential close to the MS interface, as numerical simulations have shown.

The most interesting form of SBH inhomogeneity is the presence of small regions with a low SBH, $\Phi_{B,n}^{oo} - e\Delta$, embedded in an interface with an otherwise-uniform high SBH, $\Phi_{B,n}^{oo}$. Two convenient geometries to consider for the low-SBH regions are small circular patches and narrow semi-infinite strips. For a low-SBH circular patch with a radius, R_o, small compared with W, the potential at any point (ρ, z), may be written down using Eq. (6.10). Along the z-axis ($\rho = 0$), the potential has an analytic form:

$$\varepsilon_{CBM}(\rho = 0, z) = eV_{bb}\left(1 - \frac{z}{W}\right)^2 + eV_N + eV_a - e\Delta\left[1 - \frac{z}{(z^2 + R_o^2)^{1/2}}\right]. \qquad (6.11)$$

An important aspect of the potential distribution described by Eq. (6.11) is the likelihood of potential pinch-off. "Pinch-off" is a term often used to describe the operation of a field-effect transistor. For the inhomogeneous MS interface, a low-SBH area is said to be pinched off if the majority carriers originating from outside the space charge region need to go over a potential barrier higher than the band-edge position at the MS interface in order to reach the MS interface. Potential contours near a pinched-off region (SBH = 0.5 eV) are shown in Fig. 6.2 to illustrate that the potential rises initially rather than monotonously decreases.[2] Obviously, the reason for potential pinch-off is the influence from the surrounding regions with high SBH. Note that the potential distribution is in the shape of a "saddle-point" near the location of the minimum potential barrier for electron transportation to a low-SBH patch. The potential described by Eq. (6.11) is plotted as solid lines in Fig. 6.3 for assorted values of R_o, Δ, W, and V_a. The agreement with numerical solutions obtained from solving Poisson's equation, shown as circles, is excellent. Whenever the potential on the z-axis has a positive slope at small z, the potential has to go through a maximum before descending at large z, i.e. the potential is pinched off. Differentiating Eq. (6.11) with

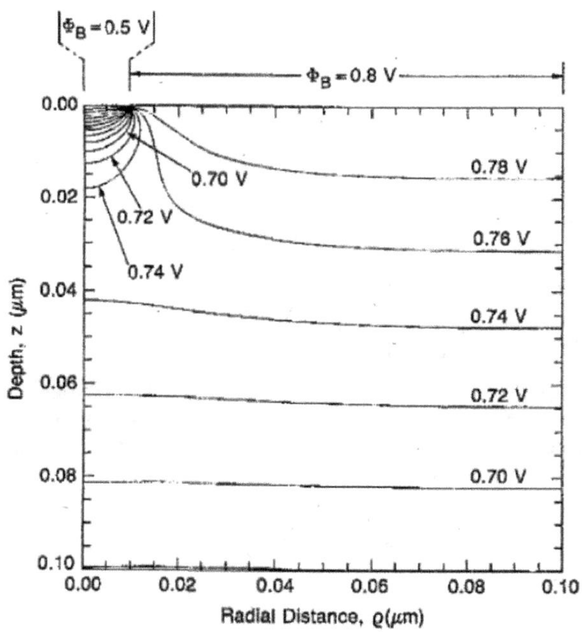

Fig. 6.2. Potential contours at zero applied bias along a slice of constant azimuthal angle for a diode with cylindrical symmetry. The semiconductor is assumed to be n-type Si with doping $1 \times 10^{15}\,\text{cm}^{-3}$. The SBH of the high barrier region is 0.8 eV and the low-SBH circular patch has a radius of $R_o = 0.01$ mm and an SBH of 0.5 eV.

Source: Reprinted from Sullivan *et al.*[2] with the permission of AIP Publishing.

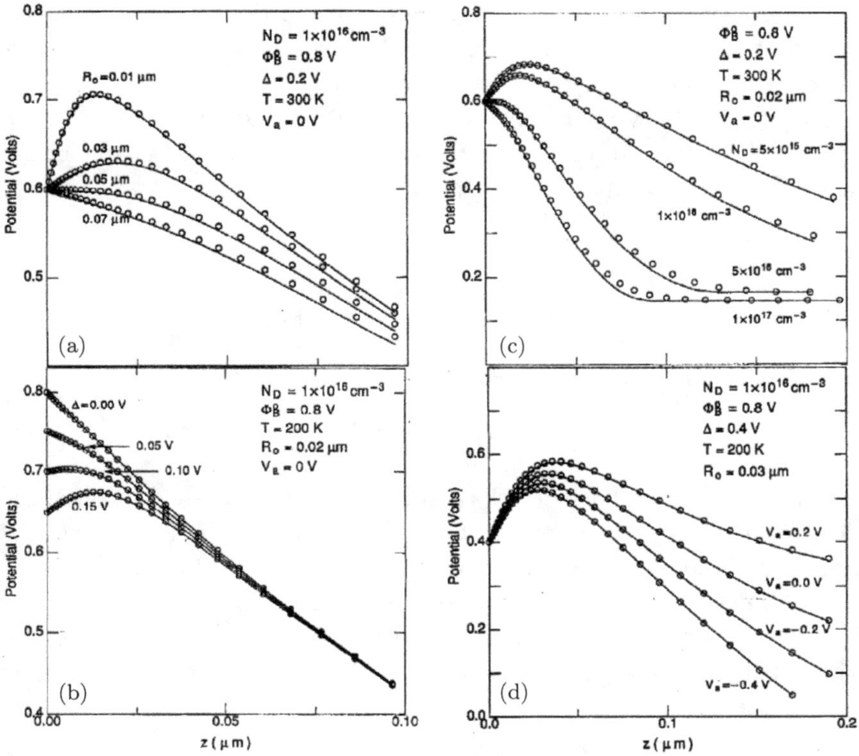

Fig. 6.3. Numerically determined potential distribution (shown in circles) along $\rho = 0$ of the low-SBH region. The potential based on the analytic theory is shown as the solid lines: (a) The radius of the low SBH patch R_o is varied; (b) the SBH difference Δ is varied. Note that with the same Δ, only the smaller patches are pinched off, whereas for a fixed R_o, only patches with larger Δ are pinched off. (c) The n-type doping level is varied, and (d) the applied bias voltage is varied.

Source: Reprinted from Sullivan *et al.*[2] with the permission of AIP Publishing.

respect to z and setting positive the derivative at $z = 0$, we arrive at the condition for pinch-off as

$$\frac{\Delta}{V_{bb}} > \frac{2R_o}{W}, \tag{6.12}$$

for a circular patch. This condition shows that potential pinch-off should be a common phenomenon when the semiconductor doping level is low and/or the low-SBH patch is small. The dependence of the saddle-point potential on the applied bias, illustrated in Fig. 6.3(d), is particularly worth noting. As discussed in the following, this dependence significantly affects the transport properties at inhomogeneous SB junctions and is responsible for a host of "anomalous" phenomena in I–V measurements. To first order, the potential does not depend on temperature.

For a strip of low-SBH, which is semi-infinite along the y-axis and has a width of L_o in the x-direction, the CBM position in the space charge region may also be expressed analytically:

$$\varepsilon_{CBM}(x,y,z) = eV_{bb}(1-\frac{z}{W})^2 + eV_N + eV_a + \frac{e\Delta}{\pi}\tan^{-1}\frac{|x|-L_o/2}{z} - \frac{e\Delta}{\pi}\tan^{-1}\frac{|x|+L_o/2}{z}. \quad (6.13)$$

As before, the condition for pinch-off may be obtained:

$$\frac{\Delta}{V_{bb}} > \frac{\pi L_o}{2W}. \quad (6.14)$$

The dipole layer approach described above gives an accurate account of the electric potential in the presence of SBH inhomogeneity, irrespective of the size of the inhomogeneity. Equation (6.13) is plotted in Fig. 6.4 to give a 2-D view of a saddle point in front of a low SBH strip.

When the dimension of low SBH patches or strips is small compared with the depletion length, one may further simplify the math by replacing

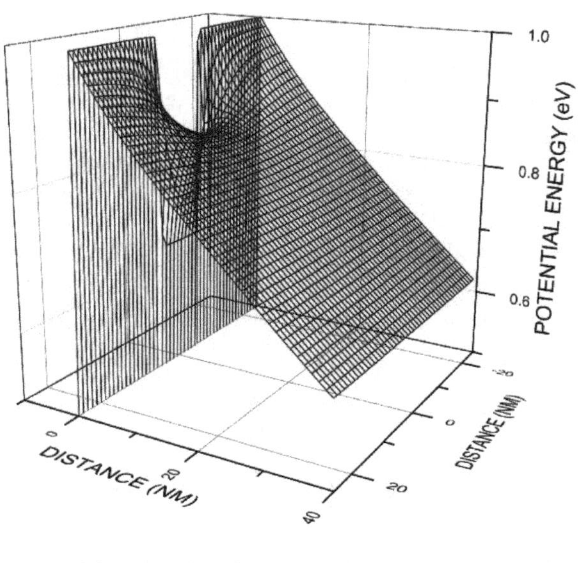

Fig. 6.4. The potential energy distribution of the space charge region near a low-SBH strip with the following characteristics: $\Phi^{oo}_{B,n} = 1.00\,\text{eV}$, $\Delta = 0.31$ eV, $L_O = 10$ nm, and W= 200 nm.

the spatially extended dipole patch/strip with a point dipole or a dipole line. For a circular patch with radius R_o, the total dipole moment is $2\varepsilon_S \pi R_o^2 \Delta$. The potential around the patch, except near the core of the dipole, is given by

$$\varepsilon_{CBM}(\rho, z) = eV_{bb}\left(1 - \frac{z}{W}\right)^2 + eV_N + eV_a - \frac{eV_{bb}\Gamma^3 z W^2}{(\rho^2 + z^2)^{3/2}}, \tag{6.15}$$

where Γ is a dimensionless quantity which measures the "strength" of a given patch and is defined by

$$\Gamma^3 = \frac{R_o^2 \Delta}{4\eta V_{bb}^2}, \tag{6.16}$$

where $\eta = \varepsilon_S/(eN_D)$. Numerical simulation of Fig. 6.5 shows that the point dipole approximation accurately describes the potential around the saddle-point when $R_o \ll W$, although it is not accurate immediately next to the MS interface.[3] This approximation is nevertheless appropriate for a description of the expected I–V behavior because only the potential distribution near the saddle-point is important for electron transport. One notes that Eqs. (6.15) and (6.16) are not limited to low-SBH patches with a circular geometry, as Γ may be defined for any small patch with an irregular

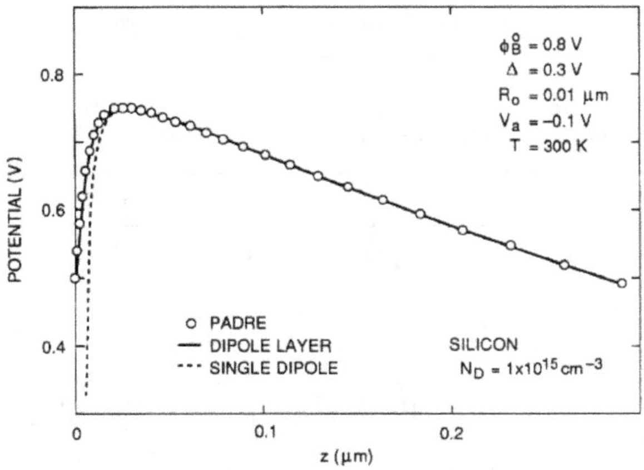

Fig. 6.5. CBM position, with respect to the FL, in front of a low-SBH circular patch. Circles are numerical solutions to Poisson's equation, while the solid curve is the analytical prediction from the dipole layer approach. The dashed curve is from the point-dipole approximation.

Source: Reprinted from Tung.[3] Copyright (1992) by the American Physical Society.

shape and/or a varying SBH by an integral:

$$\Gamma^3 = \frac{-1}{4\pi\eta V_{bb}^2} \iint_{\text{patch}} dx_1 dy_1 \delta(x_1, y_1). \tag{6.17}$$

Since the point-dipole approximation lumps the effect of a dipole patch with a finite size into a point source, it errs on slightly overestimating the depth of the potential "valley" at the saddle point and slightly underestimating the valley width.

When pinch-off occurs, the location of the saddle-point, at $(0, z_o)$, may be obtained by differentiating Eq. (6.15) along the z-axis and setting the derivative to zero. When $\Gamma \ll 1, z_o$ is found to be, approximately,

$$z_o \approx \Gamma W, \tag{6.18}$$

and the saddle-point potential is

$$\varepsilon_{CBM}^{sad-p} = \varepsilon_{CBM}(0, z_o) \approx eV_{bb}(1 - 3\Gamma) + eV_N + eV_a. \tag{6.19}$$

This represents the minimum potential barrier separating the low-SBH patch from the neutral semiconductor.

The dipole layer due to a semi-infinite strip with a width L_o, may be approximated by a "dipole line" with an infinitesimal width. The dipole moment per unit length is $2\varepsilon_S L_o \Delta$ and the potential may be written down

Table 6.1. Relevant parameters for an isolated low-SBH region with a local SBH of $\Phi_{B,n}^{oo} - e\Delta$, surrounded by regions with a SBH of $\Phi_{B,n}^{oo}$.

Geometry	Circular Patch	Semi-infinite Strip
Dimension	Radius $= R_o$	Width $= L_o$ Length $= L_{\text{strip}} \gg L_o$
Region Parameter	$\gamma = 3\left(\frac{R_o^2\Delta}{4}\right)^{1/3}$	$\omega = 2\left(\frac{\sqrt{2}L_o\Delta}{\pi}\right)^{1/2}$
Space Charge Parameter	$\Gamma = \left(\frac{R_o^2\Delta}{4W^2 V_{bb}}\right)^{1/3}$	$\Omega = \left(\frac{L_o\Delta}{2\pi W V_{bb}}\right)^{1/2}$
Saddle-Point Position	$(\rho = 0,\ z = \Gamma W)$	$(x = 0,\ y,\ z = \Omega W)$
Effective Area, A_{eff}	$\frac{4}{3}\pi\lambda_D^2\Gamma$	$\sqrt{\pi\Omega}\lambda_D L_{\text{strip}}$
Effective SBH, Φ_{eff}	$\Phi_{B,n}^{oo} - 3e\Gamma V_{bb}$	$\Phi_{B,n}^{oo} - e4\Omega V_{bb}$
Ideality Factor, n	$\approx 1 + \Gamma$	$\approx 1 + \Omega$

as follows:

$$\frac{\varepsilon_{CBM}(x,y,z)}{e} = V_{bb}\left(1-\frac{z}{W}\right)^2 + V_N + V_a - \frac{2V_{bb}\Omega^2 W z}{x^2+z^2}, \tag{6.20}$$

where Ω is a dimensionless quantity which measures the strength of the strip, defined in Table 6.1. Using procedures similar to those described above, the location and the potential of the saddle point may also be found for the strip geometry.

6.2 Electrical Measurement of Uniform SBH

Due to the apparent FLP phenomenon, SB research traditionally assumed that the SBH was uniform. Theories and techniques for SBH measurement were developed based on a uniform SBH. The experimental demonstration of the near ubiquitous inhomogeneity of SBH in MS interfaces routinely studied and the invention of spatially resolved measurement techniques only came after the measurement methodologies for homogeneous SBH had been firmly established. These traditional techniques and how SBHs are extracted from the data are discussed here. Obviously, the interpretations of experimental results from these techniques need to be modified and broadened for inhomogeneous SBH, as discussed in the following section along with spatially resolved techniques that can directly map the SBH distribution of an MS interface.

6.2.1 *Current–Voltage method*

One of the most widely used techniques to measure the SBH is the current–voltage (I–V) technique. In particular, the forward-bias portion of the I–V characteristics has been used to deduce the magnitude of the SBH. We assume that the SBH is homogeneous, which can be roughly defined as one that may vary laterally on a scale longer than the depletion region length. Under this condition, the current transport at one region of the interface is independent of that in other regions. Principally responsible for the current flow at the majority of Schottky diodes on non-degenerate semiconductors is the thermionic emission process, as found in various studies. This is a process treated in detail for the unidirectional emission of electrons from a hot metal surface to vacuum in Chapter 3, where the energy barrier is the work function. When applied to a solid interface, the thermionic emission process should be recognized as operative in both directions because carriers are present on both sides. With a uniform energy barrier of Φ_A for carrier transport in one direction, the thermionic emission in thermal equilibrium over the barrier is

$$I_{\text{uni-direc.}} = AA^*T^2 \exp\left(-\frac{\Phi_A}{k_B T}\right), \tag{6.21}$$

where A is the cross-sectional area of the barrier, A^* is the Richardson's constant, k_B is the Boltzmann constant, and T is the absolute temperature, as shown in Fig. 6.6. Applying this theory to a uniform Schottky barrier with an applied bias, one notices that the transport of electrons in the forward direction, i.e. from the n-type semiconductor to the metal, has a barrier height of $\Phi_{B,n} - eV_a$, while the electrons in the metal need to overcome a higher barrier of $\Phi_{B,n}$ in the reverse direction. Adding these two components of current, the net flow of charge carriers at the MS interface is

$$I_{SB}(V_a) = AA^*T^2 \exp\frac{-\Phi_{B,n}}{k_B T}\left[\exp\left(\frac{eV_a}{k_B T}\right) - 1\right] \tag{6.22}$$

At large forward bias, the numeral 1 in the square bracket can be ignored, and the current essentially displays an exponential dependence on the applied bias.

While the thermionic emission theory is regularly used to interpret current transport data, an implicit assumption for its applicability to the SBH should be pointed out. The energy barrier for electronic transport to the semiconductor is abrupt, and the use of Eq. (6.21) is well justified. However, the energy distribution of the electrons transported from the

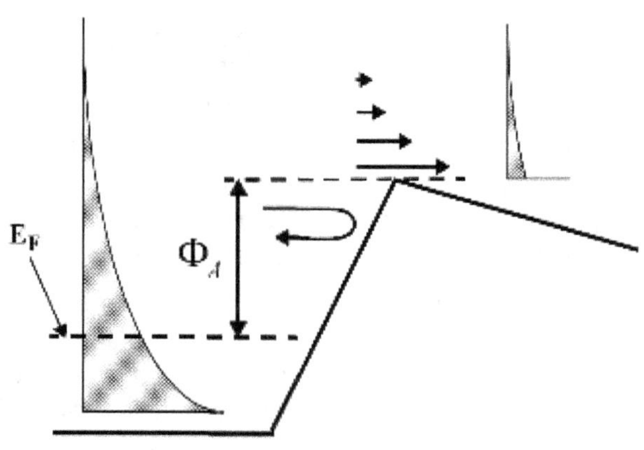

Fig. 6.6. Schematic of the thermionic emission process of electrons from a reservoir (left) over a barrier with a barrier height of Φ_A. Note that this process is independent of the shape of the barrier and that the barrier height is measured from the Fermi level.

semiconductor to the metal is based on the electrochemical potential position in the neutral semiconductor, according to Eq. (6.22). Implicit in such an equation is that the scattering of electrons in the space charge region is insignificant, so their velocity distribution was still governed by the electrochemical potential of their origin. This does not seem likely to hold for a lightly doped semiconductor with a long depletion width. However, the bottom line is that Eq. (6.22) has been found in experiment to adequately describe current transport for all kinds of MS interfaces. Therefore, it is accepted here as a working basis for our discussions of electronic transport at homogeneous and inhomogeneous SBs. The barrier height that controls thermionic emission of carriers, $\Phi_{B,n}$, is actually slightly less than the nominal SBH, $\Phi^o_{B,n}$, because of image force lowering:

$$\Phi_{B,n} = \Phi^o_{B,n} - \delta\Phi_{image} = \Phi^o_{B,n} - \left[\frac{e^3 N_D V_{bb}}{8\pi^2 \varepsilon_S^3}\right]^{\frac{1}{4}}. \tag{6.23}$$

As shown in Fig. 6.7, the maximum of the energy barrier for thermionic emission is not at the MS interface but a small distance inside the semiconductor. This "Schottky lowering" effect due to image force is already discussed at length in regard to work function formation at metal surfaces. Due to the image force lowering, even uniform $\Phi_{B,n}$ has a weak dependence on the interface electric field. An implicit assumption of Eq. (6.22) is that the entire applied bias is maintained across the MS interface. In real devices,

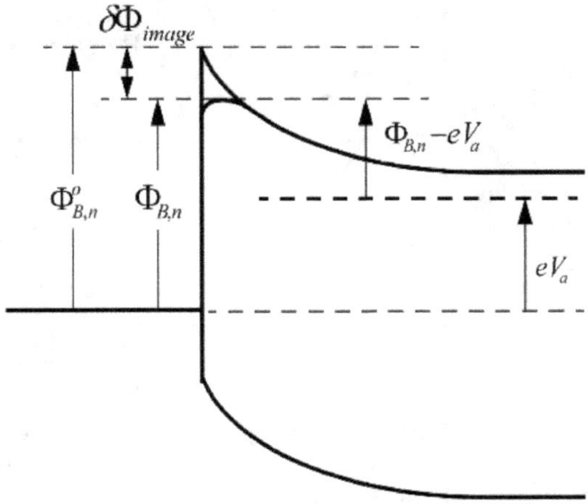

Fig. 6.7. Schematic of the image force lowering (Schottky lowering) at a Schottky barrier, drawn under an applied forward bias condition.

the applied bias dissipates in voltage drops across the Ohmic contact and the bulk semiconductor. Taking into account the effect of series resistance, an Ohmic drop of $I_{SB}R_S$ should be subtracted from the applied bias to render the actual voltage drop across the MS interface. One gets

$$I_{SB}(V_a) = AA^*T^2 \exp\left(-\frac{e\Phi_{B,n}}{k_BT}\right)\left[\exp\left(\frac{eV_a - eI_{SB}(V_a)R_S}{k_BT}\right) - 1\right], \tag{6.24}$$

where the series resistance R_S is independent of the applied bias. Under this condition, the effect of series resistance can be easily deconvolved from the experimental I–V curves by the use of various ingeniously designed plots.

On heavily doped semiconductors, tunneling of carriers at energies lower than the full barrier height can lead to a current that is comparable to, or even far exceeds, thermally activated transport of carriers over the barrier. Since the tunneling probability for a potential barrier increases sharply with electron energy, and the carrier density decreases exponentially with electron energy, the energy distribution of carriers that tunnel through the barrier is narrowly peaked, as schematically shown in Fig. 6.8. Tunneling thus reduces the effective Schottky barrier for carrier transport by an amount that depends on the shape of the energy barrier and the

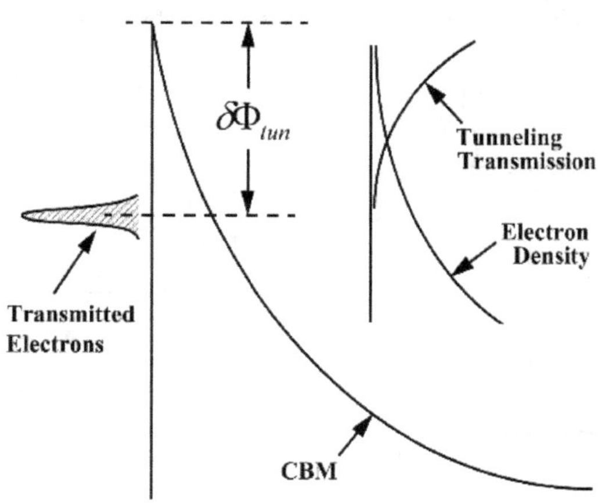

Fig. 6.8. Schematic of the thermionic–field emission (TFE) process. The product of the electron density and the transmission coefficient leads to a sharply peaked energy profile of electrons tunneling through the potential barrier. The runneling process leads to a reduction in the effective barrier height of $\delta\Phi_{\text{tun}}$.

temperature. The combined mechanism of partially thermally activation and tunneling is known as thermionic–field emission (TFE).[4,5] The total TFE current has been calculated within the WKB approximation to be

$$I = I_{S-TFE}\left[\exp\left(\frac{eV_a}{E_o}\right) - 1\right], \tag{6.25}$$

where I_{S-TFE} is the saturation current for the TFE process and E_o is given by

$$E_o = E_{oo}\coth\left(\frac{E_{oo}}{k_B T}\right). \tag{6.26}$$

The parameter E_{oo} is a property of the bulk semiconductor,

$$E_{oo} = \frac{e\hbar}{2}\left(\frac{N_D}{\varepsilon_S m^*}\right)^{\frac{1}{2}}, \tag{6.27}$$

and it is an indicator of how efficient tunneling is in a semiconductor. E_{oo} can be thought of as the amount of band bending for a semiconductor that would yield a tunneling probability of ~37% (=e^{-1}) at the base of the barrier. Since the actual band bending is much larger than E_{oo}, some degree of thermal activation is always more efficient than direct tunneling. A comparison of E_{oo} with $k_B T$ determines whether thermionic emission or tunneling will be more efficient near the very top of the energy barrier, i.e. whether to go over or go through the last few E_{oo}'s (or $k_B T$'s) of the Schottky barrier. As a rule of thumb, the effect due to tunneling is small at room temperature for common semiconductors with a doping level of 1×10^{17} cm^{-3} or less ($E_{oo} \sim 3$ meV), whereas tunneling will be efficient for semiconductors with a doping level higher than 1×10^{18} cm^{-3} ($E_{oo} \sim 10$ meV). At lower temperatures, the relative importance of tunneling increases.

Current–voltage measurements are mainly carried out to determine the magnitude of the SBH. As different current transport mechanisms predict the I–V characteristics slightly differently, experimentally observed I–V curves have also been analyzed with the intention of determining the transport mechanism responsible. One of the commonly used parameters, that seemingly reveals the transport mechanism directly, is the ideality factor of the I–V traces. This factor, n, is defined as the inverse slope of the semilogarithmic I–V curve, normalized by the thermionic emission process:

$$n = \frac{e}{k_B T}\left[\frac{\partial(\ln I_{SB})}{\partial V_a}\right]^{-1},$$
$$= \left[1 - \left(\frac{\partial \Phi_{B,n}}{e\partial V_a}\right)\right]^{-1}. \tag{6.28}$$

Even though n is named the "ideality" factor, its numerical value actually reflects the departure from ideality. The larger n is, the less "ideal" the SB is! Most I–V curves can be fitted with a single n and expressed as (with the inclusion of the series resistance effect)

$$I_{SB} = I_{Sat.}\left[\exp\left(\frac{eV_a - eI_{SB}R_S}{nk_BT}\right) - 1\right], \tag{6.29}$$

which can be used through Eq. (6.24) to deduce the magnitude of the SBH. When the effect of series resistance is small, a logarithmic current vs. voltage plot, at any fixed temperature, usually contains a wide enough linear section which can be used to deduce an ideality factor as well as extrapolate to zero bias to yield the saturation current. When the series resistance is significant, it is not always possible for a linear region to be identified from the I–V plot, especially at elevated temperatures. Various schemes have been proposed which greatly facilitate the analysis of I–V traces. The use of the small bias portion of the I–V trace,[6] the use of a special function of the current and the bias,[7] and the use of two temperature measurements[8] have all made the job easier. When the incremental conductance is available from digitized data, there are several plots which allow n and R_S to be determined graphically, as illustrated in Fig. 6.9.[9] In these plots, I_d is the diode current, G the small signal conductance ($G = \partial I_d/\partial V_a$), and R_{dr} is the differential resistance ($R_{dr} = G^{-1}$). It is important to remember the two implicit assumptions of Eq. (6.29), namely, the SBH is homogeneous and the series resistance is independent of the bias.

6.2.2 *Capacitance–Voltage technique*

A convenient technique to measure the SBH is the capacitance–voltage (C–V) technique. The charge per unit area due to exposed dopants is given by

$$Q_{SC} = eN_DW = \sqrt{2\varepsilon_S N_D(eV_{bb} - k_BT)}, \tag{6.30}$$

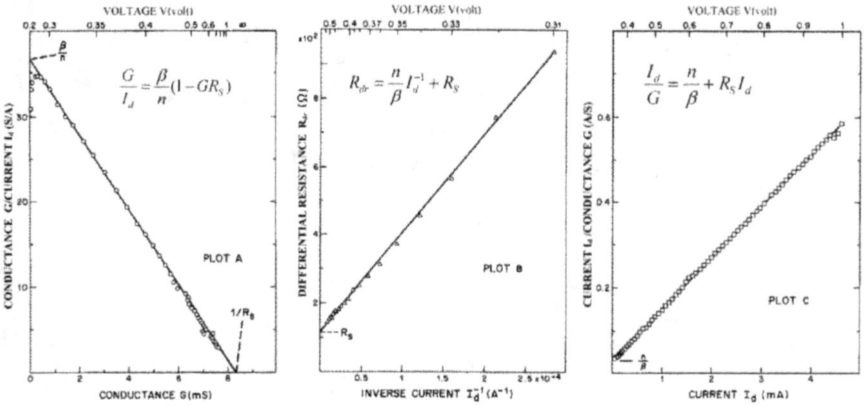

Fig. 6.9. The same set of experimental I–V data from a PtSi/Si Schottky diode, corrected for shunt resistance, is used to three different plots. The series resistance and the ideality factor of the diode can be deduced from each of these plots.

Source: Reprinted from Werner,[9] with permission from Springer Nature.

for a homogeneous SBH, when the effect of the charge distribution at the edge of the depletion region is included. The small signal capacitance for the Schottky diode is

$$C = -\frac{\partial Q_{SC}}{\partial V_a} = \sqrt{\frac{e^2 \varepsilon_S N_D}{2(eV_{bb} - k_B T)}}. \tag{6.31}$$

Equation (6.31) shows that a plot of C^{-2} as a function of the applied bias, known as a Mott–Schottky plot, can be used to deduce the doping profile of the semiconductor. In case the doping profile is homogeneous throughout the space charge region, a C–V plot should yield a straight line in the reverse bias region with a slope of $-2/e\varepsilon_S N_D$, as shown in Fig. 6.10 for epitaxial $NiSi_2$/Si(111) interfaces of different orientations and doping levels.[10] Each intercept with the bias voltage axis, known as the built-in potential eV_{bi}, is the total band bending at zero bias. The SBH can be deduced through the following relationship:

$$\Phi^o_{B,n} = eV_{bi} + eV_n + k_B T. \tag{6.32}$$

The C–V technique gives a convenient measure of the average SBH when the SBH is inhomogeneous. The capacitance signal comes from the modulation of the total space charge as a function of bias. For an inhomogeneous SB, the total space charge is equal to that for a uniform SB with a height equal

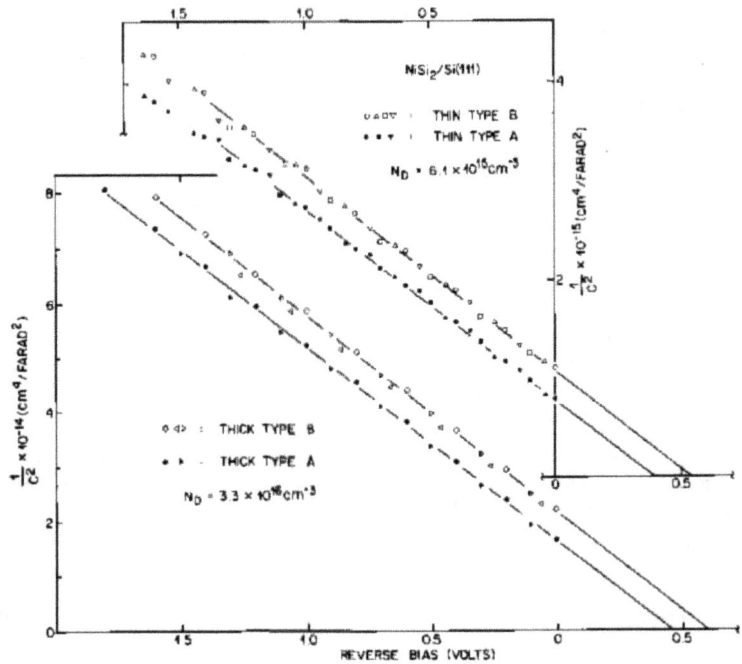

Fig. 6.10. Capacitance per unit area of $NiSi_2$-Si(111) junctions, as a function of reverse bias, measured at 1MHz.

Source: Reprinted from Tung,[10] with permission from Elsevier.

to the mathematical average of the inhomogeneous SBH. To see this result, one notes that the total space charge is equal to the charge in the metal, which can be integrated using Eq. (6.10)

$$
\begin{aligned}
Q_{SC} &= \frac{1}{A}\iint dx\,dy \left(\frac{\varepsilon_S}{e}\right) \frac{\partial \varepsilon_{CBM}(x,y,z)}{\partial z}\Bigg]_{z\to 0} \\
&= \frac{1}{A}\iint dx\,dy \sqrt{2\varepsilon_S N_D(\Phi_{B,n}^{oo} - eV_N - eV_a)} \\
&\quad + \frac{1}{A}\iint dx\,dy \iint dx_1 dy_1 \frac{\delta(x_1,y_1)}{2\pi\left[(x_1-x)^2+(y_1-y)^2\right]^{3/2}} \\
&= \sqrt{2\varepsilon_S N_D(\Phi_{B,n}^{oo} - eV_N - eV_a)}.
\end{aligned} \tag{6.33}
$$

The second term on the second line can be seen to vanish, neglecting minor edge effect, by reversing the integration sequence and making use of the fact that $\int \delta(x,y)dxdy = 0$. Therefore, the total space charge of an inhomogeneous SB is controlled only by its average SBH.

6.2.3 *Photo-response and photoemission*

As shown earlier for the heterojunctions, the photo-response technique can be used to measure the SBH. Monochromatized photons with energies greater than the SBH, and less than the band gap, can lead to photocurrent by exciting carriers in the metal from near the Fermi level to above the energy barrier. As discussed, the short-circuit current flowing in the SB diode depends on the photon energy, as governed by the Fowler equation. For photons with energies at least a few k_BT 's larger than the SBH, the photoelectric current can be shown to roughly scale with the square of the excess energy:

$$I_{ph} \propto (h\nu - \Phi_{B,n})^2. \tag{6.34}$$

A plot of the square root of the photoelectric current against the photon energy thus gives a straight line with an intercept with the energy axis at the SBH, as shown in Fig. 6.11 for diodes fabricated on *p*-type GaAs.[11] Since current flow is being monitored, the SBH measured is subject to the Schottky lowering effect. To reduce effects due to noise and leakage, PR is usually conducted in a.c. mode with a chopped light source and a lock-in amplifier. More often, radiation enters from the semiconductor side of the sample. However, for thin layers of metal (<30 nm), light can also enter from the front side with enough intensity to reach the MS interface. The magnitude of the SBH and the size of the band gap limit the photon energies to use, which also explains why PR cannot be used to measure large SBHs that approach the band gap. In cases where the MS interface is made up of areas with several different local SBHs, the PR signal may contain linear segments which may be used to identify each of these individual barrier heights.[12] For a more complicated distribution of inhomogeneous SBH, the PR signal is nonlinear and, hence, unreliable for SBH determination.

Ever since the mid-1970s, core-level photoemission spectroscopy (PES) has been a popular technique surface scientists use to study SBH formation at semiconductor surfaces. A core level is always at a fixed energy with respect to the average potential energy of a semiconductor, the same way the band edges are. Therefore, any observed movement of the core level with respect to the Fermi level also reflects an identical movement in the band edges. PES offers a unique opportunity to study the development

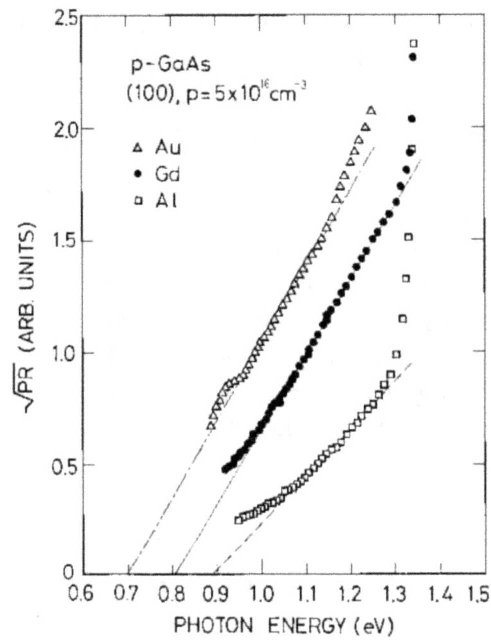

Fig. 6.11. Internal photo-response plots for Au, Gd, and Al diodes on p-type GaAs. *Source*: Reprinted from Okumura and Tu,[11] with the permission of AIP Publishing.

of the band bending as a function of the coverage of metal deposited on the semiconductor surface, all under UHV conditions. Monochromatic x-ray or soft x-ray, often from a synchrotron radiation source, excites core electrons to high enough energy to escape into the vacuum. A schematic drawing of the surface photoemission process is shown in Fig. 6.12. The thickness of the metal layer has to be kept small for photoelectrons from the semiconductor to be detected. The change in the band bending is usually a few hundred meV, which is the minimum energy resolution needed to study the formation of the SBH. Photon energy can be chosen to keep the kinetic energy of core electrons of interest low to achieve bulk sensitivity and good energy resolution at the same time. Measurements of band bending with (sub-)monolayers of metal may have issues with lateral inhomogeneity due to islanding of metal on the surface. Furthermore, intense light radiation generates electron–hole pairs that locally charge the isolated metal clusters and induce significant changes in the measured band bending. This surface photovoltage effect is most prominent for wide-gap semiconductors and/or at low temperatures.[13]

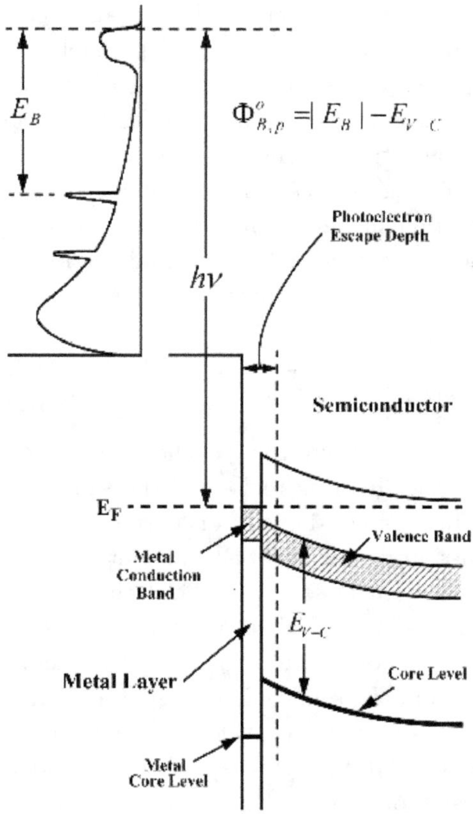

Fig. 6.12. Schematic of the principle of SBH measurement by photoemission spectroscopy. The binding energy of a core level of the semiconductor is measured at the MS interface. This, along with the E_{V-C} measured from a bulk semiconductor sample, can be used to determine the SBH.

6.3 Measurement of Inhomogeneous SBH

6.3.1 *Current–Voltage analysis: Parallel conduction model*

The junction current of an inhomogeneous SB was originally proposed to be described by a parallel conduction model.[14] In such a model, the junction current is a linear sum of the thermionically emitted contributions from all individual areas, namely,

$$I(V_a) = A^*T^2 \left[\exp\left(\frac{eV_a}{k_BT}\right) - 1\right] \sum_i A_i \exp\left(-\frac{\Phi_i}{k_BT}\right), \qquad (6.35)$$

where A_i and Φ_i are, respectively, the area and the SBH of the i-th "patch". The main assumption behind the parallel conduction model is the lack of interaction between different areas of the same MS interface. When such a condition holds, transport theories for homogeneous SBH can be applied individually to different areas of an inhomogeneous SB. From earlier discussions of the potential distributions for inhomogeneous SBH, the parallel conduction model is appropriate only when the SBH varies laterally on a scale greater than the average depletion width. This makes the parallel conduction model inappropriate for the vast majority of SBs routinely studied.

6.3.2 *Current–Voltage analysis: Saddle point model*

What has been missing from the parallel conduction concept is the interaction between neighboring regions of the same interface. When the actual band bending is considered for current transport, not only is the statistical distribution of the SBH magnitude at the MS interface important, but also the length scale with which the SBH varies. As discussed above, rapid variations of the SBH at an inhomogeneous MS interface are smoothed out in the space charge region. This masks the low-SBH areas from direct access from the semiconductor side. Potential pinch-off causes the current transport to deviate considerably from the parallel conduction model. Numerical simulations[2] gave excellent support to expressing the total current as a sum of currents flowing in each patch[3]:

$$I(V_a) = A^*T^2 \left[\exp\left(\frac{eV_a}{k_BT}\right) - 1\right] \sum_i A_{i,\text{eff}} \exp\left(-\frac{\Phi_{i.\text{eff}}}{k_BT}\right), \qquad (6.36)$$

where the effective SBH, $\Phi_{i,\text{eff}}$, and the effective area of a patch, $A_{i,\text{eff}}$, depend on whether the patch is pinched off or not. For patches that are not pinched off, the SBHs at the interface and the actual sizes are used. However, $\Phi_{i,\text{eff}}$ and $A_{i,\text{eff}}$ assume the values of the saddle-point potential and the effective area listed in Table 6.1 for a patch when pinch-off occurs. A fundamental difference between Eqs. (6.35) and (6.36) is that the $A_{i,\text{eff}}$'s and the $\Phi_{i,\text{eff}}$'s depend on the bias, the temperature, the doping level, etc., while the A_i's and the Φ_i's do not. It is these dependencies that are responsible for a wealth of interesting I–V characteristics routinely observed. Most notably, the increase of the saddle point potential with forward bias, shown in Fig. 6.3(b), leads to an ideality factor greater than 1 for the component of current which flows through this saddle-point.

Specifically, for the current flowing to a low-SBH patch,

$$n \approx 1 + \Gamma = 1 + \frac{\gamma}{3\eta^{1/3} V_{bb}^{2/3}}. \tag{6.37}$$

A diode dominated by pinched-off low-SBH patches will display an ideality factor greater than 1.

At a real MS interface, SBH variations may occur at a variety of shapes and spatial frequencies. The observed I–V behavior from an inhomogeneous diode depends very much on the SBH distribution. A random variation of the SBH at an MS interface may be simulated by a statistical distribution of the patch/strip characteristics. For instance, one may assume a Gaussian distribution of the patches:

$$P(\gamma) = \frac{c_1}{\sigma_{SBH}} \exp\left[-\frac{\gamma^2}{2\sigma_{SBH}^2}\right] \quad \gamma > 0, \tag{6.38}$$

where $P(\gamma)\mathrm{d}\gamma$ is the density of circular patches with a characteristic γ between γ and $\gamma + \mathrm{d}\gamma$. In Eq. (6.38), σ_{SBH} is the standard deviation of the normal distribution. The current flowing in such composite inhomogeneous SB diodes is made up of two components. One is the current over the entire diode, which is governed by a uniform SBH of $\Phi_{B,n}^{oo}$, and the other is an additional current due to the presence of the low-SBH patches.[3] The combined effect of all the low-SBH patches is as if there were a big low-SBH region in the diode with an effective SBH of

$$\Phi_{\mathrm{EFF}} = \Phi_{B,n}^{oo} - \frac{e\sigma_{SBH}^2 V_{bb}^{2/3}}{2\eta^{2/3} k_B T} \tag{6.39}$$

and with an "overall ideality factor" of

$$n_{\mathrm{EFF}} \approx 1 + \frac{e\sigma_{SBH}^2}{3\eta^{2/3} V_{bb}^{1/3} k_B T}. \tag{6.40}$$

Equations (6.39) and (6.40) are fascinating. They show that, even though the effective SBH and the ideality factor of each individual patch are both temperature-independent, the collective current from a distribution of patches is characterized by an effective SBH and an ideality factor that are explicitly temperature-dependent! Analytic expressions similar to those derived above have also been obtained for the strip geometry. The main difference between the two geometries is in the dependence of the effective SBHs on bias: $\Phi_{B,n}^{oo} - \Phi_{\mathrm{eff}}$ is proportional to $V_{bb}^{1/3}$ for the patch geometry and proportional to $V_{bb}^{1/4}$ for the strip geometry. Hence, the current for both patch and strip geometries may be expressed in a generalized form:

$$I_{total} = A^* A T^2 f(\beta, V_d)[\exp(\beta V_a) - 1] \times \exp(-\beta \Phi_{Bn,o}^{mean} + \beta \kappa_1 V_d^{\xi/2} + \beta^2 \kappa_2 V_d^{\xi}), \tag{6.41}$$

where β is $(k_B T/e)^{-1}$, f is a slowly varying function in comparison to the exponential function, and the exponent ξ is 2/3 and 1/2, respectively, for the patch and strip geometry. The constants, κ_1 and κ_2, depend on the doping level and the distribution of SBHs.

So far, the semiconductor is implicitly assumed to be ultrathin to avoid the effect of series resistance and keep the equations uncluttered. In real SBH measurements, the effect of ohmic drop can be significant and can vary locally for current transport at inhomogeneous SBH. Since the current density is laterally inhomogeneous, the strategy of replacing $\exp(\beta V_a)$ in the current equation with $\exp(\beta V_a - \beta I R_S)$, appropriate for uniform SBH and used in Eq. (6.24), cannot be applied to inhomogeneous SBH. In general, the current densities, and thus the Ohmic drops, to different patches may be drastically different. This effect is particularly pronounced for low-SBH patches that are pinched off, for which the spreading resistances associated with the spatial confinement of current flow are largely responsible for the Ohmic effect. Furthermore, since the current flow is most congested near the saddle point, the effective area of the saddle-point, $A_{i,\text{eff}}$, rather than the actual size, A_i, of the low-SBH patch should be used to calculate the spreading resistance for each patch. The spreading resistance of a circular disc is approximately

$$R_{sp} = \frac{\rho}{2d}, \tag{6.42}$$

when its diameter, d, is much smaller than the sample thickness.[15] In (6.42), ρ is the specific resistivity of the semiconductor. Incorporating effects due to series resistance, the current of an inhomogeneous SB can be expressed as

$$I(V_a) = A^* T^2 \sum_i^{\text{“not\_p–o”}} A_i \exp(-\beta \Phi_i) \left[\exp\left(\beta V_a - \beta I_i \frac{\rho t}{A_i} \right) - 1 \right], + A^* T^2 \sum_i^{\text{“p–o”}} A_{i,\text{eff}} \exp(-\beta \Phi_{i,\text{eff}}) \times \left[\exp\left(\beta V_a - \beta I_i \frac{\rho}{4\sqrt{A_{i,\text{eff}}/\pi}} \right) - 1 \right], \tag{6.43}$$

where the patches not pinched off (first term) are assumed to act independently, with t being the semiconductor thickness. One should note that since $A_{i,\text{eff}}$ and $\Phi_{i,\text{eff}}$ have explicit dependence on the applied bias, a change of variables from V_a to $(V_a - I_i\rho/4\sqrt{A_{i,\text{eff}}/\pi})$ has been used for each individual current contribution. Eq. (6.43) has been shown to give results in close agreement with the numerical solutions of Poisson's equation and the continuity equation.[2]

According to the analytic theory, current transport at an inhomogeneous SB is frequently dominated by potential pinch-off and saddle-point formation. Detailed and systematic simulations have been carried out to test the validity of the saddle-point concept and the various predictions from that concept.[2] As shown in Fig. 6.13, numerically calculated ideality

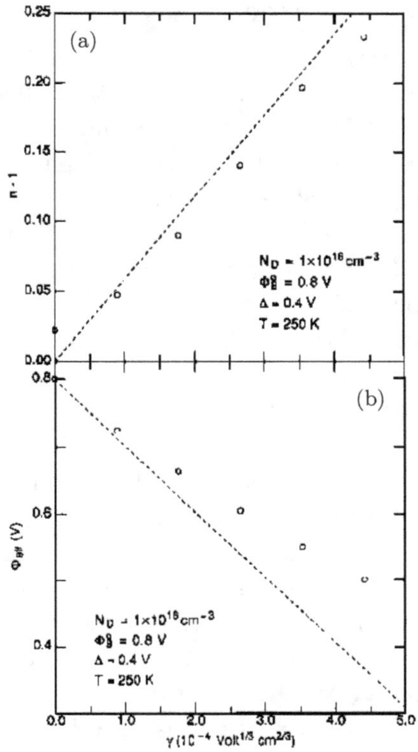

Fig. 6.13. Computer-simulated ideality factor (a) and effective SBH (b) for the current flowing to a circular patch with a local SBH of 0.4 eV at 250K, against the γ parameter (open symbols). This low-SBH patch is embedded in a background of 0.8 eV SBH on n-type Si with a 1×10^{16} cm^{-3} doping level. The dashed lines are predictions from the analytical theory.

Source: Reprinted from Sullivan *et al.*[2] with the permission of AIP Publishing.

factor and effective SBH agree with predictions of the analytic theory summarized in Table 6.1. Numerical results are in such excellent agreement with the predictions on all aspects of the analytic theory that the overall conclusion of numerical studies was essentially that numerical simulations were unnecessary! It is important to note that inhomogeneous SBs do not always behave in the same fashion. Depending on the distribution of inhomogeneous SBH, the temperature, the doping level, the bias, etc., carrier transport can look very different. Some examples of the spectrum of I–V characteristics from inhomogeneous SBH will be compared with the predictions of the analytic theory when experimental polycrystalline data are examined.

6.3.3 *Spatially resolved techniques*

The desire to correlate SBH variation with defects, structure, etc. has stimulated the development of a few techniques that allow spatially resolved determination of the SBH. A resolution at ~1 mm scale can be achieved in traditional SBH measurement techniques, such as photo-response or surface photoemission, simply by reducing the spot size of the illuminated area and scanning on an x-y stage. Electron–hole pairs generated in the space charge region of the SBH, in a scanning electron microscopy (SEM), are swept by the electric field to opposite ends of the sample, thus setting up an electron beam induced conduction (EBIC).[16] It is worth noting that the spatial resolutions of these coarse scanning techniques are not set only by the beam spot but also by the diffusion length of the generated carriers.

By far, the most powerful technique, and the one which has had the most impact on our present view of the lateral distribution of the SBH, is the ballistic electron emission microscopy (BEEM),[17] implemented in STM. Electrons (or holes) are injected into a thin metal layer in contact with a semiconductor, at energies considerably above (below) the Fermi level of the metal, as drawn in Fig. 6.14(a). A fraction of the hot electrons can traverse the thin metal layer without scattering and clear the SBH to enter the semiconductor. This BEEM current is then detected and used to map out a microscopic profile of the MS interface. Alternatively, when operating in the ballistic electron emission spectroscopy (BEES) mode, the energy dependence of the ballistic current at any specific point of the interface can be used to deduce the onset of the emission and thus the local SBH,[18] as shown in Fig. 6.14(b).[19] The thickness of the metal layer is usually kept

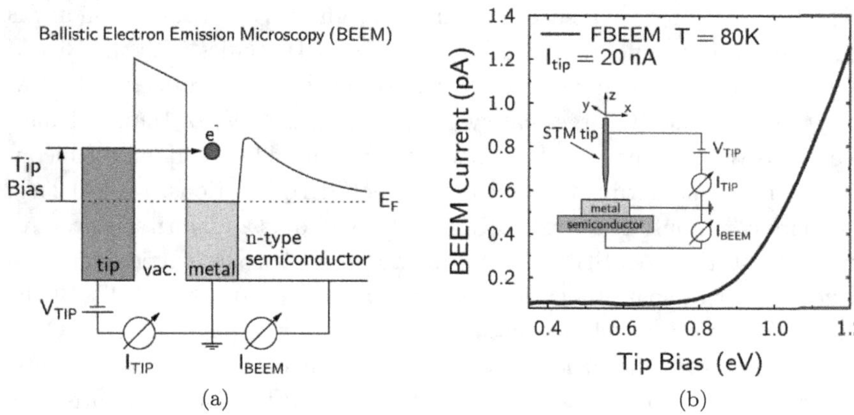

Fig. 6.14. (a) BEEM energy diagram. (b) Averaged BEEM spectra with inset showing schematic BEEM setup.

Source: Reprinted from Durcan *et al.*,[19] with the permission of AIP Publishing.

much shorter than the electron mean free path to minimize scattering in the metal film. BEEM can be conducted in air, which prompted the preferred use of noble metals (Au, Pt, etc.) to avoid oxidation. The emission process in a BEEM has been described by a simple theory, based on the free electron band, that explains some experimental data.[18] The injected hot electrons have a large forward momentum, which limits their lateral spread to a cone of a certain angle, in what has been called the "search light" effect. It has also been pointed out that the electron focusing effect, which has to do with the band structure of the metal, could lead to a spatial resolution of the BEEM technique that is better than expected from the free electron model. Even though all the steps leading to the transport of ballistic electrons into the semiconductor have excellent spatial resolutions, the potential pinch-off effect in the semiconductor can still severely limit the ability of BEEM to faithfully map the SBH profile at the MS interface. Since the BEEM current needs to not only surmount the energy barrier at the MS interface but also clear other hurdles, such as the saddle-point potential, in order to reach the neutral semiconductor and be detected, the SBH profile obtained by BEEM is subject to masking by the band bending in the space charge region. Potential pinch-off is always a factor that can mask the actual extent of SBH inhomogeneity from BEEM images. In that respect, semiconductors with high doping levels should be preferred for a more accurate map of the SBH profile.

The most direct evidence for barrier height variation in common SB diodes naturally came from spatially resolved SBH studies with the BEEM technique. The overall uniformity of an SB diode is easily revealed by BEEM images. Occasionally, large-scale variations (0.7–1.1 eV) of the SBH have been observed in compound semiconductor.[20] But on lightly doped semiconductors more susceptible to pinch off, smaller SBH variations, ~0.1–0.2 eV, are more commonly observed. BEEM can be used as a quantitative tool to study the statistics of SBH distributions. BEES scans are carried out for a large number of spatial locations of the same diode to get a profile of the SBH distribution, which is then used to calculate the expected I–V, C–V, and PR characteristics from the same diode in macroscopic measurements. Generally, the SBHs extracted from BEEM distributions compare very closely with those measured with macroscopic techniques. This agreement is expected because the SBH distribution measured by BEEM reflects the effective SBH with the effect of potential pinch-off already included. An excellent example illustrating the lateral distribution of n-type SBH at a W/Si(100) interface is shown in Fig. 6.15.[19] Direct evidence for the potential pinch-off effect has been observed by BEEM in artificially generated inhomogeneous SBs.[21] A continuous Au film was deposited over

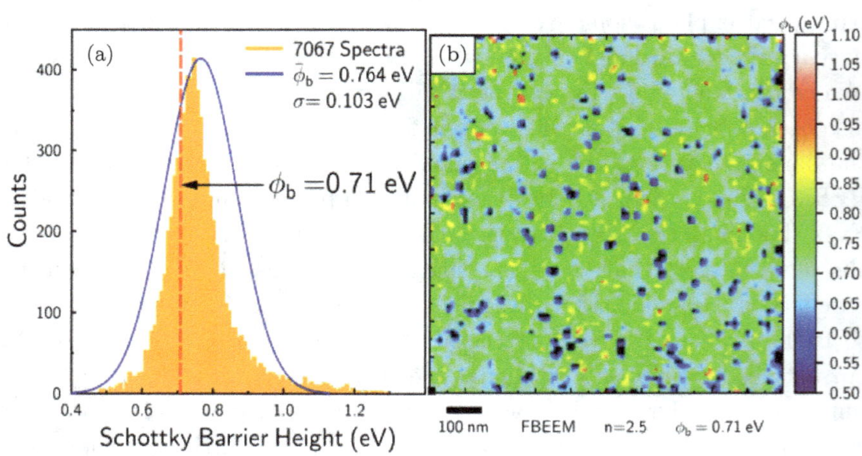

Fig. 6.15. (a) Histogram of SBHs from the 7225 spectra taken on a tungsten n-type silicon (100) Schottky junction by BEEM. The histogram has a standard deviation of 0.103 eV with an average R^2 value of 0.93. (b) Spatial map of the tungsten to n-type silicon Schottky junction in (a).

Source: Reprinted from Durcan *et al.*,[19] with the permission of AIP Publishing.

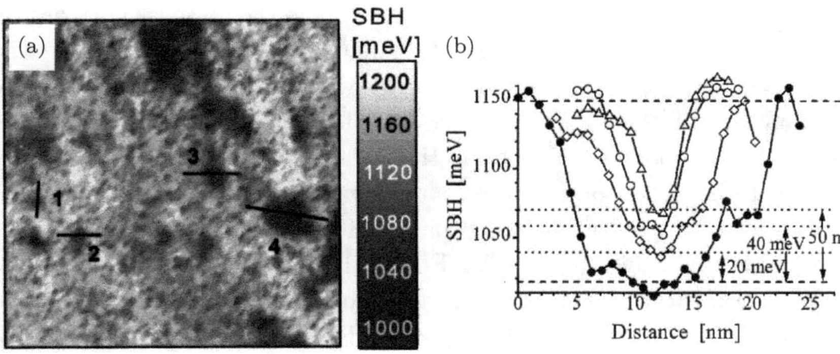

Fig. 6.16. (a) Schottky barrier map of an inhomogeneous 6 nm Au/0.5 nm Co/ GaAs0.67 P 0.33-Schottky diode (size: 114 nm × 114 nm), dark areas correspond to Co clusters at the MS interface. The black lines (1–4) indicate the linescans through Co patches of different sizes. (b) SBH profiles along the linescans in (a). The diameters for the four Co patches are 5, 7, 10, and 18 nm, respectively.

Source: Reprinted from Olbrich *et al.*,[21] with the permission of AIP Publishing.

small Co islands, of variable sizes, which were pre-deposited on $GaAs_xP_{1-x}$. Such "mixed" diodes are expected to have a background of high SBH decorated with small patches of low-SBHs because the Co/$GaAs_xP_{1-x}$ interface has a SBH that is lower than that of the Au/$GaAs_xP_{1-x}$ interface by about 0.15 eV. Such an overall BEEM pattern was indeed observed, as shown in Fig. 6.16(a). Since the sizes of the Co islands, 5–10 nm, are smaller than the depletion region width of $GaAs_xP_{1-x}$, most of the low-SBH patches should be pinched off. Furthermore, the smaller patches should be more severely pinched off than the larger ones. Line scans taken across low-SBH regions of different dimensions, shown in Fig. 6.16(b), indeed show this size dependency clearly. Saddle-point potentials calculated according to the analytic theory (Table 6.1), using the actual island sizes, were found to be in good quantitative agreement with the "valleys" measured by BEEM. This example is a vivid demonstration of the potential pinch-off effect in real space.

6.4 Polycrystalline Schottky Experiments: Evidence for Inhomogeneity

Early investigations of the SBH were conducted almost exclusively on polycrystalline MS interfaces, from which the overwhelming experimental observation was the FLP phenomenon. The interpretation then was that

FLP was a fundamental mechanism responsible for the insensitivity of the SBH to the metal. That interpretation naturally led to the automatic assumption that SBH was uniform. This was a time when researchers were unaware of the potential pinch-off effect and could not recognize signs of SBH inhomogeneity from their data. In light of the analytic theory just discussed, some experimental data from that period of time are reexamined in this section to show that SBH inhomogeneity was pervasive for polycrystalline interfaces. The FL of these MS interfaces was thus unpinned, contrary to the conclusion researchers drew at the time from the same data. The real reason for the apparent FLP phenomenon is addressed in the following chapter.

6.4.1 *Leakages and edge-related currents*

Experimentally observed I–V curves were analyzed and shown to be comprised of two or more components of current, as illustrated in Fig. 6.17(a). At small biases, the forward current is sometimes dominated by a "soft", or "leaky", component that leads to both a curvature in the I–V curve and an ideality factor much in excess of 1. As the bias increases, the I–V relationship becomes semi-logarithmic with an ideality factor not far from unity, before at very large biases, when the current turns over due to series resistance.[22] The presence of a leaky component is particularly easy to detect at lower temperatures. It was customary to attribute the linear portion of the I–V curve to the main conduction mechanism, e.g. thermionic emission, and the leakage current at low biases to a different mechanism. Generation and recombination in the space charge region and edge-related conduction were the main mechanisms invoked to explain the leakage currents. However, the fact that leakage currents are observed to dominate some diodes and are completely absent in other diodes identically processed is inconsistent with a uniform distribution of generation-recombination centers. Since edge-related currents scale with the peripheral length of the diode, and not with the area of the diode as would be expected from common junction conduction mechanisms, they may be unambiguously identified by using diodes with different sizes. Specially designed experiments showed that edge-related currents had ideality factors similar to, or smaller than, the ideality factor associated with the center portions of the diodes, as shown in Fig. 6.18.[23]

Experimentally observed leakage currents and edge-related currents are consistent with SBH inhomogeneity. The presence of a few large low-SBH

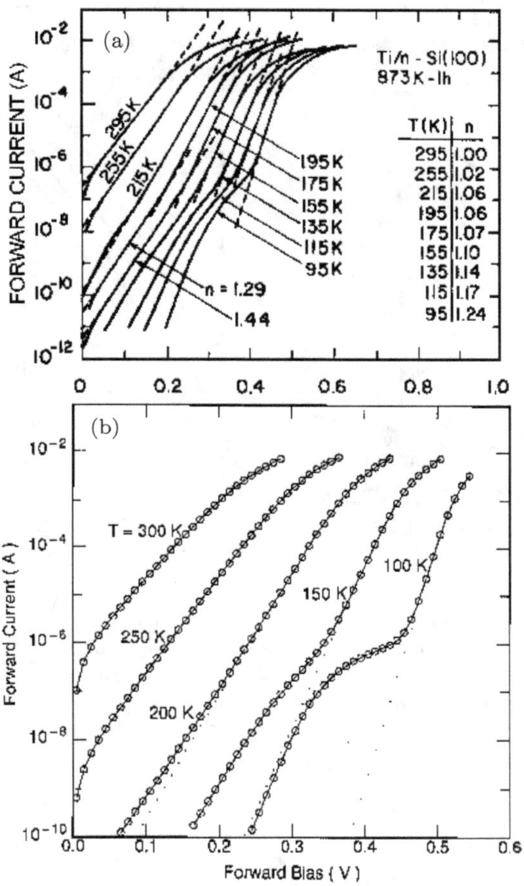

Fig. 6.17. (a) Forward current-voltage characteristics of Ti on n-type Si(100) as a function of temperature for samples annealed at 873K for 1h. Diode area is 4×10^{-4} cm^2. Reprinted from M. O. Aboelfotoh, reference 22, with the permission of AIP Publishing. (b) Calculated I–V characteristics of an inhomogeneous SB.

Source: Reprinted with permission from Tung.[3] Copyright (1992) by the American Physical Society.

regions in the SB diode can certainly lead to the observation of a leaky component in the junction current, as shown in Fig. 6.17(b).[3] The small slope of this current (high ideality factor) and effects due to series resistance limit the predominance of this leakage component to small forward biases. Due to a lower effective SBH, the presence of even a few low-SBH regions can lead to the observation of leakage current. Experimentally observed diode-to-diode variations of the leakage current are in much better

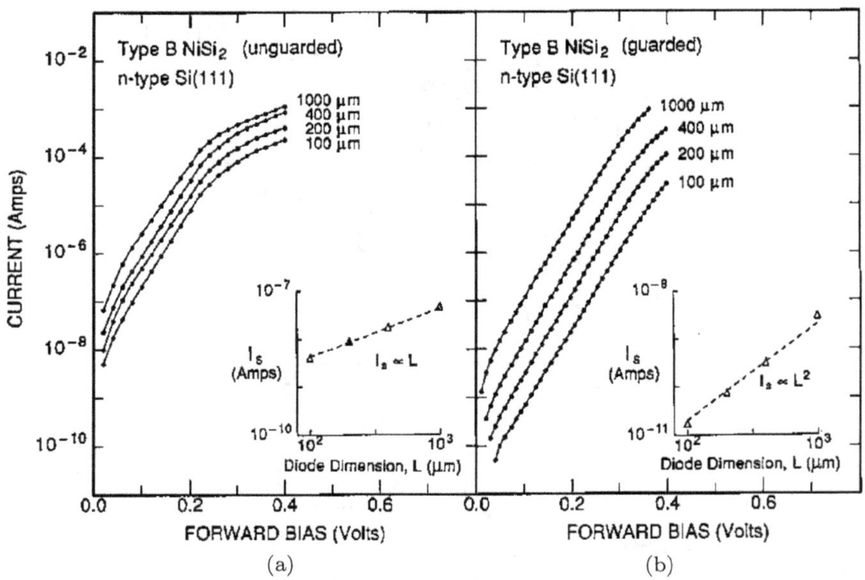

Fig. 6.18. Current–voltage characteristics of epitaxial type B $NiSi_2$/Si(111), *n*-type Schottky circular diodes with varied diameter. (a) Edges unprotected, showing edge-related current that is linearly proportional to the diode diameter, but with excellent ideality factors, ~1.01. (b) Diode edges are protected by guard rings. The current is proportional to the area of the diode (square of the diameter).

Source: Reprinted with permission from Sullivan *et al.*[23] Copyright (1992), American Vacuum Society.

agreement with isolated "leakage spots" due to low local SBH rather than with a distribution of recombination centers in the space-charge region. The existence of a current component that is proportional to the perimeter of the diode is also in good agreement with SBH inhomogeneity. As demonstrated by computer simulations, a low-SBH patch is less effectively pinched off when it is situated in close proximity to the edge of a diode than when it is at the central portion. When an SB diode contains a uniform distribution of low-SBH patches, the number of patches that are located near the edges is proportional to the peripheral length. Therefore, most edge-related currents observed in experiments are consistent with the presence of SBH inhomogeneity.

6.4.2 *Greater-than-unity ideality factor*

Even though the forward I–V plots experimentally observed from SBs are almost always semi-logarithmic, the slopes of such traces often differ from the prediction of the thermionic emission theory. An ideality factor greater than 1.03 cannot be explained by thermionic emission through the image-force lowering mechanism and generally indicates a dependence of the SBH on bias. For decades, this bias dependence was proposed to be due to interface states, similar to the way the SBH dependence on the metal work function was weakened. However, the fact that these two dependencies contradict each other shows that interface states cannot be the explanation for either dependency.[3, 24] The bias dependence of the saddle-point potential of an inhomogeneous SB can explain the observed behavior of the ideality factors. When the doping level increases, the Φ_{eff} of a low-SBH patch with a fixed γ decreases and its ideality factor increases, in good agreement with some experimental observations. Among diodes fabricated identically, diodes with larger ideality factors have frequently been measured to have lower SBHs. This correlation agrees with the expected behavior of diodes with a distribution of low-SBH patches. Plugging Eq. (6.40) into Eq. (6.39), one gets

$$\Phi_{\text{EFF}} = \Phi_{B,n}^{oo} - (n_{\text{EFF}} - 1)V_{bb}\xi^{-1}, \tag{6.44}$$

which shows the observed inverse correlation between the ideality factor n_{EFF} and the SBH Φ_{EFF}. An example of experimental I–V scans that display a correlation between the ideality factors and SBHs of the same diode obtained at different temperatures is shown in Fig. 6.19.[25] The increase in the SBH and decrease in the ideality factor with temperature are good examples of what to expect from inhomogeneous SBH, according to Eqs. (6.39) and (6.40).

6.4.3 T_o *Anomaly and other dependencies of the ideality factor*

Many different temperature dependences of the ideality factor have been experimentally observed. Frequently, the ideality factor of a diode increases when the sample temperature is lowered. At many MS interfaces, the deduced SBH and ideality factors are found to vary with the measurement temperature in a fashion generally referred to as the "T_o anomaly".[26]

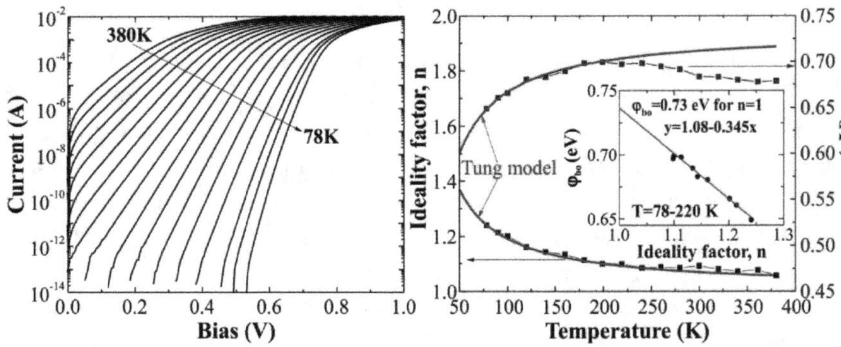

(a) (b)

Fig. 6.19. Experimental forward-bias I–V dependences of the Au/Ti/n-InAlAs SB. Layers of 500 nm lightly doped n–$In_{0.52}Al_{0.48}As$ and 300 nm heavily n+$In_{0.52}Al_{0.48}As$ are epitaxially grown on InP(100) substrate with prefabricated heterostructures designed to minimize dislocation defects. (a) Temperature scans are conducted at 78K and then from 100K to 380K with a spacing of 20K. (b) The extracted n and SBH are shown against the temperature. The inset shows the correspondence of SBH with n.

Source: Reprinted with permission from Aksenov *et al.*,[25] with the permission of AIP Publishing.

Such a behavior has been observed from all types of SBs, on elemental semiconductors and compound semiconductors alike. A diode is said to display the T_o effect if its junction current may be expressed as

$$I_{To} = A^{**}AT^2 \exp\left(\frac{-e\Phi_{To}}{k_B[T+T_o]}\right)\left[\exp\left(\frac{eV_a}{k_B[T+T_o]}\right)-1\right], \qquad (6.45)$$

where T_o, typically 10–60 Kelvin, and Φ_{To} are both constants. Demonstration of the T_o effect is usually accomplished by plotting nk_BT (the inverse slope of an I–V curve) against k_BT and observing a straight line, parallel but offset from the n = 1 line,[27] as illustrated in Fig. 6.20(b). This was a spectacular observation at a time when the pervasiveness of SBH inhomogeneity was not known. Understandably, many different transport mechanisms were proposed to explain the T_o and other observed dependences of the ideality factor on temperature. This led to contradictions and inconsistencies. Once the SBH is realized to be inhomogeneous for most polycrystalline MS interfaces, the widely different I–V characteristics experimentally observed can be coherently explained with the usual thermionic emission, without the need to invent a special transport mechanism. Shown in Fig. 6.20 are examples of ideality factor dependences numerically simulated for inhomogeneous

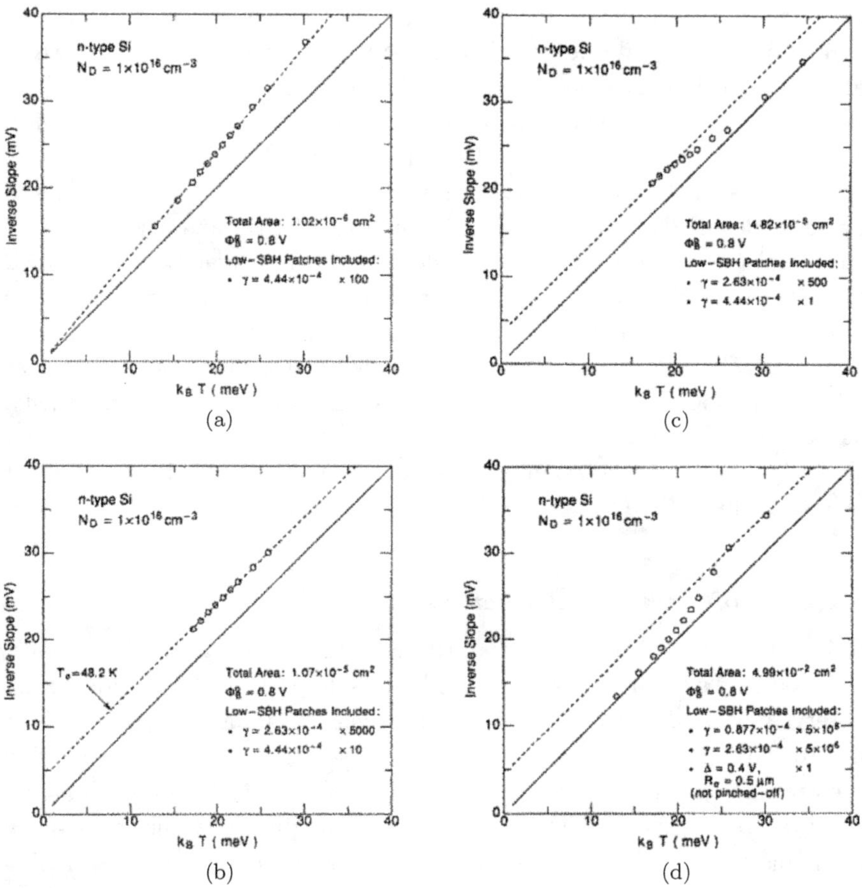

Fig. 6.20. Plots of the inverse slope of the I–V plot as a function of k_BT for inhomogeneous MS contacts composed of a distribution of low-SBH patches as described in the figures (γ is in units of $V^{1/3}\,cm^{2/3}$): (a) Plot showing a constant ideality factor in excess of unity, (b) plot showing the T_o anomaly, (c) plot showing $n \approx 1+T_o/T$ behavior at low temperature and ideal behavior at high temperature, and (d) plot showing ideal behavior at low temperature and $n \approx 1 + T_o/T$ behavior at high temperature.

Source: Reprinted from Sullivan *et al.*[2] with the permission of AIP Publishing.

SBs with different SBH distributions.[2] A temperature-independent, greater-than-unity, ideality factor, often observed experimentally,[28] is consistent with an SB diode whose current transport is dominated by low-SBH regions with narrowly distributed γ's, as shown in Fig. 6.20(a). An increase in the ideality factor with decreasing temperature, as is the case for the majority of SB diodes, is consistent with the presence of many low-SBH regions with a distribution in their low-SBH characteristics. The current of a random, inhomogeneous, SB diode is described by Eq. (6.36), which may be expressed phenomenologically in a form identical to Eq. (6.45). Therefore, SBH inhomogeneity offers the only valid explanation of the T_o anomaly: as the temperature is lowered, the junction current is dominated by fewer low-SBH regions with lower effective SBHs and larger ideality factors. As an example, the inverse slope, in an I–V simulation, of an SB diode with low-SBH patches of just two distinct γ's is plotted in Fig. 6.20(b) as a function of temperature, to illustrate the origin of the T_o mystery.[2] The apparent switchovers between different ideality factor categories observed at different temperature ranges are also consistent with SBH inhomogeneity, as shown in Fig. 6.20(c). From Fig. 6.20, there is no particular behavior that all inhomogeneous MS interfaces are expected to display. Very different behaviors may all be caused by SBH inhomogeneity.

6.4.4 *"Soft" reverse characteristic*

It is a universal observation that the current from any SB diode never truly saturates at large reverse bias. These "soft" reverse characteristics are observed even when the utmost care is taken to eliminate possible effects due to the edges of the diode. Image-force lowering is capable of explaining the reverse currents at some SB diodes. However, the majority of SB diodes show reverse currents which far exceed those predicted by the image-force mechanism alone. An early analysis of the reverse characteristics of SB's showed that, in addition to the image-force mechanism that scaled with the square root of the electric field, there needed to be an SBH reduction proportional to the electric field to explain the experimental data.[29] From the fact that polycrystalline MS interfaces are generally inhomogeneous, the reverse current is expected to be leaky. Any SBH lowering mechanism that varies with the electric field more rapidly than $E_{\max}^{1/2}$, could lead to a satisfactory fit with the experimentally observed reverse currents. As shown in Eq. (6.41), a reduction in the SBH with the reverse bias, such as proportional to the 1/2 and 2/3 powers of V_{bb}, is possible in inhomogeneous SB diodes with a random distribution of low-SBH patches. As shown in Fig. 6.21, experimental data agree with the behavior expected of inhomogeneous SBs.[3]

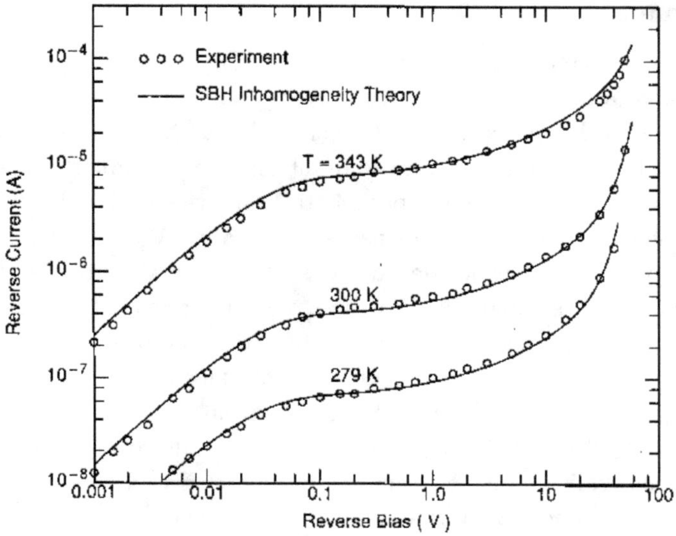

Fig. 6.21. Reverse characteristics (circles) of a ZrSi2 Schottky diode experimentally observed by Andrews and Lepselter.[29] Solid lines are calculated currents of an inhomogeneous SB diode, which has a $\Phi^{o}_{B,n}$ of 0.56 eV and contains low-SBH patches with γ's of 5.6×10^{-5} $V^{1/3}\,cm^{2/3}$ and 4×10^{-5} $V^{1/3}\,cm^{2/3}$.

Source: Reprinted from Tung.[3] Copyright (1992) by the American Physical Society.

6.4.5 *SBH dependence on measurement technique*

The SBH measured by the I–V technique, Φ_{I-V}, often decreases with increasing doping level, while the SBH measured by the C–V method, Φ_{C-V}, remains constant. Frequently, the SBHs depend on the technique of measurement, namely, Φ_{C-V} significantly exceeds Φ_{I-V} and the SBH derived from PR techniques, Φ_{PR}. While this SBH dependence on the measuring technique was puzzling decades ago and elicited explanations based on interface states, it is now clear that SBH inhomogeneity is its explanation. Current transport at inhomogeneous SBs is dominated by low-SBH patches, leading to the deduction by I–V and PR techniques of apparent SBHs which are lower than the arithmetic average of the entire diode. Since, according to Eq. (6.33), the C–V technique yields an average SBH for the whole diode,[3] the experimentally observed dependence of SBH on the technique of measurement is consistent with SBH inhomogeneity.

6.5 Summary

Traditional concepts and theories on the SBH were largely developed decades ago from experiments with polycrystalline MS interfaces. In these studies, the SBH was always assumed to be uniform across the entire diode, and the experimental data were interpreted accordingly. There were always some aspects of the experimental data that did not fit together or remained unable to be explained scientifically. A mystery that defied satisfactory explanations for decades was the "ideality factor" that needed to be included in the analysis of carrier transport at Schottky barriers. This artificial parameter led to the proposal of various empirical models, many of which were based on the assumption of interface states, which had since been ruled out because of a contradiction it would lead to.

All the mysteries with experimental SBH results disappeared quickly when it was realized and experimentally demonstrated that the SBH for most polycrystalline MS interfaces was inhomogeneous. An important effect that sets inhomogeneous MS interfaces apart from ones with uniform SBH is the likelihood of potential pinch-off in the space charge region of the semiconductor. Since the potential distribution in the space charge region has to satisfy Poisson's equation, any rapid changes in the potential profile required by the inhomogeneous SBH at the interface, e.g. near $z = 0$ in Fig. 6.22(a), need to be smoothed out. As illustrated in this figure, the potential distribution near a low-SBH strip in a high-SBH background is in the shape of a saddle (marked by an arrow), being a local maximum along the z direction yet a local minimum parallel to the interface. The potential at the saddle point is higher than that at the low-SBH strip, which means that carrier transport from the semiconductor to the low-SBH strip has to reach a potential at least as high as this in order to succeed.[3] This saddle-point potential thus becomes the de facto SBH for the low-SBH strip and is the one that will be measured experimentally rather than the actual SBH at the interface! The conduction path of the low-SBH is said to be pinched off by surrounding high-SBH regions. Potential pinch-off fundamentally modifies current transport at an SB and changes the way experimental current-voltage data should be interpreted. A natural explanation of the ideality factor is offered through Figs. 6.22(b–d), depicting the potential distribution near a saddle point. When a positive voltage (forward bias) is applied, the saddle-point potential is raised slightly as shown in Fig. 6.22(b). And in Fig. 6.22(c), the saddle-point potential is seen to be lowered with a reverse negative voltage. This dependence of the saddle potential on

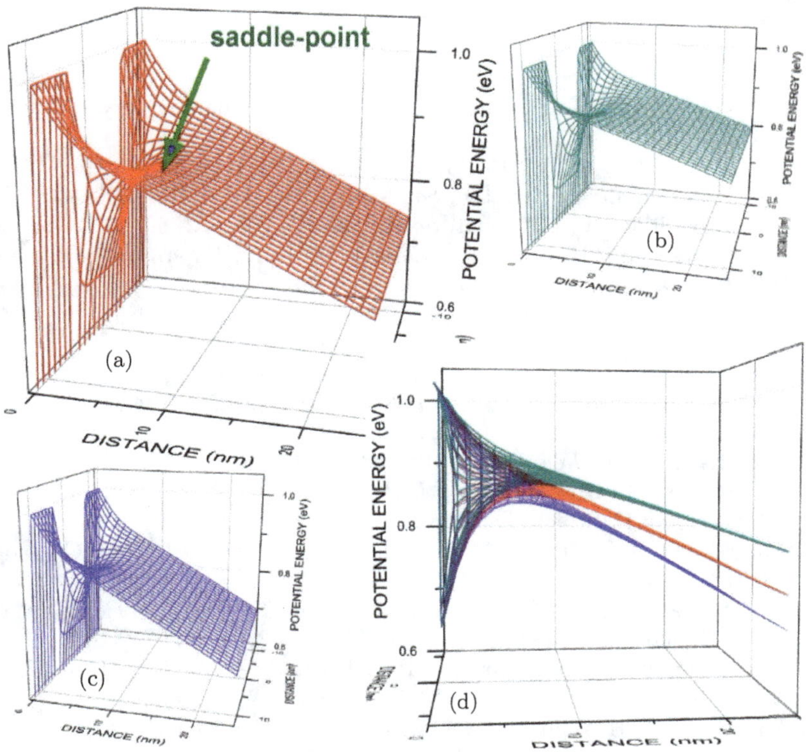

Fig. 6.22. The potential energy distribution of the space charge region near a low-SBH strip with the following characteristics: $\Phi_{B,n}^{oo} = 1.00\,\text{eV}$, $\Delta = 0.31\,\text{eV}$ and $L_0 = 10$ nm. (a) Without applied potential. The location of the saddle point is indicated. (b) With a bias of +0.36 V. (c) With a bias of −0.36 V. (d) A superposition of the potential distribution under different biasing conditions. Viewed from the direction parallel to the interface in (d), the increase and decrease of the saddle-point potential, respectively, with positive and negative applied potential are clearly on display. This dependence of the saddle-point potential on the applied bias voltage is the origin of a host of different effects experimentally observed from current–voltage characteristics of polycrystalline MS interfaces.

the applied potential, displayed more clearly in the combined image of Fig. 6.22(d), is the reason for a greater-than-unity ideality factor. With the knowledge of how potential pinch-off affects carrier transport in hand, a re-examination of the experimental data from polycrystalline MS interfaces taken decades ago left little doubt that the SBHs at these interfaces were inhomogeneous. With inhomogeneity SBH, the FL at these interfaces could not have been "pinned" as hastily concluded at the time.

References

1. R.T. Tung, *Appl. Phys. Lett.* **58**, 2821–2823 (1991).
2. J.P. Sullivan, R.T. Tung, M.R. Pinto, and W.R. Graham, *J. Appl. Phys.* **70**, 7403–7424 (1991).
3. R.T. Tung, *Phys. Rev. B* **45**, 13509–13523 (1992).
4. F.A. Padovani and R. Stratton, *Solid-State Electron.* **9**, 695 (1966).
5. C.R. Crowell and V.L. Rideout, *Solid-State Electron.* **12**, 89–105 (1969).
6. M. Missous and E.H. Rhoderick, *Electron. Lett.* **22**, 477–478 (1986).
7. H. Norde, *J. Appl. Phys.* **50**, 5052–5053 (1979).
8. K. Sato and Y. Yasumura, *J. Appl. Phys.* **58**, 3655–3657 (1985).
9. J.H. Werner, *Appl. Phys. A* **47**, 291–300 (1988).
10. R.T. Tung, *Mater. Sci. Eng. Rep.* **35**, 1–138 (2001).
11. T. Okumura and K.N. Tu, *J. Appl. Phys.* **61**, 2955–2961 (1987).
12. T. Okumura and K.N. Tu, *J. Appl. Phys.* **54**, 922–927 (1983).
13. M.H. Hecht, *Phys. Rev. B* **41**, 7918–7921 (1990).
14. I. Ohdomari and K.N. Tu, *J. Appl. Phys.* **51**, 3735–3739 (1980).
15. R.H. Cox and H. Strack, *Solid-State Electron.* **10**, 1213 (1967).
16. J.M. Gibson, R.T. Tung, C.A. Pimentel, and D.C. Joy, *Inst. Phys. Conf. Ser.* **76**, 173–181 (1985).
17. W.J. Kaiser and L.D. Bell, *Phys. Rev. Lett.* **60**, 1406–1409 (1988).
18. L.D. Bell and W.J. Kaiser, *Phys. Rev. Lett.* **61**, 2368–2371 (1988).
19. C.A. Durcan, R. Balsano, and V.P. Labella, *J. Appl. Phys.* **116**, 023705 (2014).
20. A.E. Fowell, R.H. Williams, B.E. Richardson, A.A. Cafolla, D.I. Westwood, and D.A. Woolf, *J. Vac. Sci. Technol. B* **9**, 581–584 (1991).
21. A. Olbrich, J. Vancea, F. Kreupl, and H. Hoffmann, *Appl. Phys. Lett.* **70**, 2559–2561 (1997).
22. M.O. Aboelfotoh, *J. Appl. Phys.* **64**, 4046–4055 (1988).
23. J.P. Sullivan, R.T. Tung, F. Schrey, and W.R. Graham, *J. Vac. Sci. Technol. A* **10**, 1959–1964 (1992).
24. J.L. Freeouf, *Appl. Phys. Lett.* **41**, 285 (1982).
25. M.S. Aksenov, N.A. Valisheva, I.B. Chistokhin, D.V. Dmitriev, A.S. Kozhukhov, and K.S. Zhuravlev, *Appl. Phys. Lett.* **114**, 221602 (2019).
26. F.A. Padovani, *J. Appl. Phys.* **38**, 891 (1967).
27. A.N. Saxena, *Surf. Sci.* **13**, 151–171 (1969).
28. M.O. Aboelfotoh, *J. Appl. Phys.* 66, 262–272 (1989).
29. J.M. Andrews and M.P. Lepselter, *Solid-State Electron.* 13, 1011–1023 (1970).

Chapter 7

SCHOTTKY BARRIER HEIGHT EXPLAINED WITH BULK CHARGE DENSITY

The explanation of the formation mechanism of the SBH at MS interfaces has occupied the attention of scientists for decades. Analyses of band alignment problem in general, including discussions in Chapter 4, tie the magnitude of the SBH to the charge distribution of the interface. Accordingly, the SBH at any MS interface is predictable if its charge distribution can be approximately modeled. Thanks to the nearsightedness principle of electron density, this is actually feasible, which forms the main topic of this chapter. Even though the vast majority of MS interfaces employed in actual devices are polycrystalline and the development of traditional theories on SBH formation was based almost exclusively on results from polycrystalline interfaces, it is not practical to delve directly into the charge distribution at polycrystalline interfaces. A more efficient way to understand SBH formation is still through studies of epitaxial MS interfaces. The uniqueness in the atomic structure of the epitaxial interface avoids the problem of electrical inhomogeneity at polycrystalline MS interfaces and allows the experimentally measured SBH to be directly compared with the atomic structure. With each interface studied in isolation, a comparison of results from different epitaxial interfaces most readily reveals the dependence of the SBH on the specifics of the interface. Epitaxial interfaces are also important for the reason that theoretical investigations can reveal not only the magnitude of their SBHs but also the energetic and spatial distributions of their electronic states. These details, essential for a complete understanding of the SBH problem, are available only from epitaxial MS interfaces. Furthermore, the charge density modeling planned for this chapter can be demonstrated explicitly only with epitaxial interfaces of known atomic structures. Therefore, this chapter relies heavily on epitaxial MS interfaces to illustrate how SBH magnitudes can be quantitatively explained and approximately predicted simply from a knowledge of the charge distributions in the bulk crystals of metals and semiconductors. These procedures, once mastered and validated through the study of epitaxial interfaces, can be applied to polycrystalline interfaces to quantitatively model and predict the SBH, as demonstrated later in the chapter. The reason for the sensational Fermi level pinning phenomenon that mystified and fascinated researchers for decades will also be explained. The cause of this notorious mystery turns out to be alarmingly trivial: when the interface structure responsible for the SBH is essentially independent of the metal (!), the SBH is naturally insensitive to the metal.

7.1 Epitaxial MS Interfaces: Experiments

7.1.1 *Lattice-matched MS systems*

Epitaxial MS interfaces are ideal model systems for the investigation of the formation mechanism of the SBH. The initial SBH result experimentally obtained from epitaxial MS interfaces was so different from polycrystalline interfaces that it caused quite a stir in the community and prompted further searches for epitaxial systems to investigate.[1] A necessary condition for an MS interface to be fabricated epitaxially is the matching in both size and shape of the two-dimensional unit cells of two crystals along the interface direction. For this reason, lattice matching conditions found between metals and semiconductors with cubic crystal structures are particularly treasured, as epitaxy is then possible along all major interface orientations between the two crystals. Only with a small lattice mismatch, e.g. $<2\%$, is it possible for defects associated with misfit strain in the system to be avoided or reduced to a level low enough for epitaxial structures with high structural perfection to be produced. The crystal structures and lattice parameters of a few metals and semiconductors that meet these requirements and have been successfully fabricated into epitaxial interfaces with high quality are listed in the upper part of Table 7.1. Among these, $NiSi_2$/Si and $CoSi_2$/Si interfaces have demonstrated nearly perfect epitaxial quality.[2] A schematic of the cubie fluorite crystal structure, which both $NiSi_2$ and $CoSi_2$ occupy, is shown in Fig. 7.1 along with the diamond lattice structure of Si. Single-crystal MS interfaces have also been successfully fabricated on GaAs and AlAs substrates, using intermetallic group III alloys with the CsCl structure[3] and rare-earth monopnictides with rock salt lattice structure.[4] In addition to the MS epitaxial systems that have been experimentally fabricated and studied, there are numerous other metal–semiconductor combinations with excellent lattice matching conditions that either have not

Table 7.1. Structure and lattice constant of some metals suitable for epitaxial growth on semiconductors.

Metal	Structure	Lattice (Å)	Semiconductor
$NiSi_2$	CaF_2	5.41	Si
$CoSi_2$	CaF_2	5.37	
NiAl	CsCl	2.86	GaAs, AlAs
NiGa	CsCl	2.88	
CoGa	CsCl	2.85	
ErAs	NaCl	5.75	
ScAs	NaCl	5.48	
Ca	fcc	5.59	Si, GaP, ZnS
Rh	fcc	3.8	
Ir	fcc	3.84	
Rb	bcc	5.7	Ge, GaAs, ZnSe
Al	fcc	4.05	
Pt	fcc	3.924	
Cr	bcc	2.885	
Au	fcc	4.08	InP, CdS
Ag	fcc	4.086	
Na	bcc	4.29	

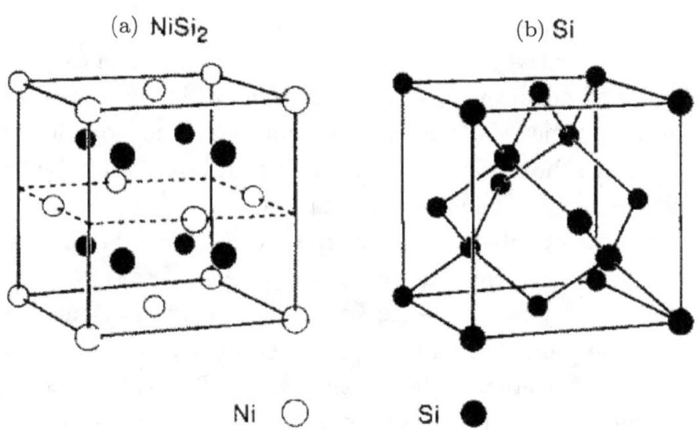

Fig. 7.1. Ball-and-stick models of the crystal structures of (a) NiSi2 and (b) Si.

been explored experimentally or have only been studied briefly for epitaxial growth. The existence of these potential epitaxial systems, also listed in Table 7.1, is important to SBH research, as they represent opportunities for theoretical investigations to be broadly conducted and the detailed electronic structures to be systematically compared.

7.1.2 *Fabrication and atomic structure of epitaxial silicide interfaces*

Since a semiconducting material has to be of nearly perfect structural quality for the transport properties and the energy band positions experimentally measured to be meaningful and useful, epitaxial MS interfaces are generally fabricated by depositing a metallic material onto the surface of a pristine semiconductor crystal. To prevent unwanted impurities from being incorporated during fabrication and affecting the quality of epitaxial structures, both the fabrication environment and the surface of the semiconductor are required to be atomically clean. Advanced fabrication facilities such as molecular beam epitaxy (MBE) and atomic layer epitaxy (ALE) have been employed to deposit metallic materials with atomic-level control over the growth process. Growth processes are typically conducted at elevated temperatures and usually involve the supply of all the elements needed for the composition of the overlayer. For example, both Ni and Al are deposited in an optimized ratio in order to grow an epitaxial layer of NiAl on a GaAs crystal. On the other hand, it is not always necessary to deposit Si in order to grow silicides on a Si substrate, e.g. $NiSi_2$/Si and $CoSi_2$/Si interfaces, because the Si for the overgrowth can also come from the substrate. Indeed, the use of the reaction between the deposited metal, Ni or Co, with the Si crystal toward the formation of the disilicide phase has been essential in achieving epitaxial interfaces of extremely high quality. An added advantage to the utilization of Si from the substrate for silicide reaction is that the eventual location of the silicide/Si interface is buried underneath the original surface and is thus less affected by impurities present on the original Si surface inadvertently.

There are two possible epitaxial orientations for $NiSi_2$ and $CoSi_2$ on Si(111) that have been named simply "type A" and "type B". Type A oriented disilicide layers have exactly the same crystal orientation as the underlying Si. The surface normal [111] direction is also shared by a type B oriented disilicide layer and the Si substrate. However, there is a 180° rotation of the type B disilicide crystal about that axis with respect to the Si. Thick $NiSi_2$ layers routinely grown by solid phase reaction on Si(111) contain areas with both orientations. An intriguing discovery from the reaction between deposited Ni and Si(111) is the complete control over the orientation of epitaxial $NiSi_2$ thin films.[5] As shown in Fig. 7.2, the orientation of the $NiSi_2$ grown on Si(111) by Ni deposition at room temperature, followed by moderate thermal annealing, depends sensitively on the thickness of Ni deposition. Spectacularly, $NiSi_2$ thin films with

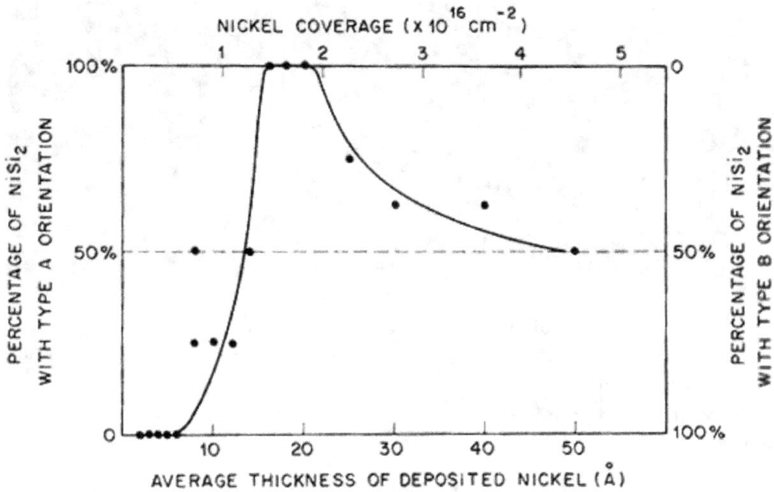

Fig. 7.2. Orientation of thin $NiSi_2$ layers grown on Si(111) by the deposition of nickel at room temperature and annealing to ~450° C.

Source: Reprinted with permission from Tung *et al.*[50] Copyright (1983) by the American Physical Society.

entirely type A or entirely type B orientation can be formed, respectively, with 1.6–2.0 nm and 0.1–0.4 nm of deposited Ni. Alternative methods to grow purely type B oriented $NiSi_2$ layers, involving the deposition or co-deposition of Si, have also been successful. These ultrathin $NiSi_2$ layers can be used as a "template" to increase the film thickness to a more workable range without affecting the layer orientation.[6] Various advanced characterization techniques have been used to study both type A and type B oriented $NiSi_2$ thin layers and have verified the single orientation and the near structural perfection of these films. High-resolution electron microscopy investigations of these interfaces, such as those shown in Fig. 7.3, conclude that both of these interfaces have the 7-fold coordinated atomic structures, models of which are shown in Fig. 7.4.[7,8] In a bulk fluorite lattice, each Ni (or Co) atom is called 8-fold coordinated because it has eight Si nearest neighbors. At either type A or type B $NiSi_2$/Si(111) interface, each interface Ni atom has only seven Si nearest neighbors, hence its name. We should note that the two available $NiSi_2$/Si(111) interfaces represent a unique opportunity to study the effect of an azimuthal rotation on the SBH, while every other aspect of the interface, such as crystal orientation and atomic structure, is kept unchanged.

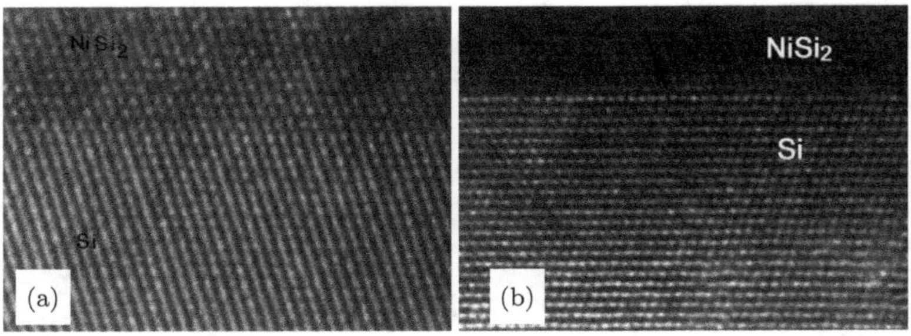

Fig. 7.3. Cross-sectional high-resolution transmission electron micrographs (HREM) of the interfaces of thin $NiSi_2$ layers grown on Si(111) with (a) type A orientation and (b) type B orientation.

Source: Reprinted with permission from Tung and Gibson.[43] Copyright (1985) American Vacuum Society.

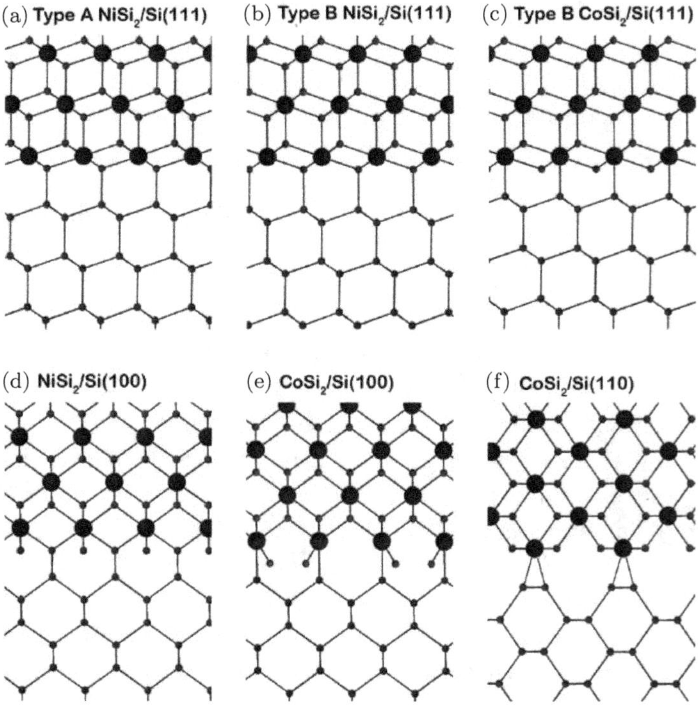

Fig. 7.4. Ball-and-stick models of epitaxial silicide/Si interfaces, viewed in the [110] direction, (a) the 7-fold type A $NiSi_2$/Si(111) interface, (b) the 7-fold type B $NiSi_2$/Si(111) interface, (c) the 8-fold type B $CoSi_2$/Si(111), (d) the 8-fold $NiSi_2$/Si(100) interface; (e) the reconstructed 7-fold $CoSi_2$/Si(100), and (f) the laterally shifted 8-fold $CoSi_2$/Si(110). Large spheres indicate either Co or Ni, while small spheres represent Si.

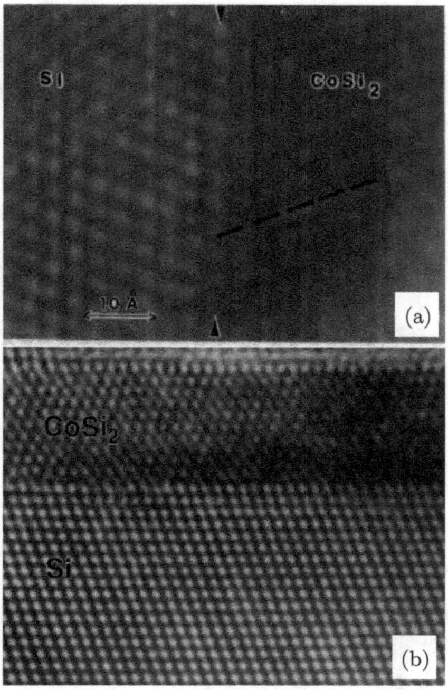

Fig. 7.5. Axial bright-field electron micrographs of (a) high-temperature reacted type B CoSi2/Si(111) interface. Reprinted from J. M. Gibson *et al.*, Applied Physics Letters 41, 818 (1982), with the permission of AIP Publishing. (b) Co-rich type B "X" interface, reacted at low temperatures.

Source: Reprinted with permission from Sullivan *et al.*[13] Copyright (1993) American Vacuum Society.

For reasons completely different from those just described for the $NiSi_2/Si(111)$, the interfaces of $NiSi_2/Si(100)$ also offer a unique opportunity to understand SBH formation. Layer nonuniformity has been a common problem for reacted $NiSi_2$ layers on Si(100), which frequently contain islands, pinholes, twins, and interface facets. Moderate annealing of deposited monolayers of Ni can lead to the growth of continuous and pinhole-free epitaxial $NiSi_2$ template layers. However, the interfaces of thin $NiSi_2$ layers formed at moderate temperatures are not completely uniform as they invariably contain "facet bars", which are trough-shaped protrusions of $NiSi_2$ into Si bounded by $\{111\}$ planes, an example of which is shown in Fig. 7.6.[9] Typically, facet bars are 50–100 Å in width and a few hundred Å in length along either the $[011]$ or the $[1\bar{1}0]$ direction. With the use of thicker $NiSi_2$ layers prepared from optimized thin templates, facet bars can be completely eliminated by high-temperature anneals to

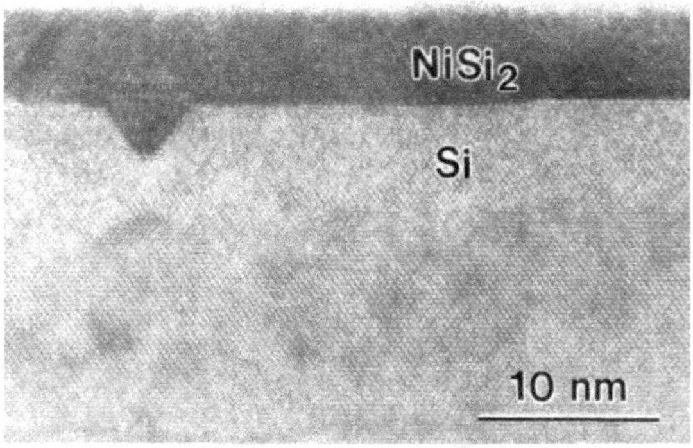

Fig. 7.6. Bright-field high-resolution transmission electron microscopy image of the cross-section of an 80-Å-thick $NiSi_2$ layer grown on Si(100). On the left side of the micrograph, the protrusion of $NiSi_2$ into the Si, bound by inclined {111} planes at the interface, is a facet bar.

Source: Reprinted from Batstone *et al.*,[9] with the permission of AIP Publishing.

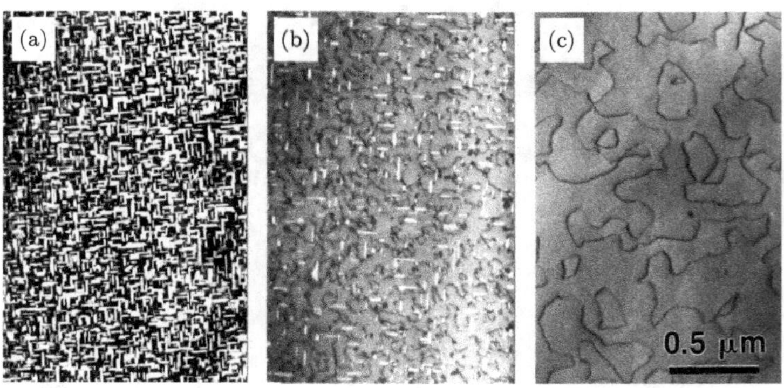

Fig. 7.7. Plan-view, (001) dark-field, TEM images of ~80-Å-thick $NiSi_2$ layers grown on Si(100). A facet bar shows up as a bright streak under this imaging condition. Dark lines are defects with 1/4(111) characters, which decorate steps with an odd number of atomic planes at the interface. (a) A layer which is nearly completely faceted. (b) A layer with mixed morphology. (c) A uniform layer.

Source: Reprinted with permission from Tung *et al.*[10] Copyright (1991) by the American Physical Society.

allow atomically flat $NiSi_2$/Si(100) interfaces to be fabricated. On the flip side, facet bars can also be added to $NiSi_2$ layers by additional Ni deposition and moderate annealing. It is thus possible to adjust the density of "facet bars" at a $NiSi_2$/Si(100) interface over a wide range through these procedures. Examples of $NiSi_2$(100) layers that are uniform, with a low density of facet bars, and completely faceted are shown in Fig. 7.7.[10]

Uniform $NiSi_2(100)$ layers have been found to be of high structural quality by standard characterization techniques. HREM images[8,11] are consistent with the $NiSi_2/Si(100)$ interface being 8-fold coordinated, a schematic of which is shown in Fig. 7.2(d). The inclined facets at a $NiSi_2/Si(100)$ interface have the type A 7-fold (111) structure locally, different from the structure of the flat area of the same interface. A spectrum of $NiSi_2/Si(100)$ interfaces with controllable mixtures of two different atomic structures is thus available experimentally. Controllable interface morphology makes the epitaxial $NiSi_2/Si(100)$ interfaces ideally suited for the investigation of the effect of structural inhomogeneity on the magnitude of the SBH.

Although with a lattice structure identical to the $NiSi_2$, thermally reacted $CoSi_2$ only forms with the type B orientation on Si(111). With the co-deposition of Co and Si, type B $CoSi_2$ layers can be grown at room temperature on clean Si(111) without annealing. Experimental studies have indicated that the 8-fold model, shown in Fig. 7.4(c), is most likely the structure for well-annealed, type B $CoSi_2/Si(111)$ interface.[12] However, it is also known that the interfacial structure of type B $CoSi_2/Si(111)$ may vary according to preparation. Evidence for a 7-fold coordinated structure has been obtained from layers that have only been annealed to lower temperatures. The atomic structure of a particular type B $CoSi_2/Si(111)$ interface, known as the "X" and shown in cross-sectional HREM in Fig. 7.5(b), cannot be conclusively determined, as it appears to be defective and nonuniform.[13] Study of the growth of $CoSi_2$ layers on Si(100) by solid phase reaction finds several competing epitaxial orientations. However, conditions can be optimized for (100)-oriented single-crystal $CoSi_2$ layers of high structural quality to be fabricated. At the well-annealed interfaces of (100)-oriented epitaxial $CoSi_2/Si(100)$ layers, half-order spots are observed in transmission electron diffraction studies. Therefore, interface reconstruction has to be part of the structure of this interface. After considerable work and some debate, a 7-fold structure of the interface that consists of dimerized Si atoms, shown in Fig. 7.2, is now understood to be responsible for the extra diffraction spots experimentally observed.[14,15] Although the unreconstructed 8-fold structure has also been identified in some experiments.

An unusual substrate for the epitaxial growth of disilicides is the Si(110), on which single-crystal $NiSi_2$ and $CoSi_2$ can be grown with the same orientation as the Si. However, no part of the $NiSi_2/Si(110)$ interface is actually parallel to the surface-normal (110) plane, as the entire interface is made up of inclined {111}-oriented facets, running long in the $[1\bar{1}0]$ direction with alternating (111) and $(11\bar{1})$ orientations.[16] Therefore, the interfaces of epitaxial $NiSi_2$ layers fabricated on Si(110) are actually isostructured with the 7-fold type A $NiSi_2$/Si(111) interfaces! For $CoSi_2$, layers with uniform planar interfaces may be grown with single-crystal orientation on Si(110). A unique feature of the atomic structure of epitaxial $CoSi_2$/Si(110), not found at other epitaxial silicide interfaces, is that the Co at the interface does not reside at an extended lattice site of the Si. Co or Ni atoms at the interface layer of all other epitaxial silicide interfaces are positioned close to a lattice site, or a twin-related site, with respect to the Si crystal. Two possible rigid shifts of the $CoSi_2$ crystal with respect to the Si(110) crystal have been found, $\pm\frac{1}{8}[1\bar{1}0]$. Either one brings the interface Co to a bridge site of Si atoms at the interface to achieve full 8-fold coordination. These rigid shifts are likely driven by the energetic advantages of the 8-fold coordinated structure. As a result of the lateral shifts, the atomic rows of the Si crystal along the surface normal [110]

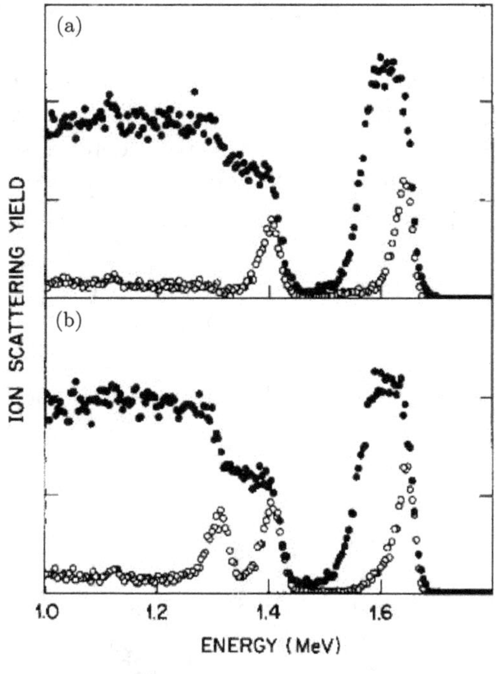

Fig. 7.8. Channeling (open circles) and random (closed circles) ion scattering spectra, obtained with a glancing exit angle, of thin $CoSi_2$ layers. (a) A 100 Å thick layer on Si(100) and (b) a 110 Å thick layer on Si(110).

Source: Reprinted from Tung.[2] Copyright (1992), with permission from Elsevier.

direction are brought into the channels (hollows) of the $CoSi_2$ crystal in the ion beam scattering and channeling experiment. A significant interface peak in the scattering spectrum, due to scattering off of only Si atoms, is observed for the $CoSi_2$/Si(110) system.[2] Such peaks are absent from all other epitaxial $CoSi_2$/Si and $NiSi_2$/Si interfaces studied, none of which includes a lateral shift in the structure.

7.1.3 *SBH Measurements of epitaxial silicide interfaces*

Whether type A and type B $NiSi_2$/Si(111) interfaces have the same or different SBHs was a question of great scientific interest and considerable initial debate on the experimental side.[1,17] These two interfaces have since been experimentally studied by many research groups, and it can be safely concluded that the n-type SBH at the type B $NiSi_2$/Si(111) interface is higher by about 0.14 eV over that of type A interface.[18–20] Furthermore, the planar $NiSi_2$/Si(100) interface was found to have an n-type SBH much smaller than either of the (111) interfaces.[10] Current-voltage characteristics obtained from these three single-crystal $NiSi_2$ interfaces are shown in Fig. 7.9.[21] The huge difference in SBHs of ~0.40 eV, or over one-third of the Si band gap, between the type B (111) interface and $NiSi_2$/Si(100) interface is in sharp contrast to the apparently "pinned" behavior of SBHs routinely found for polycrystalline interfaces between common metals or silicides and Si. This dependence of the SBH observed at epitaxial $NiSi_2$ interfaces on the atomic structure of the interface illustrates the importance and necessity of relying on MS interfaces with homogeneous structure to definitively test the validity of SBH concepts and models. Since the SBHs of type A and type B oriented $NiSi_2$ are different, it may be asked whether this SBH difference carries over locally to $NiSi_2$ diodes with mixed A and B orientations. BEEM has confirmed the presence of local SBH variations at the interfaces of routinely formed $NiSi_2$/Si(111) with mixed orientations. The formation of SBH is indeed a local effect, having to do only with the local atomic structure and not the larger structures on macroscopic scale.

As may be expected from the possible presence of facet bars with a different SBH, the current transport of $NiSi_2$/Si(100) interfaces depends sensitively on the morphology of the interface. Interfaces that are completely faceted, such as that shown in Fig. 7.7(a), yield an SBH of ~0.65 eV, in agreement with that observed at 7-fold type A $NiSi_2$/Si(111) interfaces, while the SBH of uniform $NiSi_2$/Si(100) layers is ~0.40 eV, characteristic of the 8-fold structure. For $NiSi_2$/Si(100) layers with a low density of facet

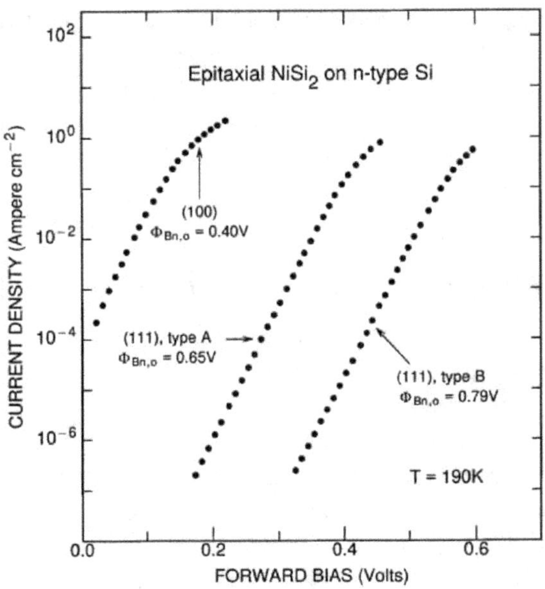

Fig. 7.9. I–V characteristics of single-crystal $NiSi_2$ Schottky diodes on n-type Si, recorded at 190 K.
Source: Reprinted with permission from Tung.[21] Copyright (1993) American Vacuum Society.

bars but otherwise uniform, such as the one shown in Fig. 7.7(b), electronic transport at n-type interfaces is not significantly affected by facet bars because of the higher local SBH for facet bars. The SBH measured by the C–V technique is also little affected by the presence of facet bars because of the small areal fraction. However, when a layer with the same morphology is formed on p-type Si(100), the locally lower SBH of the facet bars now offers an easier conduction path for hole transport. The current flow to a small low-SBH patch is controlled by the saddle-point potential in the p-type Si, which depends on many factors, including the size of the facet bar and the doping level.[10] Experimentally measured SBHs from $NiSi_2$ with similar morphologies grown on p-type Si(100) with different doping levels are shown in Table 7.3. These interfaces are clearly inhomogeneous electrically as the SBH depends on the measurement technique. While the C–V result is not affected by the presence of facet bars, I–V measured SBH decreases with doping level in agreement with a reduction in the pinch-off of the conduction paths to the facet bars. For comparison, the saddle-point potentials from

Table 6.1 based on the known SBHs of the faceted and flat regions are also listed.

The SBHs of well-annealed type B $CoSi_2$/Si(111) interfaces as well as those grown at room temperature are in the range 0.65–0.70 eV on *n*-type Si, which may be attributed to the observed 8-fold structure. On the other hand, $CoSi_2$ layers grown by solid phase reaction at lower temperatures show considerable variations in their SBH. In particular, the Co-rich "X" interface has a much lower SBH on *n*-type Si(111) and a correspondingly higher *p*-type SBH, as illustrated in Fig. 7.10 by the widely different I–V characteristics obtained from different interfaces with the same type B orientation.[13] The SBHs for the epitaxial $CoSi_2$/Si(100) and $CoSi_2$/Si(110) interfaces have been found experimentally to be similar to that for the 8-fold type B $CoSi_2$/Si(111) interface.

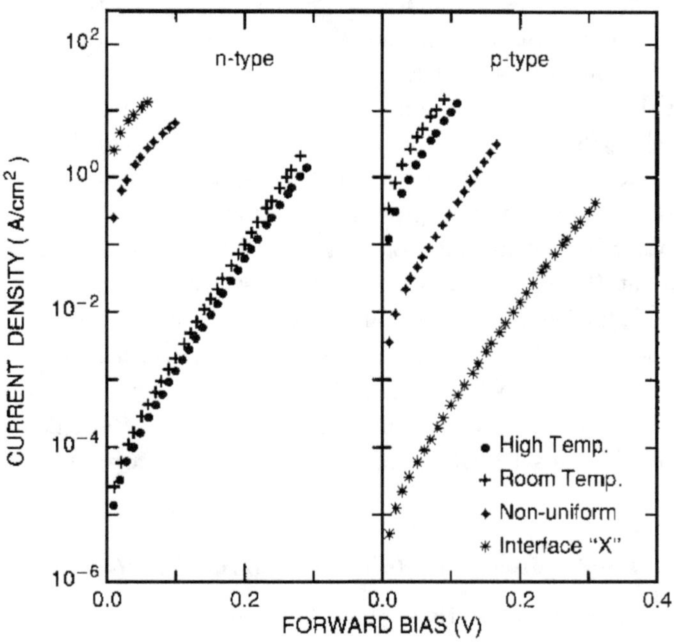

Fig. 7.10. I–V plots for room temperature measurements of $CoSi_2$ interfaces with *n*- and *p*-type Si(111). "High Temp." and "Room Temp" both indicate single-crystal type B $CoSi_2$ interfaces with 8-fold coordination. "X" marks a Co-rich interface fabricated at low temperature and "non-uniform" indicates low temperature fabricated interface with a low density of dislocations.

Source: Reprinted with permission from Sullivan *et al.* [13]. Copyright (1993) American Vacuum Society.

Table 7.2. Experimentally measured SBH of epitaxial silicide/Si interfaces. Calculated SBHs are listed in comparison.

			SBH Experiment (eV)		SBH Theory (eV)
Silicide	Substrate	Interface Structure	n-type	p-type	p-type
$NiSi_2$	Si(111)	7-fold type A	0.65	0.47	0.49
$NiSi_2$	Si(111)	7-fold type B	0.79	0.33	0.35
$NiSi_2$	Si(100)	8-fold	0.40	0.73	0.73
$NiSi_2$	Si(100)	7-fold type A {111}	0.65	0.45	0.49
$NiSi_2$	Si(110)	7-fold type A {111}	0.65	0.45	0.49
$CoSi_2$	Si(111)	8-fold type B	0.67	0.44	0.52
$CoSi_2$	Si(100)	8-fold and 7-fold 2×1	0.71	0.41	0.59
$CoSi_2$	Si(110)	8-fold	0.70	0.42	0.57

Even though $CoSi_2$ and $NiSi_2$ have identical crystal lattices, seemingly opposite behaviors are found for the SBHs of their epitaxial interfaces. Based only on the SBH magnitudes of the $CoSi_2$/Si epitaxial interfaces, one would have suspected that the FL were pinned because the SBH is largely independent of orientation. On the other hand, the SBH of epitaxial $NiSi_2$/Si interfaces would have led to the easy conclusion that the SBH depended sharply on the orientation! Epitaxial silicide interfaces thus illustrate the general inappropriateness in drawing conclusions about the fundamental SBH mechanism from only the SBH magnitudes. Each epitaxial silicide interface offers an opportunity to study the connection between its SBH and the distribution of charge associated with its atomic structure, which is not feasible for polycrystalline interfaces. Ultimately, the true merit of the epitaxial $NiSi_2$/Si and $CoSi_2$/Si interfaces is the opportunity they provide, through experimental results and *ab initio* calculations to understand the causal relationship between the atomic structure and the SBH.

7.1.4 *SBH of epitaxial metallic compounds on GaAs*

Crystalline metallic materials that have been grown epitaxially on zincblende compound semiconductors include elemental metals such as Al and Fe, and two groups of cubic solids with the CsCl and the rock salt structures. The interfaces between GaAs(AlAs) and a number of epitaxial intermetallic compounds have been examined by HREM, although without conclusive determination of their atomic structure. Evidence for a reconstruction has been observed at the interface between NiAl and overgrown GaAs. The SBH of various structures involving epitaxial metallic compounds has been studied in detail. It was found that buried NiAl has an SBH with respect to the overgrown (AlAs)GaAs different from that for NiAl overlayers on (AlAs)GaAs.[22] Even though the interfaces of NiAl and CoAl with GaAs were epitaxial, evidence for SBH inhomogeneity was found. The observed SBH also showed a dependence on the thickness of the NiAl and CoAl layers. The most remarkable result from these epitaxial metallic compound interfaces was the observation of a dependence of the SBH on epitaxial orientation for lattice-matched $Sc_{1-x}Er_xAs/GaAs$ interfaces, shown in Fig. 7.11.[4] It showed that annealing lead to significant changes in the observed SBH. The SBH of the ternary compound $Sc_{0.32}Er_{0.68}As$ varied by as much as 0.4 eV when the n-type GaAs orientation was changed from (100) to $(1\bar{1}\bar{1})$. It should be noted that multiple atomic structures may be possible for these epitaxial interfaces because of the alloy or compound nature of the conducting materials used. The electronic structures of various

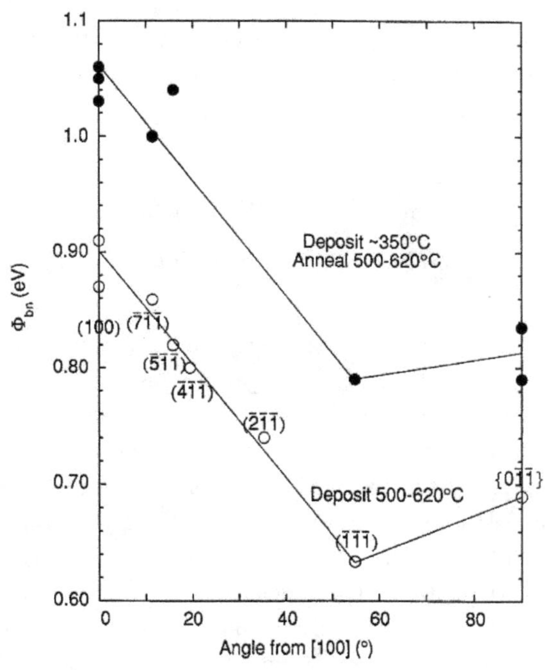

Fig. 7.11. Schottky barrier height versus substrate orientation for lattice-matched $Sc_{1-x}Er_xAs/GaAs\{h11\}$ diodes.

Source: Reprinted with permission from Palmstrom *et al.*[4] Copyright (1992) American Vacuum Society.

Table 7.3. SBH of $NiSi_2$ layers on Si(100) with different p-type doping concentrations measured experimentally with the C–V and I–V techniques. These layers have similar interface morphologies, with facet bars occupying ~10% of the total area. Calculated saddle-point potentials for these doping levels are listed for comparison.

	Expt. p-SBH (eV)		Theory (eV)
Doping (cm^{-3})	C–V	I–V	Saddle-Point
1×10^{17}	0.73	0.50	0.52
3.6×10^{16}	0.74	0.60	0.59
5×10^{15}	0.73	0.64	0.68

ErAs/GaAs interfaces have been investigated theoretically. The buckling of atoms in the mixed ErAs plane was found responsible for a very large adjustment (>1 eV) in the interface dipole during lattice relaxation. The p-type SBH calculated for an energetically favored (100) interface structure was lower than that calculated for the ErAs/GaAs(110) interface by about 0.2 eV,[23] in agreement with the experimental result shown in Fig. 7.11 for $Sc_{1-x}Er_xAs$.

7.2 First Principles Studies of MS Interfaces

7.2.1 *Calculated SBHs of epitaxial silicide interfaces*

The sharp dependence of the SBH measured from epitaxial $NiSi_2$/Si interface on its orientation and the availability of the atomic structures of these interfaces from HREM investigations present an opportunity and a challenge for theoretical calculations to explain and explore. Numerical calculations are conveniently carried out in the "supercell" geometry, with silicide and silicon crystals occupying opposite ends of the cell. Identical structures are constructed for the two back-to-back MS interfaces, as drawn in the example of Fig. 7.12, to avoid long-range electric fields in the semiconductor and allow its band positions to be established accurately with respect to the metal FL. As shown in Fig. 4.2, the SBH of an MS interface is determined from the distribution of potential energy. The planar-averaged potential energy distribution calculated for the type A $NiSi_2$/Si(111) interface, plotted as a thin line in Fig. 7.13, is used as an example to illustrate this process. The macroscopically averaged potential energy[24] (thick broken line) settles at constant local values separately for the metal portion and the semiconductor portion of the supercell. This

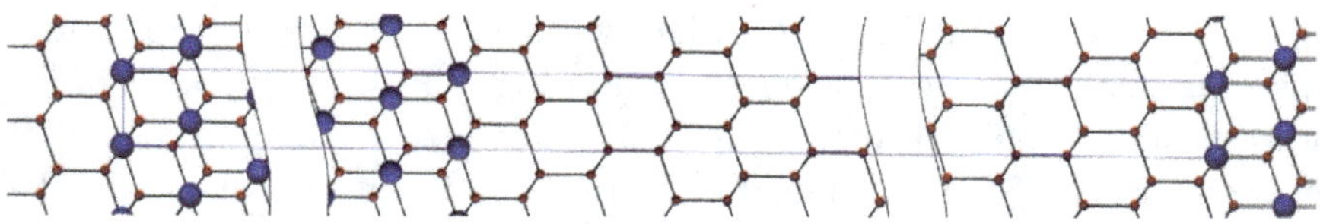

Fig. 7.12. Model of a supercell employed for the calculation of an epitaxial type A NiSi2/Si(111) interface. Thin lines mark the boundary of the unit cell used for the calculation, which contains two back-to-back interfaces with identical atomic structures.

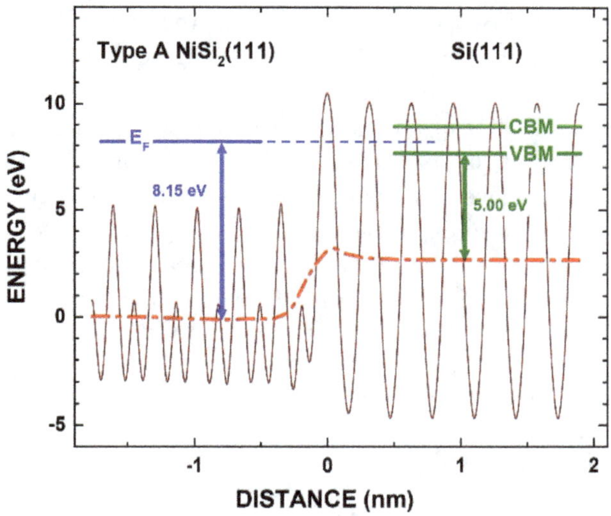

Fig. 7.13. Planar-averaged (pseudo-)potential energy (thin line) calculated for a relaxed type-A-oriented $NiSi_2$/Si(111) interface. The macroscopically averaged potential energy is shown as thick broken line, with respect to which the Fermi level of $NiSi_2$ and the band edges of Si may be identified.

difference in the average potential energies is the $e\Delta_{ISR}$ term needed for SBH determination. Following Eq. (4.2), the internal chemical potential energies for the metal, μ_M and the semiconductor, μ_{SC}, are now added to the average potential energy curve, respectively, to establish the projected VBM position of Si with respect to the FL of $NiSi_2$. As the band diagram of Fig. 7.13 illustrates, the FL of $NiSi_2$ is positioned at 0.48 eV (=8.15 − 5.00 − 2.62) above the VBM of the Si, which is the calculated *p*-type SBH of the type A $NiSi_2$/Si(111) interface. To deduce the *n*-type SBH for this interface, the experimentally measured band gap is used to arrive at ~0.64 eV as the calculated *n*-type SBH for the type A $NiSi_2$/Si(111) interface. Note that the supercell calculation also produces an FL that may be used directly for SBH determination, without making use of the potential distributions in the metal, as illustrated in Fig. 7.13. Both methods of FL identification should yield the same SBH. The use of the entire potential energy curve for SBH determination in Fig. 7.13 is a demonstration of the explicit dependence of the SBH on the charge distribution, which is the basis of SBH modeling to be discussed.

First-principles calculation was able to quantitatively explain the SBHs experimentally observed at epitaxial $NiSi_2$/Si interfaces. Due to the use of LDA, early DFT calculations could only approximately reproduce the experimental SBH difference between the type A and type B $NiSi_2$/Si(111).[25] With the addition of non-local interactions in advanced functionals, more recent calculations[26] were able to reproduce numerically the experimental SBHs of epitaxial silicides, as shown in Fig. 7.14. The sharp dependence of the experimental SBH on the orientation of the epitaxial $NiSi_2$/Si interface was quantitatively reproduced theoretically. Total energy calculations also explained the lack of experimental observation of any flat $NiSi_2$/Si(110) interface, as it was unstable against breaking up into inclined type A $NiSi_2$/Si(111) facets even at the cost of an increased

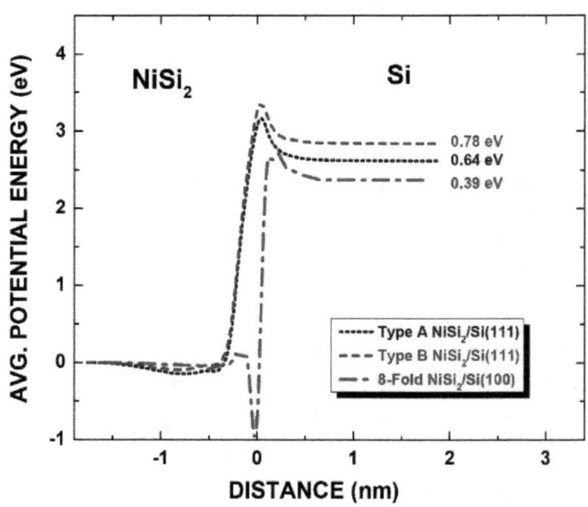

Fig. 7.14. Macroscopically averaged potential energy distributions calculated with hybrid functionals for relaxed epitaxial $NiSi_2$/Si interfaces with different orientations and interface structures. For each interface, zero distance marks the mid-point between the last atomic layer of the $NiSi_2$/Si and the first layer of Si crystal. The n-type SBH that may be derived is marked for each calculated potential energy curve.

total interface area! These $NiSi_2$ results, along with the calculated SBHs for epitaxial $CoSi_2$/Si interfaces, are summarized in Table 7.2. The lack of a strong dependence of the SBH of well-annealed epitaxial $CoSi_2$/Si interface on the orientation, which is the main experimental result for these interfaces, was well reproduced by theoretical calculations. However, quantitative agreement individually between theory and experiment was only adequate. At present, it is not clear why the calculated *p*-type SBHs for epitaxial $CoSi_2$ interfaces are consistently ~0.15 eV higher than the experimental results. Importantly, experimentally determined atomic structures of various epitaxial $CoSi_2$/Si interfaces were verified by total energy calculations. For example, the 8-fold type B $CoSi_2$/Si(111) interface was calculated to be favored over both the 7-fold type B and the 8-fold type A interfaces, as found experimentally. In general, calculations confirmed that bond formation can significantly affect the net electric dipole at a single-crystal silicide interface. Both the sharp dependence of the experimental $NiSi_2$ SBH on interface orientation and the lack of a strong dependence for the $CoSi_2$ interfaces were reproduced by theoretical calculations.

7.2.2 *Calculations of Metal/Zincblende interfaces*

Despite the presence of a rather large number of metals and semiconductors that have similar lattice structures and excellent matching conditions in lattice parameters, only a small number of epitaxial MS interfaces have been fabricated experimentally with high structural quality. Information gained experimentally from epitaxial interfaces alone is insufficient for an in-depth understanding of the SBH mechanism. The scarcity of available interfaces for experimental investigation need not impede the logical path to unravel the SBH mystery through epitaxial MS interfaces, as any lattice-matched metal–semiconductor combination is amenable to theoretical investigation of epitaxial interface. For such an epitaxial MS interface with a specific orientation, theoretical investigations first identify through interface energy analysis the most likely atomic structure of the interface were it to be fabricated in reality. Full-scale supercell calculation of its electronic structure then follows. Epitaxial MS interfaces between lattice-matched elemental metals and zincblende semiconductors have been systematically investigated, with the results of some calculations conducted for the non-polar (110) interfaces with GaAs and ZnSe summarized in Table 7.4. Following tradition, calculated *p*-type SBHs of epitaxial (110) interfaces between metals and zincblende semiconductors, including those listed in

Table 7.4. Calculated *p*-type SBHs for the relaxed non-polar (110) interfaces between metals and two semiconductors, GaAs and ZnSe, are listed under "Calc.", in comparison with the charge relaxed NPT predictions. Hybrid functionals are employed for DFT calculations.

	p-SBH GaAs(110) (eV)		*p*-SBH ZnSe(110) (eV)	
Metal	Calc.	NPT	Calc.	NPT
Rb	0.70	0.73	1.30	1.40
Al	0.84	1.47	2.05	2.23
Cr	1.23	1.28	2.46	2.05
Pt	0.95	0.45	1.73	1.25
Ca	0.25	0.46	1.53	1.31
Rh	0.76	0.61	1.39	1.45
Ir	0.86	0.83	1.58	1.63
Na	0.41	0.80	1.78	1.65
Ag	0.52	0.84	1.88	1.65
V	1.14	1.20	2.41	2.02
Au	0.69	1.09	1.68	1.92
Pd	0.62	0.37	1.27	1.28
CrFe	1.09	0.83	2.06	1.72
TiAl mix	1.10	1.17	2.42	2.05
TiAl_Al	0.89	0.85	1.88	1.68
TiAl_Ti	0.75	0.97	1.96	1.82

Table 7.4, are plotted against the SMT prediction in Fig. 7.15, with the straight line indicating perfect agreement.[27] There is little correlation between the calculated SBH and the SMT. Some of these discrepancies may be attributed to the surface dipole problem inherent in the SMT. However, it should be noted that the SBHs of single-crystal (110) interfaces do not display the apparently pinned behavior experimentally found for polycrystalline MS interfaces! For each semiconductor, the downward trend in the SBH observed is opposite to the small yet positive S parameters observed experimentally from polycrystalline interfaces. The epitaxial interfaces of Fig. 7.15 and Table 7.4 include not only those with excellent lattice-matching conditions, but also some with more significant lattice mismatches to broaden the scope of the study and reveal possible systematics. Some simple alloys are also included.

Since the atomic planes of a zincblende crystal alternate between cations and anions in the (100) direction, it is possible to have the metal bonded entirely to either the cation or the anion at a planar (100) interface. These two different epitaxial (100) interfaces have been studied theoretically,

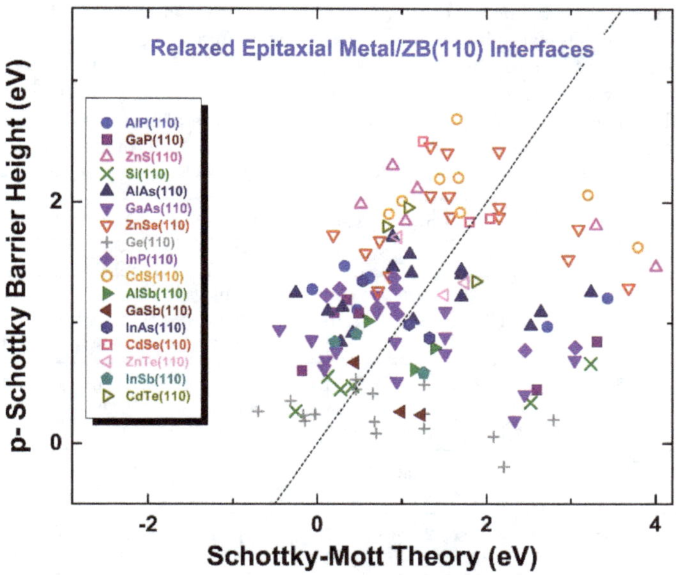

Fig. 7.15. Theoretically calculated *p*-type Schottky barrier height of non-polar (110) interfaces between elementals and zincblende semiconductors, plotted against the prediction of the Schottky-Mott Theory. No obvious correlations are detected. The DFT calculations are conducted with hybrid functionals to avoid band gap issues with the determination of Schottky barrier height.

Source: Reprinted with permission from Tung and Kronik.[27] Copyright (2021) by the American Physical Society.

but only for closely lattice-matched systems.[27] Some examples of these SBHs are listed in Table 7.5. Within the same MS system, the SBHs are generally different for the metal–cation and the metal–anion interfaces. For metals with larger WFs, the *p*-type SBH for the metal–anion interface exceeds that of the metal–cation interface, but the reverse is generally true for electropositive metals with smaller work functions. When SBHs are compared with the predictions of the SMT, the metal–cation interfaces and the metal–anion interfaces behave differently. As shown in Fig. 7.16, neither group agrees with the SMT, due partially to the surface dipole problem with SMT. Like the (110) interfaces just described, the SBHs of epitaxial metal–anion interfaces exhibit a systematic trend opposite to that predicted by the SMT. On the other hand, the SBHs of the (100)-oriented metal–cation epitaxial interfaces show a small, but positive, slope in Fig. 7.16(b), in agreement with the "pinned" behavior experimentally

Table 7.5. *P*-type Schottky barrier heights calculated with hybrid functional for relaxed epitaxial interfaces between elemental metals and zincblende semiconductors, with three different orientations or interface structures. SBH values produced by NPT model solids, with charge density equilibration and interface bond formation, are also listed. Consistent with the specific location of the metal–semiconductor interface concluded from analyses explained in the text, a different NPT model solid, involving embedded atoms, is employed to represent the metal for the metal–cation interfaces.

		(110) *p*-SBH		M-cation (100) *p*-SBH		M-anion (100) *p*-SBH	
Metal	Semi-conductor	Calc.	NPT	Calc.	NPT	Calc.	NPT
Na	InP	0.78	0.79	0.99	1.05	0.48	0.43
V		1.38	1.16	1.10	0.93	1.52	1.45
Au		1.23	1.11	1.00	0.97	1.29	1.57
Na	CdS	2.06	1.84	1.74	2.11	0.76	1.12
V		2.69	2.24	1.88	1.93	2.28	2.18
Au		1.90	2.22	1.77	1.95	1.87	2.30
Rh	GaP	1.08	1.02	0.98	0.77	1.47	1.44
Ir		1.09	1.05	0.83	0.80	1.56	1.95
Rh	ZnS	2.12	2.18	1.77	1.87	2.72	2.21
Ir		2.30	2.21	1.54	1.89	2.94	2.71

observed for polycrystalline MS interfaces. A comparison of Fig. 7.15 and the two panels of Fig. 7.16 leads to the odd conclusion that the metal–cation (100) interfaces are different from the other two types of interfaces. There is thus a suspicion that the SBHs at the metal–cation interfaces may require an explanation different from that for the other two!

7.2.3 *Calculations of MIGS distributions*

The magnitude of the SBH represents only a small piece of the information that can be extracted from theoretical calculations of epitaxial MS interfaces. Details of the electronic states and the spatial distribution of charge, which are responsible for the observed SBH, may all be examined. There is a particular interest in the distribution of MIGSs at MS interfaces because the CNL of MIGS was widely discussed as responsible for the experimentally observed SBHs for decades. Whether various assumptions about the MIGS are right or wrong can be judged from the results of these ab initio calculations. Out of the several hypotheses of empirical CNL models, the most important one is that the formation of MIGS is a bulk property of the semiconductor that does not depend significantly on the metal.[28,29] This assumption, which is necessary for the MIGS to be able to "pin the FL", has been shown to be incorrect repeatedly.[27,30,31] Discussions

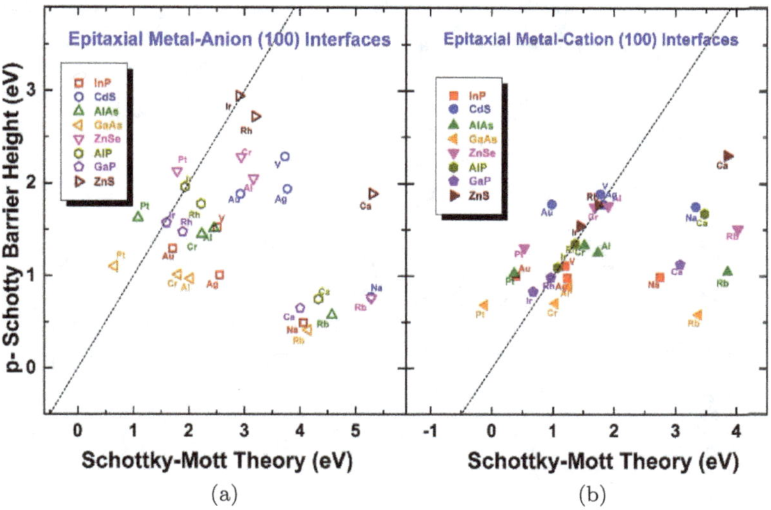

Fig. 7.16. Theoretically calculated *p*-type Schottky barrier heights for (100) oriented epitaxial interfaces between elemental metals and zincblende semiconductors with (a) metal–anion bonds and (b) metal–cation bonds, plotted against the prediction of the Schottky–Mott theory. These DFT calculations are conducted with hybrid functionals. It should be noted that the overall dependence of SBH on the metal is opposite to the trend predicted by the SMT for metal–anion bonded interfaces in (a). For metal–cation interfaces, the systematic trend is in general agreement with the SMT prediction, although the dependence on metal is much greater. In other words, apparent Fermi level pinning is on display for (b).

Source: Reprinted with permission from Tung and Kronik.[27] Copyright (2021) by the American Physical Society.

in Sec. 4.4 also showed that the formation of any MIGS is a direct result of the bulk metal state at that energy and $\vec{k}_{||}$, and therefore, it must depend on the metal! This is shown explicitly for metal/GaAs(100) interfaces in Fig. 7.17. There is no similarity in the densities of states in the projected band gap on the GaAs side of the interface, with different metals.[27,31] With the same metal, different MIGS distributions are also found for different atomic structures. What can be concluded from theoretical studies is that the distribution of MIGS is not a bulk property of the semiconductor.

Another major assumption of MIGS models is the existence of a unique CNL in the bandgap of each semiconductor that separates MIGSs with valence band character below and conduction band character above. The validity of this assumption is directly evaluated by examining the characters of MIGS states computed from ab initio calculations. It is known that

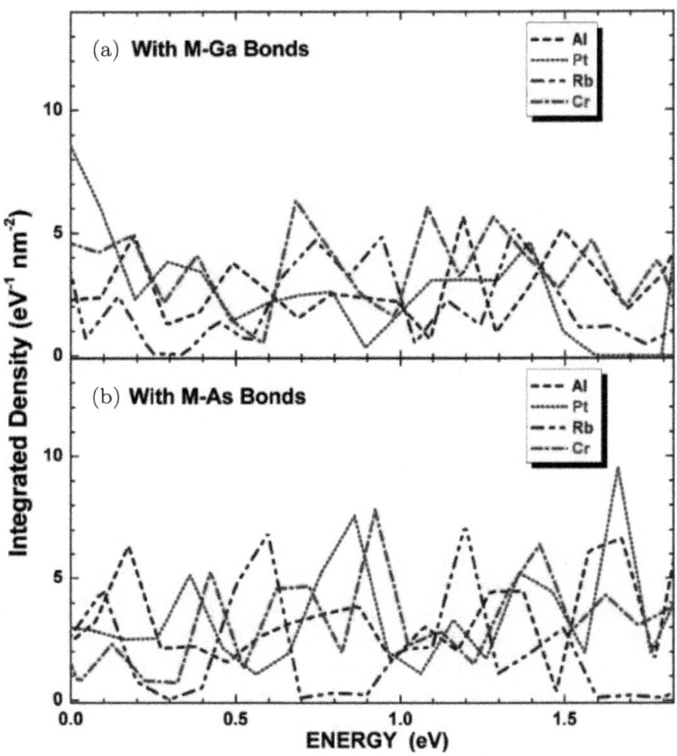

Fig. 7.17. Integrated electronic density within the projected band gap of GaAs for metal/GaAs(100) interface (a) with metal–Ga bonds and (b) with metal–As bonds.

the top part of the valence band of a bulk zincblende semiconductor has predominantly *p*- character as the sp^3 bonding orbitals are close to being filled. Strong s- character dominates states at the bottom of the conduction band, as the anti-bonding orbitals begin to fill. Symmetry-projected densities of states of two epitaxial metal–GaAs interfaces, plotted in Fig. 7.18,[27,31] indeed change from mainly *p*- character in the valence band to more s- in the conduction band of GaAs. However, this transition does not take place sharply, or even monotonically, in the projected band gap. Furthermore, the approximate location at which the character of MIGS changes is different for the two interfaces shown and is uncorrelated to the location of the FL, marked by an arrow, for either interface. These results disagree with the scenario assumed in order for CNL models to pin the Fermi level.[28,29] Metal always plays a crucial role in the formation of MIGS! There is perhaps no better place to illustrate that the interface

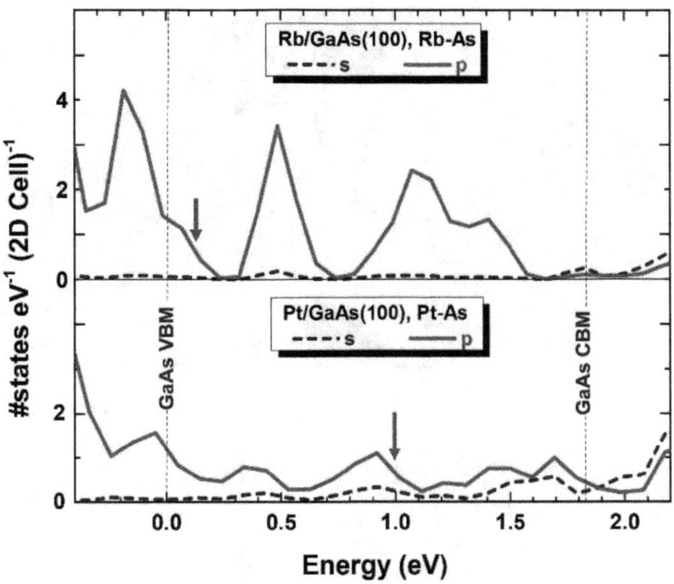

Fig. 7.18. The calculated numbers of states derived from the atomic orbitals of Ga and As at the Rb/GaAs(100) interface with Rb-As bonds (upper) and the Pt/GaAs(100) interface with Pt-As bonds (lower) are shown as a function of energy. Atomic orbitals of different symmetries are distinguished. The location of the FL for each supercell is marked by arrow and the projected band edges of GasAs are indicated by vertical lines.

structure also significantly affects the distribution of MIGS than at the type A and type B $NiSi_2/Si(111)$ interfaces. As shown by the energy-resolved and spatially resolved charge density maps of Fig. 7.19, the distributions of MIGS in the projected bandgap of Si are significantly different from a subtle change in the atomic structure.

7.2.4 *Spatial location of interface charge transfer*

In the original surface states pinning model, the interface dipole was assumed to arise from a charge transfer across a physical (oxide) gap between the metal and the semiconductor. With the proposal of the CNL concept, the interface dipole is envisioned to extend some distance into the semiconductor to be consistent with the idea that MIGS's responsible for FL pinning are properties of only the semiconductor. These conjectures disagree with what first-principles calculations actually show about the spatial location of the charge transfer at MS interfaces.[32] The spatial extent of charge transfer is directly revealed through the local net charge

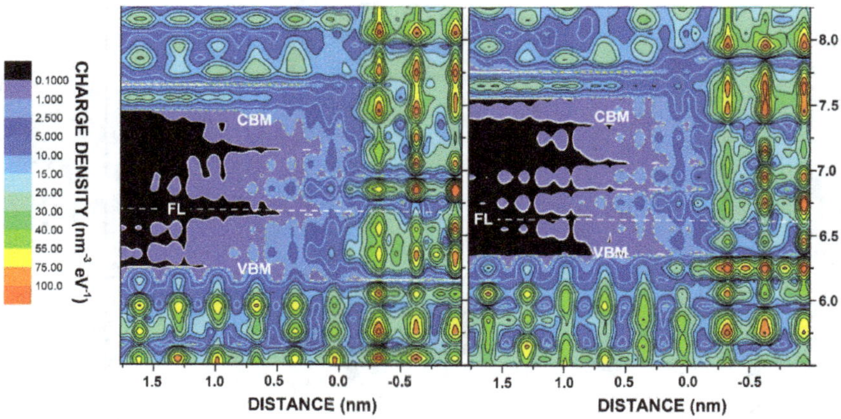

Fig. 7.19. Calculated density of electronic states as a function of the energy and the spatial location for (left) type A $NiSi_2$/Si(111) and (right) type B $NiSi_2$/Si(111) interfaces. The zero on the horizontal scale marks the mid-point location between the $NiSi_2$ ($z < 0$) and the Si crystals ($z > 0$). The Fermi level position and the approximate locations of the projected Si band edges are marked for each interface. In the band gap of the Si, the distributions of charge from MIGS (between VBM and CBM) are different for type A and type B interfaces.

of an MS interface. In a bulk zincblende crystal, the net density between consecutive planes (NDBP) vanishes along the (100) or (110) major direction. NDBP is defined as the number per unit area of all electrons between consecutive planes, subtracted by 1/2 of the nuclear charges on the two boundary planes. For bulk elemental metal, the NDBP also vanishes. A study of the NDBP through an MS interface thus reveals which parts of the interface gain and lose electrons, compared to the bulk. Shown in Fig. 7.20 are such plots from epitaxial interfaces with GaAs(100) and ZnSe(100) of four different metals, each with metal–cation and metal–anion interface structures.[27,31] Marked as "interface" here is the volume between the last metal plane and the first semiconductor plane, which largely houses the interface bonds and cannot be assigned to either the metal or the semiconductor. This region is set at zero on the x-axis. Vertical dashed lines mark the approximate locations of the first metal and semiconductor planes and partition each supercell into three regions: the metal, the semiconductor, and the interface. As shown in Fig. 7.20, the NDBP in the interior of the semiconductor is usually much smaller than those in the metal and the interface regions, suggesting that the main part of the charge transfer does not extend into the semiconductor. There are

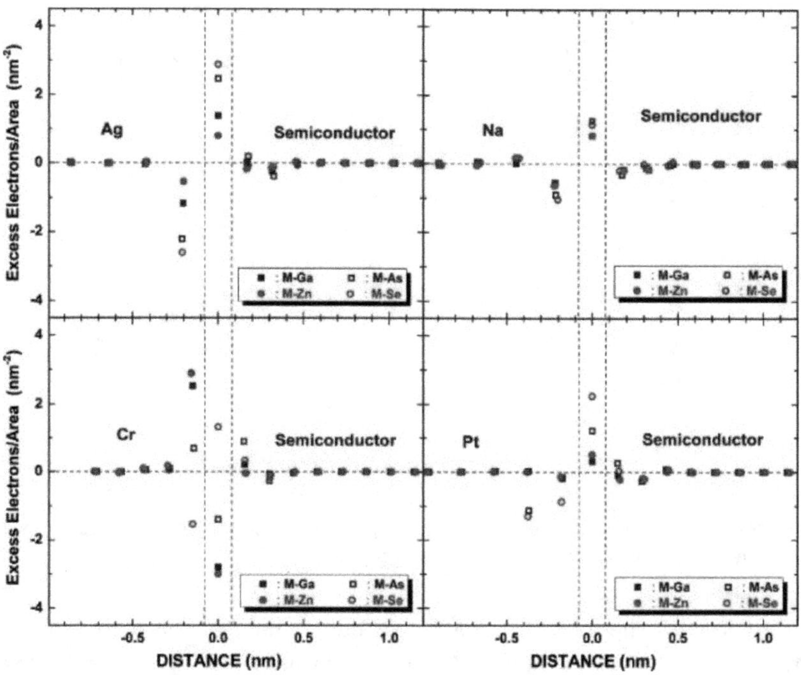

Fig. 7.20. The net density between planes (NDBP), which is the total number of electrons per area between consecutive atomic planes at MS interfaces, subtracted by one-half of the total positive ion charge on these two atomic planes, is plotted as a function of the distance from the interface. Vertical lines mark the approximate locations of the first metal and the first semiconductor planes. Data is shown for the metal–cation (solid symbols) and metal–anion (open symbols) interfaces between four metals, Ag, Na, Cr, and Pt, and two semiconductors, GaAs(100) and ZnSe(100).

significant net charges in the first one or two planes inside the metal. We caution that the charge transfers implied by Fig. 7.20 are not necessarily responsible for the entire overall dipole, as asymmetry in charge distribution within each layer could also contribute. However, the lack of significant charge transfers inside the semiconductor is consistent with, and suggestive that, the interface dipole being mainly located elsewhere, i.e. the interface region and the metal. An excellent correlation was found between the net charge in the metal and that in the interface region, as shown in Fig. 7.21. The closeness of points to the straight line, which has a slope of -1 and is not a fit, suggests that the charge transfer process takes place largely outside the first semiconductor atomic plane. Therefore, charges residing inside the semiconductor, including those associated with the MIGS, have

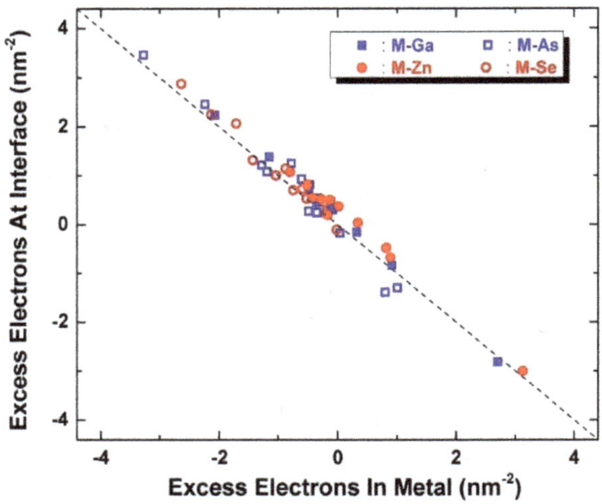

Fig. 7.21. Total number of excess electrons per area in the interface region is plotted against the density of excess electrons in the metal.

little to do with the interface dipole! Available analyses conducted on other epitaxial MS interfaces, such as those performed on type A and type B $NiSi_2/Si(111)$ interfaces shown in Fig. 7.22, confirm the main location of charge transfer being outside the semiconductor. Since the atomic planes are not equally spaced in the [111] direction for either Si or $NiSi_2$, the data shown in Fig. 7.22 have been filtered by taking centralized averages over the repeating unit of the lattice: 3 planes for $NiSi_2$ and 2 for Si. This filtering leaves a residual as the charge of the "interface", between the last $NiSi_2$ plane and the first Si plane, to ensure the neutrality of the entire interface. Similar to the metal/ZB interfaces, charge transfer toward the formation of an interface dipole at the $NiSi_2/Si$ interfaces is essentially outside Si.

7.2.5 *Length of interface dipole and nearsightedness*

In addition to the interface charge transfer revealed coarsely by NDBP, the length of the interface dipole can be defined more closely through charge distribution at an MS interface. Essentially, the question here is exactly where the two boundaries in Fig. 4.2 should be drawn to partition the charge distribution of an MS interface into an ISR and two bulk crystals? Since the charge distributions on both sides of the ISR should be essentially identical to those of the respective bulk crystals, the ISR includes the entire volume where the charge distribution is modified from

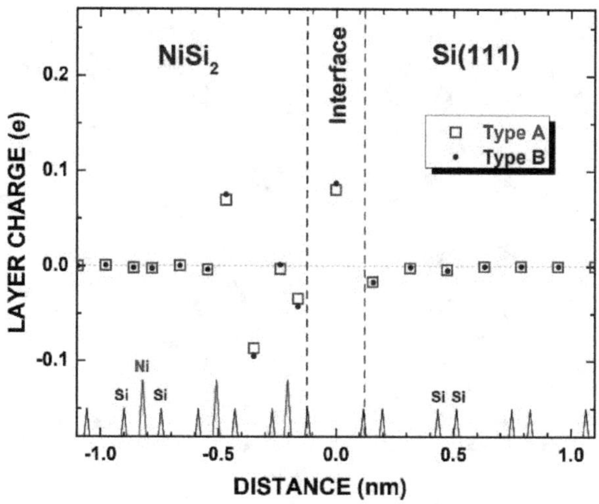

Fig. 7.22. The total number of electrons per 2D unit cell between consecutive atomic planes at type A and type B epitaxial $NiSi_2$/Si(111) interfaces, subtracted by one-half of the total positive ion charge on these two atomic planes, is plotted as discrete symbols as a function of the distance from the interface. The locations of atomic planes are marked on the bottom. As explained in the text, averaging over two or three planes is used in this graph. Vertical lines mark boundaries of the "interface region" at the locations of the last plane of the $NiSi_2$ tri-layers and the first plane of the Si bilayers.

the bulk. The minimum length of an ISR may be defined naturally as the length of the interface dipole. This length determines whether it is feasible to predict/explain magnitudes of the SBH with charge distribution modeling. If the ISR contained many atomic planes, the SBH for this interface would be essentially beyond the reach of simple chemical theories. Fortunately, however, the spatial extents of excess charges just discussed through Figs. 7.20–7.22 seem to suggest that dipole formation at an MS interface involves essentially only the first semiconductor layer. It still remains to be discovered whether the interface dipole is short enough for chemical modeling to be successful. Since earlier discussions pointed out the different systematic trends for metal–cation and metal–anion interfaces, these two types of interfaces are studied separately for the length of the interface dipole.

Generally speaking, the full spatial extent of charge distribution that has been modified from bulk density defines the length of the interface dipole. From the charge distribution near an interface, the present task is to determine the range where the local density deviates from "bulk-like".

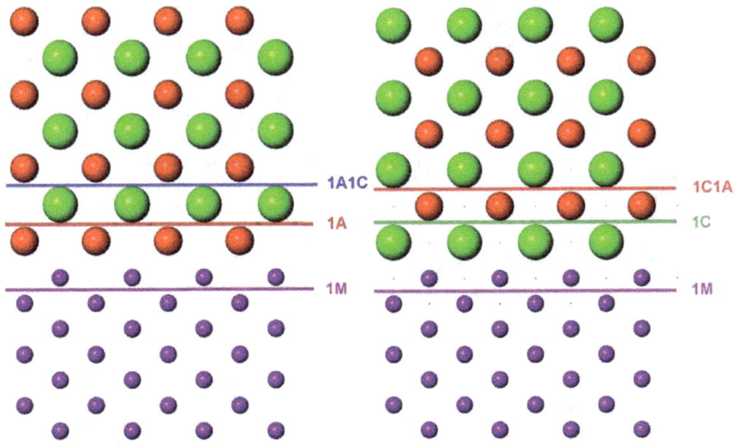

Fig. 7.23. Models of the stacking sequence at epitaxial metal–ZB (100) interface: small sphere: metal; medium sphere: anion; large sphere: cation. Neither the drawing on the left for metal–anion (100) interface, nor that on the right: metal–cation (100) interface necessarily corresponds to the actual atomic structure used in the calculations. Lines mark locations where the MS structure may be cut for the determination of the range of interface dipole.

Source: Reprinted with permission from Tung and Kronik.[31] Copyright (2021) by the American Physical Society.

Schematic diagrams of the metal–anion and metal–cation (100) interfaces are drawn in Fig. 7.23. To determine the length of charge transfer at each interface, we draw two trial demarcation lines that define a neutral ISR that is as small as possible. When the charge transfer is indeed contained within the ISR, the locations of its boundaries would be neutral or "bulk-like". Conversely, locations that are found not to be bulk-like are still part of the ISR. To define whether a location is already bulk-like or not, we first look for a salient property of an actual bulk crystal that makes it bulk-like. A bulk metal crystal can be cut in half, and the two exposed surfaces would be neutral and identical in every aspect. The calculated potentials of the two half crystals, with relaxation in electron density but not atomic positions, can then be added together, or "stitched", leaving no overall (long-range) potential shift. A bulk ZB semiconductor crystal behaves the same way if the cut is made along a (110) direction. Cutting along the polar (100) plane, however, exposes the inequivalent cation-terminated and anion-terminated surfaces on the two sides, which can no longer be stitched together without an overall potential shift. Nevertheless, a cation-terminated (100) surface can always be stitched with another cation-terminated (100) surface, and anion-terminated with anion-terminated, without long-range potential shift. The ability to be cut and then stitched together without a net potential drop is thus a hallmark of the bulk crystal that has been used to test if any location near an interface is already bulk-like.[31] To use this criterion, the structure shown in Fig. 7.23 can be cut at any chosen location into two halves. This location is already bulk-like if the potential distributions calculated separately for the two halves can be added to approximately reproduce the overall potential shift of the original (uncut) supercell, and the location is still part of the ISR if stitching fails.

This approach is demonstrated in Fig. 7.24, which shows macroscopically averaged potential energy distributions arising from various components of the Rh/ZnS(100) interface with Rh-S bonds.[31] To find out whether the interface dipole is completely formed by the first S layer at the interface, a cut is made there (line marked "1A" in Fig. 7.23), to leave bulk ZnS on one side and the Rh metal with a monolayer of S adsorbates on the other. "Monolayer" is defined here as that for a ZB (100) plane and "M+1A" is short for "metal with 1 monolayer of anion adsorbates". Calculations are then done separately for the surfaces of the two severed parts, with electronic relaxation but without lattice relaxation, leading to the potential energy distributions marked as "Rh+1S" and "ZnS(100)". The two potential energy curves are then stitched to form the curve labeled "Rh+1S Stitched". The overall potential energy shift across the interface region so stitched is nearly identical to that from supercell calculation (discrepancy ~0.04 eV), showing that the semiconductor is already bulk-like at the "1S" location and that the interface dipole is contained entirely on the "Rh+1S" side of the cut. The redundant counting of the sulfur plane does not affect the overall potential shift because the double-counted charge is symmetrically distributed about the S plane, when the location of the cut is bulk-like.

Since the stitching of the two surface potential energy curves is equivalent to superimposing their surface charge distributions without relaxation, the *p*-type SBH of the stitched structure is simply the difference between the IP of the semiconductor and the WF of the metal with an anion adsorbate. The excellent agreement between SBHs stitched in this manner and those calculated from supercells for almost all MS interfaces is shown in Fig. 7.25(a). This shows that the interface dipole is largely contained within the metal and the first anion plane. The cation does not need to be specified for the interface dipole to be calculated! The extremely

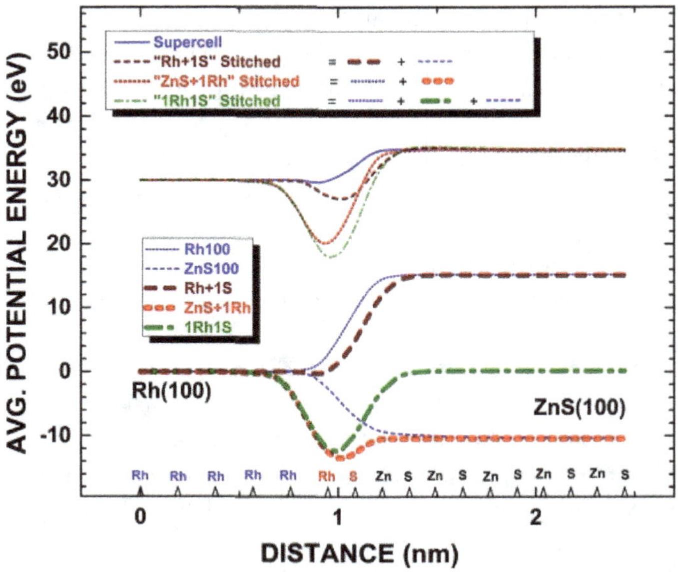

Fig. 7.24. Macroscopically averaged potential energy profiles of various structures associated with the Rh/ZnS(100) interface with Rh-S bonds. All curves are averaged over the repeating distances in both the ZnS(100) and the strained Rh(100) crystals. The profile marked "supercell" is obtained from the calculation of the entire structure. Locations of the atomic planes are marked on the bottom. Other calculations are based on the same (super) unit cell but with specified layers of atoms removed. "Rh+1S" is calculated with all semiconductor planes, marked in black on bottom, removed. "ZnS+1Rh" is calculated with all Rh planes marked in blue removed. "1Rh1S" is obtained with all planes removed, except the two interface planes marked in red. Lines marked "stitched" are the result of "stitching" two or more lines together, as illustrated symbolically in the legend box. The upper curves have been shifted vertically for clarity. *Source*: Reprinted with permission from Tung and Kronik.[31] Copyright (2021) by the American Physical Society.

short dipole length on the semiconductor side just found for metal–anion interfaces is not observed from metal–cation interfaces. A similar analysis carried out on metal–cation bonded (100) interfaces at the M+1C location leads to significant discrepancies between the calculated and the stitched SBHs for the majority of these interfaces. This suggests that the formation of the interface dipole is incomplete and extends beyond the interface cation layer for these interfaces. Enlarging the ISR to the M+1C1A line (Fig. 7.22) significantly reduces the error involved with stitching, as shown in Fig. 7.25(b). Therefore, the extent of the interface bond is indeed different for the metal–cation and the metal–anion interfaces, as speculated. This

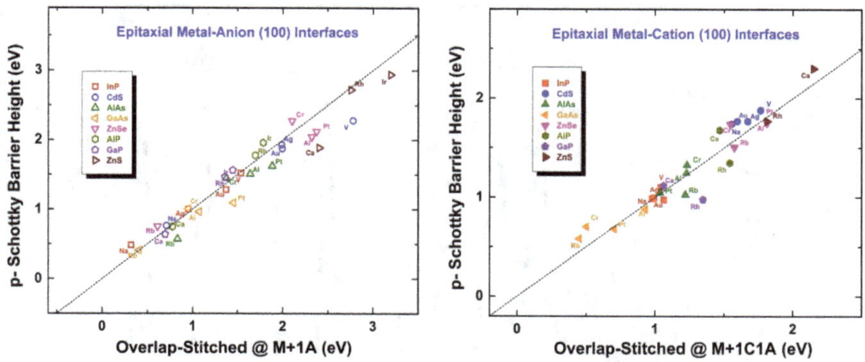

(a) (b)

Fig. 7.25. (a) Calculated p-type SBH of relaxed epitaxial metal–anion (100) interfaces, shown against the SBH stitched at the M+1A location, which is the difference between the IP of the semiconductor and the WF of the metal covered with a monolayer adsorbate of anion. (b) Calculated p-type SBHs of metal–cation (100) interfaces, plotted against the SBH stitched at the M+1C1A location.

Source: Reprinted with permission from Tung and Kronik.[31] Copyright (2021) by the American Physical Society.

difference necessitates separate approaches toward modeling the SBHs for these interfaces.

An alternative way to determine the dipole length is through a study of the minimum size of the oligo-cell needed to fully capture the transition between the metal and the semiconductor at an MS interface. One-half of the calculated charge distribution of an oligo-cell containing back-to-back interfaces can be used as the charge distribution of the ISR. When the oligo-cell is long enough, the entire charge distribution of the interface can be recreated by stitching with the charge distributions of the metal and the semiconductor on the two sides. Errors can be significant when the oligo-cell does not contain the entire interface dipole. This is the exact reversal of how the charge distribution in Fig. 4.3 was partitioned into three parts, and obviously Eq. (4.2) would hold for the stitched density. Through such a trial-and-error process, the length of the interface dipole may also be determined.

The plane-averaged charge density at the epitaxial Rh/ZnS(100) interface, with Rh-S bonds, is shown as thin solid lines in Fig. 7.26. The stacking sequence at this interface, -Rh-Rh-Rh-S-Zn-S-Zn-, seems to suggest that at least a sequence of -Rh-S-Zn- should be part of the oligo-cell to reproduce the chemistry and the charge distribution at the interface. The charge

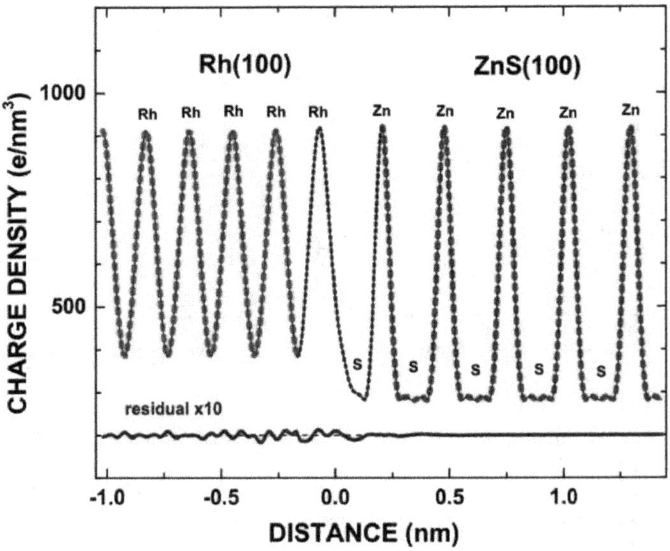

Fig. 7.26. Charge density calculated for Rh/ZnS(100) interface with Rh-S bond is shown as a thin solid line. Thick dotted lines are from calculations for the bulk crystals of Rh and ZnS, whereas the thin dotted line is the density of a "2 3" oligo-cell. The difference between the calculated and the stitched charge densities is shown with a 10x magnification and vertically shifted.

distribution from a "2 3" oligo-cell, with a unit cell of "Rh-Rh-S-Zn-S-", can provide such a charge distribution. Due to symmetry, the total charge of one-half of the oligo-cell, with a stacking sequence of "-Rh-S-½Zn", is overall neutral and therefore can be used in stitching trials. Indeed, this charge distribution, shown as a fine dotted line in Fig. 7.26, can be seen to appropriately reproduce the charge distribution at the actual interface. Also plotted to reveal the quality of agreement is the insignificant discrepancy between these two charge distributions even after a 10-fold magnification. Besides illustrating the nearsightedness of electron density generally at MS interfaces, Fig. 7.26 shows that the charge transfer at this interface involves few atomic layers at the interface.

7.2.6 *Determination of metallicity boundary*

Discussions so far have touched several times on the different behaviors exhibited by the metal–cation and the metal–anion (100) interfaces between metals and ZB semiconductors. These differences can be traced to the chemical species of the cations and the anions. The cations of zincblende semiconductors, such as Ga of GaAs and Zn of ZnSe, are ordinarily metals on their own while the anions, e.g. As and Se, are not. When a cation is fully bonded tetrahedrally to anions in a zincblende crystal, it no longer has mobile electrons. However, at a metal–cation (100) interface, cations on the interface plane are only partially bonded to anions. Therefore, they may provide metallic screening at the interface and hence become part of the metal rather than the semiconductor. Whether cations at the interface behave as metal or semiconductor is an issue of considerable importance. If they belonged to the metal side rather than were part of the semiconductor, much of the difference just described between metal–cation interfaces and metal–anion interfaces might be explained. Furthermore, metal–cation interfaces would need to be analyzed and modeled differently from the metal–anion interfaces, if that were the case. We therefore examine the issue of metallicity boundary.

Figures 7.24 and 7.25 suggest that the electrical properties of metal surfaces with a layer of cation or anion adsorbates should be investigated, as they lead to different stitching results. The WFs of the surfaces of different metals covered with cation and anion adsorbates are calculated systematically and shown against the WF of the underlying metal in Fig. 7.27.[31] For these calculations, adsorbate atoms are placed at the next lattice site of the metal (100) surface and no relaxations were allowed. Without relaxation, these results do not necessarily reflect the experimental WF. Nevertheless, they directly address the issue of metallicity boundary. Different trends are observed in the figure: the WF of the anion-covered surface decreases with the WF of the underlying metal, whereas the opposite is generally true for the cation-covered surface. These opposing trends cannot both be explained with electronegativity differences and point to a fundamental difference between the metal–cation and the metal–anion bonds. Cation adsorbates indeed behave as metal, as the WFs of cation-covered metal surfaces in Fig. 7.27 are close to the horizontal line marking the WF of the cation as a metal. In contrast, anion adsorbates form polar bonds correlated with electronegativities.

The metal–cation and metal–anion interfaces seem to differ mainly in the location where "the metal ends and the semiconductor begins" at the interface. Due to tunneling of MIGS into the semiconductor, interface calculations do not show a sharp boundary for the metal, which presumably is where electronic states near the Fermi level are confined spatially. Nevertheless, spatial variations in the density of states near the Fermi

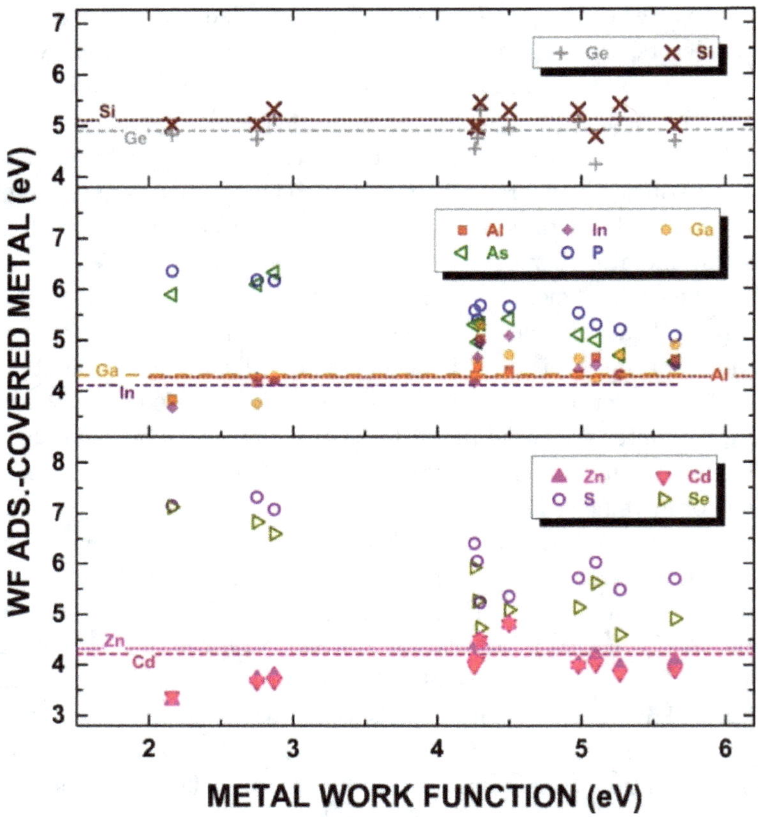

Fig. 7.27. Work function of various metal (100) surfaces covered with one monolayer of adsorbate, plotted against the work function of the bare metal. (a) Group IV adsorbates, (b) group III and V adsorbates, (c) group II and VI adsorbates. Horizontal lines mark the work functions of the metallic surface of elemental group II, III, and IV elements. *Source*: Reprinted with permission from Tung and Kronik.[31] Copyright (2021) by the American Physical Society.

level offer important clues. A comparison of the extent of metallic state decaying into the semiconductor gap is given in Fig. 7.28 for three metals with widely different WFs and two representative semiconductors. In all cases, a clear separation between charge density tails at the metal–cation interfaces and the metal–anion interfaces is observed. For each metal–semiconductor system, the exponential decay length (i.e., slope of the semi-logarithmic curves in the figure) is comparable for metal–cation and metal–anion interfaces. However, there is approximately a rigid shift of the

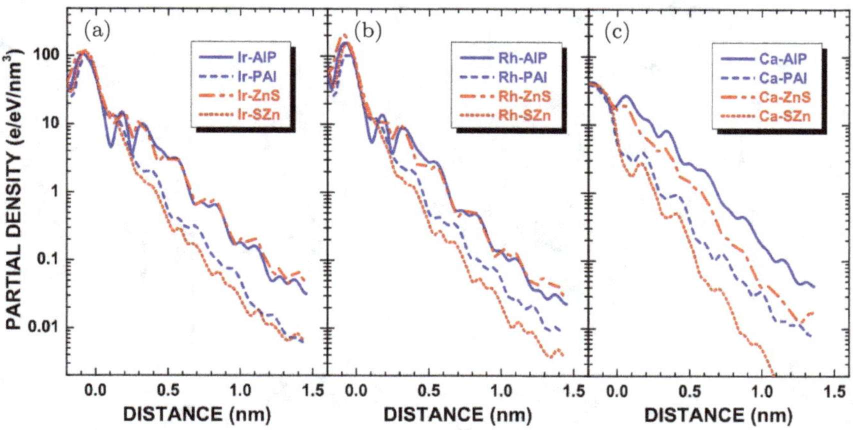

Fig. 7.28. Plane-averaged charge density of interface states with energy in the band gap of the semiconductor at the epitaxial (100) interfaces of AlP and ZnS with (a) Ir, (b) Rh, and (c) Ca, for metal–cation (solid lines) and metal–anion (dashed lines) interfaces. Zeros on the horizontal axis mark the locations midway between the last metal plane on the left and the first semiconductor plane, either cation or anion, on the right. The charge density at metal–cation interfaces extends further into the semiconductor than that at metal–anion interfaces.
Source: Reprinted with permission from Tung and Kronik.[31] Copyright (2021) by the American Physical Society.

metal–cation curves, by ~0.2 nm, toward the semiconductor, suggesting that "metallicity" extends deeper into the semiconductor at the metal–cation interface than at the metal–anion interface. The different systematic trends in SBH for these two types of interfaces can thus be attributed to a difference in the locations of the metallicity boundary.[31]

7.3 Quantitative Explanation of Epitaxial SBH

As discussed in Chapter 4, the p-type SBH of an epitaxial MS interface with known atomic structure should be modeled with the bulk densities of the metal and the semiconductor as

$$\Phi_{B,p}^{Modeled} = E_F^{ModSol} - E_{VBM}^{ModSol} + e\Delta V_{\delta\rho} + e\Delta V_{bond}, \tag{7.1}$$

with the last two terms of the RHS given by Eqs. (4.21) and (4.22). While unnecessary for the heterojunctions earlier, the $e\Delta V_{bond}$ correction is needed here for MS interfaces because MS interface bonds are generally different from that in either of the bulk crystals. Even though the metal–anion (100) and metal–cation (100) interfaces exhibit opposite systematic trends,

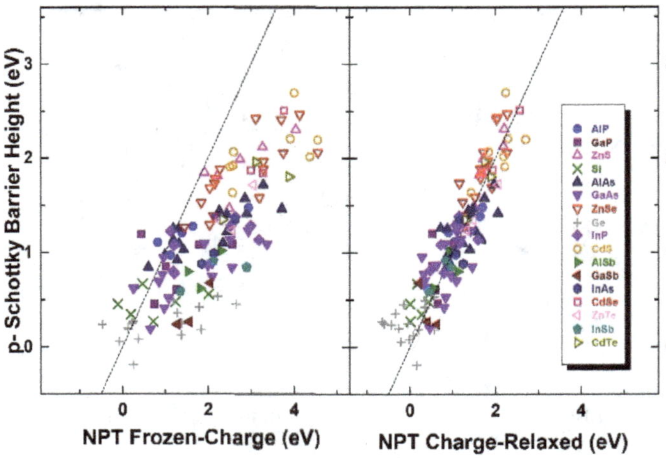

(a) (b)

Fig. 7.29. *p*-type SBHs, calculated using supercells for relaxed epitaxial metal/ZBD (110) interfaces, as a function of the prediction of NPT (a) without correction and (b) with corrections for density difference and interface dipole. Lines denote perfect agreement.

Source: Reprinted with permission from Tung and Kronik.[27] Copyright (2021) by the American Physical Society.

all SBH values and trends should be explained by charge distribution, as long as the model solids chosen adequately represent the metal and the semiconductor. Standard NPT model solids are appropriate for the metal–anion (100) and the (110) interfaces because the material interfaces are also the metallicity boundaries. When the data from Figs. 7.15 and 7.16(a) are replotted against the frozen-charge NPT model solid predictions, i.e. the first two terms on the RHS of Eq. (7.1), systematic trends emerge in Figs. 7.29(a) and 7.30(a).[31] After the corrections for density difference and bond are included, the calculated SBHs of these interfaces are in excellent agreement with those modeled with NPT, as shown in Figs. 7.29(b) and 7.30(b). In computing the bond corrections, the difference in the electronegativities of the metal and the anion of the semiconductor is used as ΔX for metal–anion (100) interfaces, while the averaged electronegativity of the cation and the anion is used for the semiconductor for the non-polar (110) interfaces. Since neither the exact amount of charge transfer nor the strength of screening is accurately known, the constant c of Eq. (4.22) used for the plots is found through a linear fit to be $\sim 0.04\,\text{eV} \cdot \text{nm}$.

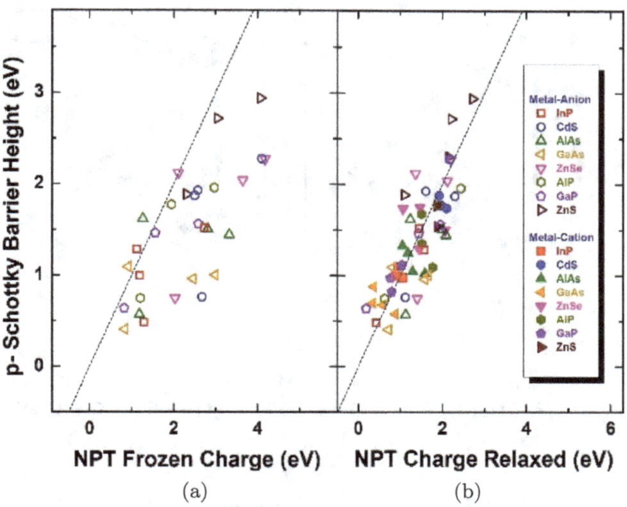

Fig. 7.30. *p*-type SBHs, calculated using supercells for relaxed epitaxial metal/ZBD (100) interfaces, as a function of the prediction of NPT (a) without correction and (b) with corrections for density difference and interface dipole. Lines denote perfect agreement.

Source: Reprinted with permission from Tung and Kronik.[27] Copyright (2021) by the American Physical Society.

A different strategy is required for the metal–cation (100) interfaces between metals and zincblende semiconductors because the metallurgical interfaces here are not the metallicity boundaries. If the standard model solids were used to plot the SBHs of these interfaces, large discrepancies would be observed as shown in Fig. 7.31(a). Since the metallicity for the metal–cation interface ends at the first cation layer and the "semiconductor" begins with the next anion plane, charge distribution modeling should be consistent with the actual location of the electrical interface. The NPT model solid still remains appropriate to represent the semiconductor. However, the proper model solid to represent the metallic side should be a metal that ends with a cation layer.[27] Specifically, we need "metal crystal covered with monolayer of cations but without surface dipole" to be the model solid for the metal side. In the spirit of the NPT, such a reference can be extracted from a metal crystal embedded with two monolayers of cations, as drawn in Fig. 7.32. The calculated charge distribution of this structure can be separated along proximity-cell boundaries between the two embedded monolayers, leading to a neutral surface of cation-covered metal that ends on polyhedron facets. Such a model solid provides

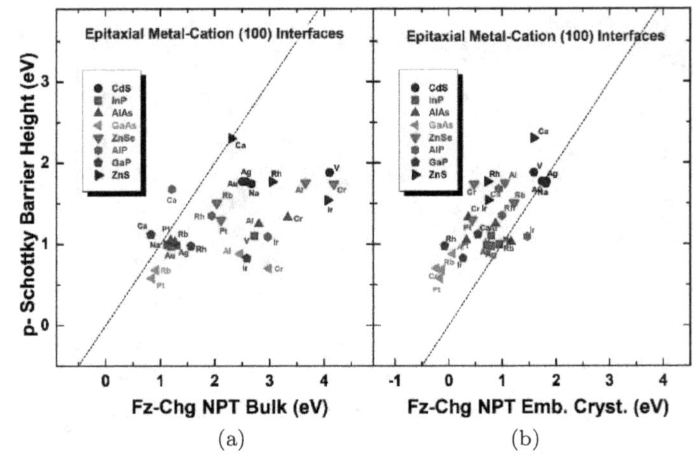

Fig. 7.31. Calculated p-type SBHs of relaxed epitaxial metal/ZBD (100) interfaces with metal–cation bonds, as a function of the frozen-charge NPT prediction, based on (a) model solids of bulk crystals and (b) model solid of "metal with cation adsorbate layer". Lines denote perfect agreement. See Fig. 7.30(b) for final result of modeling.

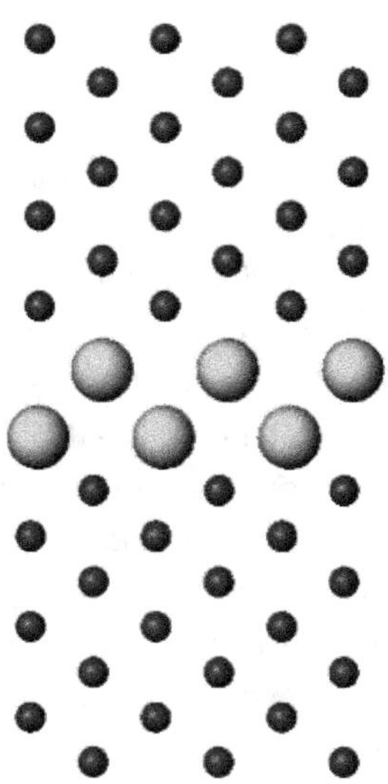

Fig. 7.32. Schematic representation of the structure used to generate an approximate bulk reference for metal covered with a monolayer of cations. Two cation layers (large symbols) are embedded in a crystal of metal (small symbols).

the appropriate ϕ^{ModSol} to use in the analysis of metal–cation interface via Eq. (7.1). With the use of such model solids for metals,[27] a tighter distribution of data is seen in Fig. 7.31(b) than Fig. 7.31(a). The small and one-sided SBH differences between unrelaxed model solids and calculation are the result of bond formation between the semiconductor's own cation and anion. With this contribution included the calculated *p*-type SBHs of the metal–cation (100) interfaces are in good agreement with the NPT model, as shown in Fig. 7.30(b). A comparison of seemingly random fashion the metal/zincblende data are plotted in the traditional way in Figs. 7.13 and 7.14 with how they closely agree with the NPT predictions in Figs. 7.29(b) and 7.30(b) confirms that the charge distribution is the foundation of the SBH at MS interfaces. Opposite systematic trends are found for different types of interfaces initially. At the end, the entire data set from metal/zincblende interfaces is quantitatively explained by the same procedures, after the issue of metallicity boundary is recognized and taken into consideration.

7.4 Polycrystalline MS Interfaces and FL Pinning

7.4.1 *Binary Oxide Schottky barriers*

The SBH of a polycrystalline MS interface is likely inhomogeneous. Areas of an interface with different atomic structures generally have different local SBHs, as found for metal/zincblende interfaces. Experimentally measured SBHs of polycrystalline interfaces are statistical averages over the range of inhomogeneous SBH and thus appear difficult to model. However, as pointed out in Chapter 5, polycrystallinity does not necessarily prohibit a density-based approach from SBH modeling. A class of polycrystalline MS interfaces with favorable conditions, those between metallic binary oxides (M_aO_b) and semiconducting binary oxides (Z_cO_d), is discussed here. Since oxygen is the only anion for these MS systems, the network of polar bonds continues through the interface with only one possible stacking sequence of atoms, – O – M – O – Z – O –. The same stacking sequence should lead to a similar SBH for areas of a polycrystalline interface with different orientations and detailed atomic structures. A uniformity in the SBH of binary oxide interfaces is expected here because the effect from interface bonds can be ignored when bulk bonds continue at the interface. In other words, the SBHs for these interfaces should be bulk-like. Experimentally measured SBHs of binary oxide interfaces can be modeled

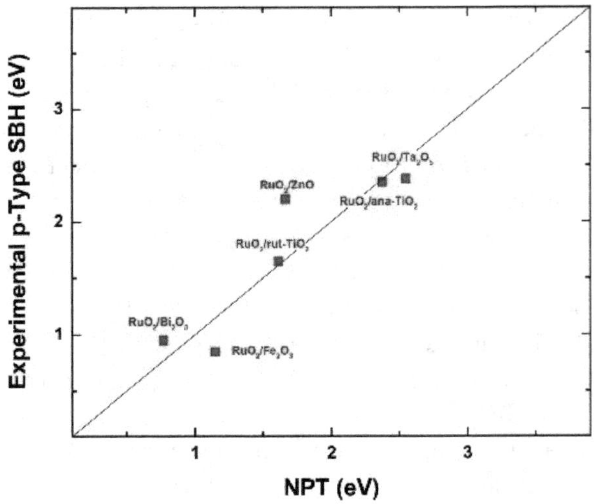

Fig. 7.33. Experimentally measured p-type SBHs from polycrystalline interfaces between conducting and semiconducting binary oxides are shown against the predictions from NPT with corrections for charge density differences. Data shown in Table 5.5 are used in this figure.

Source: Reprinted from open-access article Tung.[33]

with the NPT model solids. As shown in Fig. 7.33, the experimental SBHs for polycrystalline MS interfaces between binary oxides agree well with the differences between the NPT bulk references for the metal and he semiconductor (Table 2.8).[33] This agreement confirms that the effect of interface bonds can be ignored under the condition of a continuous network of bonds through the interface. Without the need to correct for interface bonds, NPT model solids can quantitatively model the SBH of even polycrystalline MS interfaces.

7.4.2 *The FLP myth and what it really represents*

Early investigations of the SBH dealt almost exclusively with polycrystalline MS interfaces, from which the overwhelming experimental observation was the FLP phenomenon. This unfortunately led to the opinion that, on a fundamental level, the SBH was insensitive to the metal and interface specifics. The dominance of the FLP phenomenon also led to the proposal of various models with built-in insensitivity to metal. Reexaminations of old (and new) SBH measurement data show that the SBHs of polycrystalline MS interfaces were pervasively inhomogeneous and the SBHs measured were averages of different local SBHs. Therefore, the interfaces used decades ago to demonstrate the FLP effect in fact contained irrefutable evidence for spatially inhomogeneous and therefore "unpinned" FL position in the band gap! With the issue of SBH inhomogeneity coming to light, the original experimental observations on FLP should be clarified as that "the AVERAGE SBHs of polycrystalline interfaces were insensitive to the metal and the interface structure". In other words, FLP is not a phenomenon at a fundamental level but rather a macroscopic observation on the experimental SBH averages. Even though non-fundamental, FLP is still a puzzling phenomenon. Why should the SBH, which for single-crystal interfaces depends sensitively on the metal and the structure, average out for polycrystalline interfaces to values that do not depend much on the metal?

7.4.3 *FLP mystery explained with charge density*

The FLP phenomenon was discovered from SBH studies of zincblende/diamond semiconductors. It was also for these semiconductors that the FL appeared to be most "tightly pinned", meaning that the experimental SBHs tended to fall in a very narrow range about an assumed pinning level. Therefore, a satisfactory explanation of the apparent FLP phenomenon for zincblende semiconductors would solve the mystery for all. The mystery is that the average SBH on these semiconductors appears to be nearly constant for each semiconductor and to be largely independent of the metal used. Before diving into the specifics of polycrystalline interfaces of zincblende semiconductors, we first consider possible scenarios that could make the average of spatially varying SBH appear independent of the metal. Due to the dependence of the SBH on atomic structure, any apparent similarity in SBH is likely related to similarity in atomic structures. Therefore, a possible scenario to explain the FLP phenomenon is that the polycrystalline interfaces of different metals on the same semiconductor are dominated by a specific structure that is largely independent of the metal used. Since the SBH is a direct result of the charge distribution associated with the interface atomic structure, the search for an explanation of the FLP phenomenon becomes a search for the structure(s) responsible for the average SBH of each semiconductor, i.e. the experimentally observed pinning level.

At an interface between a metal and a zincblende semiconductor, structures with metal–cation bonds and those with metal–anion bonds are clearly distinctive because of the large difference in electronegativities of the cation and the anion. Accordingly, the SBHs associated with these two structures are at the opposite ends of the range of local SBHs for polycrystalline metal/zincblende semiconductor interfaces. It seems likely that other structures of the interface could be viewed and analyzed as containing different mixtures of the metal–cation and the metal–anion bonds. For example, an examination of Table 7.5 shows that the SBH of the non-polar metal/ZB(110) interface, which contains equal densities of metal–cation and metal–anion bonds, usually lies between those for the metal–cation (100) and the metal–anion (100) interfaces. Therefore, the metal–cation and the metal–anion structures are representative of the range of disparate structures that may exist between metals and zincblende semiconductors.

In order to understand the average SBH of a polycrystalline MS interface with unknown structures, the SBHs associated with the metal–cation (100) and metal–anion (100) interface structures should provide valuable guidance. One begins by asking whether one of these structures, if actually occupying the majority of the polycrystalline interface area, would lead to the experimental observation? It is known that cations at the interface behave as a metal and significantly screen the effect of the additional metal applied. Out of the three epitaxial interface atomic structures, only the SBH of metal–cation interfaces has the same systematic trend (small positive S-parameter) as the polycrystalline FLP data. Therefore, if polycrystalline interfaces were indeed dominated by metal–cation bonds, the FLP phenomenon would be observed. But what would the pinning level be for interfaces dominated by metal–cation bonds? Clearly, the answer is that it would be the SBH between a semiconductor and its own cation (in elemental metal form). This is a quantity that has been theoretically calculated for each semiconductor.[31] Except for the Al/AlAs system, there is no lattice matching condition between a semiconductor and its own cation. These interfaces are nevertheless calculated epitaxially, with the metal form of the cation in strained f.c.c. structure. For elemental Si and Ge, there is no bulk metallic form *per se*, but the free surfaces are metallic and have WFs, toward which the underlying metals with adsorbed Si or Ge converge, as shown in Fig. 7.28. Therefore, the difference between the Fermi level and the VBM at the surface of Si and Ge may be regarded as the *p*-type SBH between the semiconductor and the metallic form of itself. It should be noted that, similar to Si and Ge, the free surfaces of cation-terminated ZB (100) are also metallic and have Fermi level positions that

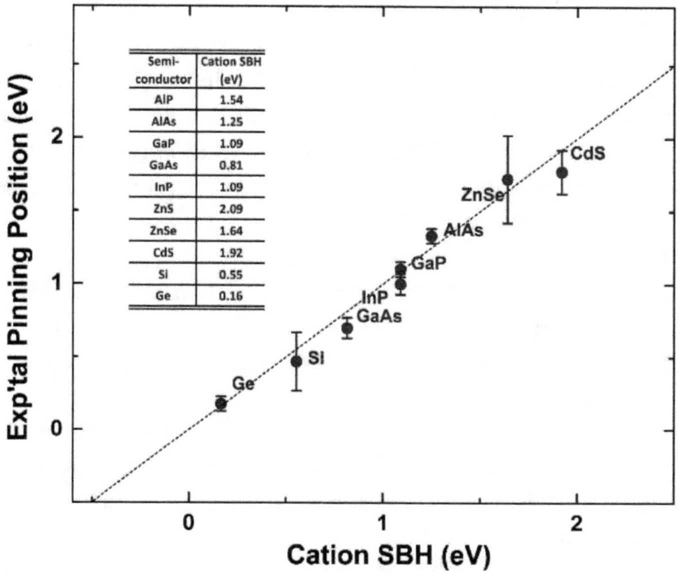

Fig. 7.34. Experimentally measured average p-type SBH, shown against the p-type SBH calculated for each semiconductor with its own cation. See text for Si and Ge. Experimental results are adopted from the literature.

Source: Reprinted with permission from Tung and Kronik.[31] Copyright (2021) by the American Physical Society.

are in general agreement with those calculated from supercells of ZB and cation. Experimentally observed pinning level is plotted against the SBH calculated for each semiconductor with its own cation in Fig. 7.34, with perfect agreement indicated by the dashed line, which is not a fit.

The excellent agreement suggests that cation-rich interface structures, including but not limited to the metal–cation (100) structure, dominate polycrystalline ZBD interfaces and are responsible for the observed FLP phenomenon. Besides the strong evidence from the agreement with the pinning level, are other experimental data in support of, or in conflict with, a prevalence of cation-rich structures at polycrystalline MS interfaces? Calculations actually show that, for the vast majority of systems, the metal–anion (100) interfaces are slightly favored energetically over the metal–cation (100) interfaces.[27] However, this preference does not indicate that the common polycrystalline interfaces would be dominated by metal–anion bonds because the kinetic path toward the formation of the physical interface depends on many other factors. MS interfaces are usually formed by the deposition of metal onto a semiconductor surface. A beam of metal atoms/molecules emitted from a hot furnace or an e-gun source carries significant kinetic energy as it impinges on a semiconductor. Decomposition of the semiconductor surface from metal deposition is quite common. Also common are the spontaneous formation of metallic compounds during metal deposition or as a result of post-deposition anneals. These are the kinetic processes that actually determine the atomic structures of the eventual MS interfaces. In controlled clean experiments, both cations and anions are observed to be released from the surface upon metal deposition and anions have a persistent tendency to segregate to the external metal surface.[34–37] Anion segregation presumably leaves the interface cation-rich. Another point worth noting is that the heats of formation for metal–anion compounds are generally larger than metal–cation compounds,[38] thus providing a driving force for cations to remain at the interface. These known facts about MS interface formation also support the idea that polycrystalline interfaces are cation-rich in general.

In the massive number of SBH investigations conducted under various experimental conditions, there were notable exceptions to the dominant FLP phenomenon. The particular processing conditions under which the measured SBHs were found to be different from the normal SBH also offer clues as to whether FLP and cation-rich structure are related. To analyze these data, we begin by asking ourselves a simple question "if cation-rich structure indeed caused the FLP, what could we do to make the SBH different from the usual, pinned SBH?". The answer is obvious: preservation of anion at the MS interface should reduce the dominance of cation-rich structures and allow the SBH to depend more on the metal or the structure. From the perspective of keeping anions in place during the formation of an MS interface, two approaches seem likely to work. Anion loss should be reduced or prevented by either "softening" the metal deposition process to minimize decomposition/damage of the semiconductor surface or by a simultaneous supply of anions during metal deposition. A number of studies have already adopted soft landing approaches for metal deposition. Buffered deposition at cryogenic temperatures,[39] indirect deposition,[40] and electrochemical deposition[41,42] have all been employed to make the metal deposition process "softer". Interestingly, these studies found that the SBH of some softly formed interfaces was significantly different from the pinned SBH usually observed from interfaces formed with conventional metal deposition.[39,41,42] We also note that arsenic was co-deposited during the growth of epitaxial $Er_xSc_{1-x}As$ layers on GaAs, shown in Fig. 7.14. Clearly, the FL was not pinned for these $Er_xSc_{1-x}As/GaAs$ interfaces as the SBH varied significantly with the orientation of the interface.[4] Therefore, some of the interfaces that did not show FLP were fabricated under conditions that avoided or reduced anion loss from the semiconductor surface. This finding also supports the attribution of FLP to interface cation richness and identifies anion preservation as a viable strategy to tune the SBH.

A strong case is thus built for the abundance of cation-rich structures at the metal–zincblende semiconductor interfaces as the reason for the FLP phenomenon. This answer is so trivial that it conceivably might have been guessed from the very beginning. At the end, the reason for the observation of the FLP phenomenon is the same as that which explains all other SBH dependencies: charge distribution.

7.5 Summary

The magnitude of the SBH is the direct result of the charge distribution of its interface, exactly the same way the VBO of a heterojunction is. Not surprisingly, the SBH may be modeled essentially the same way as VBOs were in Chapter 5. However, there is an important point that needs to be considered for SBH modeling: the location of the metallicity boundary. A knowledge of where metallic screening ends and dielectric screening begins at an MS interface is crucial not only for the charge distribution of the interface to be reasonably modeled but also at the very heart of the Fermi level pinning mystery. The important role played by the metallicity boundary came to light when the SBHs at epitaxial metal semiconductor interfaces were modeled. There are two possible planar structures in the polar (100) direction with the metal bonded either to the anion as drawn in Fig. 7.35(a), or to the cation as shown in Fig. 7.35(b). These two structures look innocuously similar as they seem to differ only in which part of the semiconductor the metal makes contact to. Even though zincblende semiconductors are insulators, they are compounds made up of cations and anions. Furthermore, the cations, e.g. Ga of GaAs or Zn of ZnSe, are metals on their own. Results show that the cations in contact with the metal at metal–cation interfaces actually behave as part of the metal, rather than the semiconductor. The metallicity boundary for the metal–cation interface is between the interface cation plane and the next anion plane, as marked on Fig. 7.35(b). For a metal–anion bonded interface, the anion is part of the semiconductor and the metallicity boundary is between the metal and the anion, as usual. Accordingly, the SBH may be successfully modeled

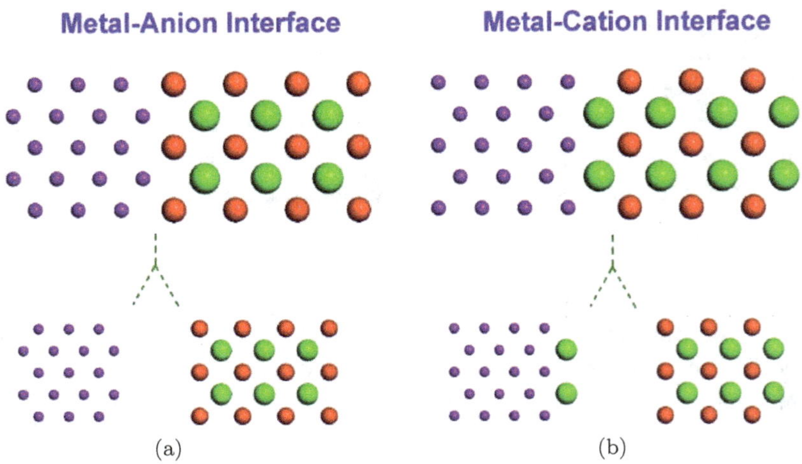

Fig. 7.35. Models of (a) metal–anion bonded and (b) metal–cation bonded interfaces between elemental metal and zincblende semiconductor with the (100) orientation. Small spheres: metal atoms, medium spheres: anion atoms, large spheres: cation atoms. In the lower part, each interface is conceptually separated along the metallicity boundary of that interface into structures that represent the metal and the semiconductor. The metal for the metal–cation interface is actually covered by a layer of cations. The bottom right configuration here is the basis for the modeling of the SBH in Fig. 7.31(b).

by starting with NPT model solids for the metal and the semiconductor, as schematically drawn in Fig. 7.35(a). Following the same rules, the metal–cation interface should be modeled with an NPT model solid for the semiconductor and a model solid that represents "a metal covered with a monolayer of cations" because that is what "the metal" is for this interface. A strategy to create a model solid for the cation-covered metal, as shown at the bottom of Fig. 7.35(b), is found that successfully modeled the charge distribution at metal–cation interfaces. The realization that interface cations in contact with the metal behave as part of the metal offers an excellent explanation of the FLP mystery that decades ago caused a sensational frenzy in this field. Cation-rich structures at the interface are the cause of FLP. Since the "metal" for these interfaces is covered with a layer of cations, as schematically drawn in Fig. 7.35(b), the Fermi level of the model solid for such a "metal" does not depend significantly on the applied metal but on the cation monolayer! The observed pinning levels of polycrystalline MS interfaces agree very well with those calculated for interfaces between semiconductors and their own cations in metallic form. Strong evidence for cation-richness at metal/zincblende interfaces is also found from a careful re-examination of availablespectroscopic data on the fabrication process of polycrystalline MS interfaces and the observed variation of the SBH with the prepration method of the interface. All experimental and theoretical results are in agreement with a cation-rich interface being responsible for the observation of the FLP phenomenon.

References

1. R.T. Tung, *Phys. Rev. Lett.* **52**, 461–464 (1984).
2. R.T. Tung, *Mater. Chem. Phys.* **32**, 107–133 (1992).
3. J.P. Harbison, T. Sands, N. Tabatabaie, W.K. Chan, L.T. Florez, and V.G. Keramidas, *Appl. Phys. Lett.* **53**, 1717–1719 (1988).
4. C.J. Palmstrom, T.L. Cheeks, H.L. Gilchrist, J.G. Zhu, C.B. Carter, B.J. Wilkens, and R. Martin, *J. Vac. Sci. Technol. A* **10**, 1946–1953 (1992).
5. R.T. Tung, J.M. Gibson, and J.M. Poate, *Phys. Rev. Lett.* **50**, 429–432 (1983).
6. R.T. Tung, J.M. Gibson, and J.M. Poate, *Appl. Phys. Lett.* **42**, 888–890 (1983).
7. D. Cherns, G.R. Anstis, J.L. Hutchison, and J.C.H. Spence, *Philos. Mag. A* **46**, 849–862 (1982).
8. J.M. Gibson, R.T. Tung, and J.M. Poate, *Mater. Res. Soc. Symp. Proc.* **14**, 395–409 (1983).
9. J.L. Batstone, J.M. Gibson, R.T. Tung, and A.F.J. Levi, *Appl. Phys. Lett.* **52**, 828–830 (1988).
10. R.T. Tung, A.F.J. Levi, J.P. Sullivan, and F. Schrey, *Phys. Rev. Lett.* **66**, 72–75 (1991).
11. D. Cherns, C.J.D. Hetherington, and C.J. Humphreys, *Philos. Mag. A* **49**, 165–177 (1984).
12. J.M. Gibson, J.C. Bean, J.M. Poate, and R.T. Tung, *Appl. Phys. Lett.* **41**, 818 (1982).
13. J.P. Sullivan, R.T. Tung, D.J. Eaglesham, F. Schrey, and W.R. Graham, *J. Vac. Sci. Technol. B* **11**, 1564–1570 (1993).
14. M. Falke, U. Falke, A. Bleloch, S. Teichert, G. Beddies, and H.J. Hinneberg, **86**, 1–3 (2005).
15. F.F. Zhao, Y.P. Feng, Y.F. Dong, and J.Z. Zheng, *Phys. Rev. B Condens. Matter Mater. Phys.* **74**, 033301 (2006).
16. R.T. Tung, S. Nakahara, and T. Boone, *Appl. Phys. Lett.* **46**, 895 (1985).
17. M. Liehr, P.E. Schmid, F.K. LeGoues, and P.S. Ho, *Phys. Rev. Lett.* **54**, 2139–2142 (1985).
18. R.J. Hauenstein, T.E. Schlesinger, T.C. McGill, B.D. Hunt, and L.J. Schowalter, *Appl. Phys. Lett.* **47**, 853–855 (1985).
19. M. Ospelt, J. Henz, L. Flepp, and H. von Kanel, *Appl. Phys. Lett.* **52**, 227–229 (1988).

20. R.T. Tung, K.K. Ng, J.M. Gibson, and A.F.J. Levi, *Phys. Rev. B* **33**, 7077–7090 (1986).
21. R.T. Tung, *J. Vac. Sci. Technol. B* **11**, 1546–1552 (1993).
22. T. Sands, C.J. Palmstrom, J.P. Harbison, V.G. Keramidas, N. Tabatabaie, T.L. Cheeks, R. Ramesh, and Y. Silberberg, *Mater. Sci. Rep.* **5**, 99–170 (1990).
23. K.T. Delaney, N.A. Spaldin, and C.G. Van De Walle, *Phys. Rev. B* **81**, 165312 (2010).
24. M. Peressi, N. Binggeli, and A. Baldereschi, *J. Phys. D* **31**, 1273–1299 (1998).
25. H. Fujitani and S. Asano, *Phys. Rev. B* **42**, 1696–1704 (1990).
26. J. Kim, B. Lee, Y. Park, K.V.R.M. Murali, and F. Benistant, IEEE-SISPAD October 9-11, 2015, 226–229 (2015).
27. R.T. Tung and L. Kronik, *Phys. Rev. B* **103**, 035304 (2021).
28. J. Tersoff, *Phys. Rev. Lett.* **52**, 465–468 (1984).
29. J. Tersoff, *Phys. Rev. B* **32**, 6968–6971 (1985).
30. R.G. Dandrea and C.B. Duke, *J. Vac. Sci. Technol. B* **11**, 1553–1558 (1993).
31. R.T. Tung and L. Kronik, *Phys. Rev. B* **103**, 085301 (2021).
32. R.T. Tung, *Appl. Phys. Rev.* **1**, 011304 (2014).
33. R.T. Tung, *J. Appl. Phys.* **136**, 165301 (2024).
34. I.M. Vitomirov, C.M. Aldao, Z. Lin, Y. Gao, B.M. Trafas, and J.H. Weaver, *Phys. Rev. B* **38**, 10776–10786 (1988).
35. F. Xu, J.J. Joyce, M.W. Ruckman, H.W. Chen, F. Boscherini, D.M. Hill, S.A. Chambers, and J.H. Weaver, *Phys. Rev. B* **35**, 2375–2384 (1987).
36. C. Bourgognon, S. Tatarenko, J. Cibert, L. Carbonell, V.H. Etgens, M. Eddrief, B. Gilles, A. Marty, and Y. Samson, *Appl. Phys. Lett.* **76**, 1455–1457 (2000).
37. C. Heske, U. Winkler, R. Fink, E. Umbach, C. Jung, and P. Bressler, *Phys. Rev. B* **56**, 2085–2093 (1997).
38. Z. Lin, F. Xu, and J.H. Weaver, *Phys. Rev. B* **36**, 5777–5783 (1987).
39. C.M. Aldao, G.D. Waddill, S.G. Anderson, and J.H. Weaver, *Phys. Rev. B* **40**, 2932–2939 (1989).
40. H. Haick, M. Ambrico, J. Ghabboun, T. Ligonzo, and D. Cahen, *Phys. Chem. Chem. Phys.* **6**, 4538–4541 (2004).
41. H. Hasegawa, T. Sato, and T. Hashizume, *J. Vac. Sci. Technol. B* **15**, 1227–1235 (1997).
42. H. Hasegawa, T. Sato, and S. Kasai, *Appl. Surf. Sci.* **166**, 92–96 (2000).
43. R. T. Tung and J. M. Gibson, *J. Vacuum Sci. Technol. A* **3**, 987 (1985).

Please view the accompanying lecture videos for this book by scanning the QR below:

www.ingramcontent.com/pod-product-compliance
Lightning Source LLC
LaVergne TN
LVHW010333220926
848484LV00004BB/892